Introductory Algebra

Derek I. Bloomfield

Orange County Community College
Middletown, New York

Reston Publishing Company, Inc.
Reston, Virginia
A Prentice-Hall Company

To my mother.

Library of Congress Cataloging in Publication Data

Bloomfield, Derek I.
Introductory algebra.

1. Algebra. I. Title.
QA152.2.B59 1983 512.9 82-13272
ISBN 0-8359-3268-0

10 9 8 7 6 5 4 3 2 1

Printed in the United States of America

Contents

Preface

This book is written for anyone who needs to learn the fundamental concepts of algebra. It assumes little or no previous knowledge of algebra. The main objective of this book is to provide a clear explanation of the basic ideas of algebra with an emphasis on *understanding* rather than formality or abstraction. In addition to these much-needed algebraic skills, the student will also develop confidence in his ability to learn mathematics —perhaps for the first time in his life.

The book is intended for use in any of three instructional modes: (1) a conventional lecture-type class; (2) a mathematics laboratory along with video tapes, film loops, or slide materials; or (3) a self-study program in which the student works at his own pace.

The book consists of 12 chapters, and material in each chapter is presented in the following manner: A concise explanation of the fundamental concepts for the particular topic is given; examples illustrating these fundamental concepts are worked out step by step; and an ample number of similar problems are given in the exercise sets enabling the student to master the concepts involved. The exercises progress from simple to more difficult in an effort to give the student confidence in his ability to solve problems. The answers to odd-numbered exercises are given at the end of the text.

The major goal of this method of presentation is a correct understanding of the topics and maximum skill in performing the mathematical operations. An achievement test is given at the end of each chapter with answers to all questions given in the answer section.

A summary of arithmetic is given in the Appendix for those who need an intensive review of the basic fundamentals.

Outstanding Features

- The material is presented in measured amounts so that the student can complete the topic before moving on to the next concept.
- Explanations are short, direct, and written to be understood.
- Rules are highlighted in boxes.
- Over 600 worked-out examples with step-by-step explanations are presented with important steps highlighted with boxes.
- Over 3,000 exercises have been carefully chosen to clarify explanations and provide necessary drill.
- STOP warnings are given to students about common errors.
- Sufficient space is provided for working out the exercises. This also provides a good reference when it comes time to review.
- Word problems covering a wide variety of applications are found throughout the text. These show the power of algebra in real-world situations.
- Chapter summaries are provided with definitions and rules restated, along with an example of each concept for quick review.
- Achievement tests at the end of each chapter examine the student's mastery of the material in that chapter.
- Answers to odd-numbered exercises and all achievement test questions are given in the answer section. Page numbers for the answers are given after each exercise set for easy reference.
- An arithmetic summary is given in the appendix for those who need a refresher.

An Instructor's Manual Is Available Containing:

- An arithmetic diagnostic test to be given before starting Chapter 1.
- Alternate forms of achievement tests for each chapter.
- A bank of test questions that may be tailored to the instructor's needs to make up a comprehensive final exam.
- Answers to the even-numbered exercises.

ACKNOWLEDGMENTS

I want to thank everyone who assisted with this book. Special thanks go to Ted Buchholz, my editor, and the staff of Reston Publishing Company for guidance and encouragement through the many steps along the way.

Many valuable suggestions and recommendations were made by the reviewers of the manuscript:

Thomas Ray Hamel	Austin Peay State University
Evelyn T. Bickford	Macon Junior College
Vincent Koehler	Howard Community College
Michael Klembara	Northern Kentucky University
Judy E. Ackerman	Montgomery College

I am grateful to my students and my colleagues at Orange County Community College for providing a stimulating atmosphere in which to work. Regina Westeris, Catherine Marrone, and my wife, Marcella, all helped with the typing. Thelma Oliver did an excellent job of checking the exercises and proofreading the manuscript.

My wife, Marcella, and my children, Jennifer, Max, Derek, Jr., and David, gave up many hours of their time. I thank them for their encouragement, their understanding, and their love.

A Note From the Author to the Student

I have written this book with the belief that every student can learn algebra if he wants to. If you have decided that you would like to learn algebra, here are some suggestions to help you attain success.

1. Attend all of your classes. You can't hope to learn what goes on in class unless you are there. Even if you don't understand everything that's being taught, you'll pick up a lot.
2. Work lots of problems. Even if you don't have a thorough understanding of what you're doing, if you do enough problems, the concepts eventually filter through.
3. Use your book. The worked-out examples are there to help you do the exercises. Go over the examples several times until you can repeat the procedures. Understanding will eventually follow.
4. Work on a regular basis. Try not to get behind. Solving today's problems usually depends on yesterday's results, so it's very hard to catch up once you're behind.
5. Try to remain confident. Even if you've never had success at algebra before, I'm convinced that you can succeed if you keep trying.

1
Signed Numbers

1.1 INTRODUCTION TO SIGNED NUMBERS

The first numbers seen in arithmetic are the counting numbers 1, 2, 3, 4, 5, 6, et cetera. When we add two counting numbers we always get a counting number for an answer. For example, $6 + 2 = 8$, $15 + 12 = 27$, and so on.

Now let's see what happens when we subtract one whole number from another.

$15 - 7 = 8$ a whole number

$10 - 6 = 4$ a whole number

$3 - 8 = ?$

Can such a subtraction be performed and, if so, what does the answer look like? To answer the question, consider the case where the temperature is 3 degrees above zero and then drops 8 degrees. The new temperature is 5 degrees below zero, or negative 5 degrees.

This example of temperature above and below zero nicely illustrates the concept of positive and negative numbers. These positive and negative numbers, along with zero, are called the **signed numbers,** or **integers.** This set, or group, of signed numbers is often looked at in terms of the number line. To make a number line, select a point on a line and call it zero. Then place the positive numbers equally spaced to the right of zero and place a positive sign (+) in front of each of them. The negative numbers are equally spaced to the left of zero and a negative sign (−) is placed in front of each.

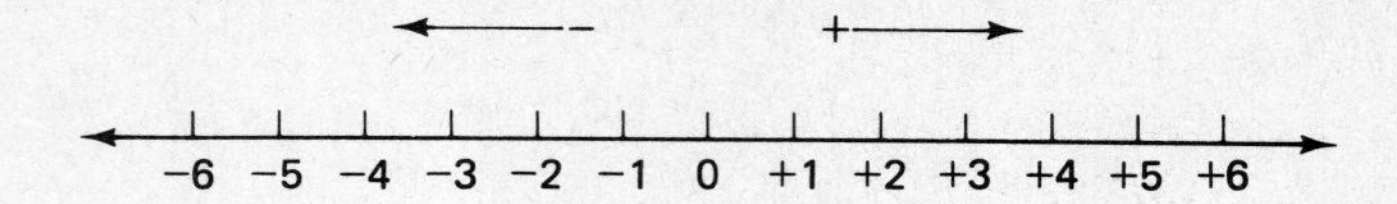

It is common practice to omit the + sign in front of positive numbers; thus a number with neither a + sign nor a − sign in front of it is positive. For example, 7 is + 7 and 34 is +34.

Signed numbers are the language of algebra, and we must become fluent in their use.

EXERCISE 1.1

Locate each of the following numbers on the number line:

1. −3
2. −1
3. +5
4. +6
5. −4
6. 5

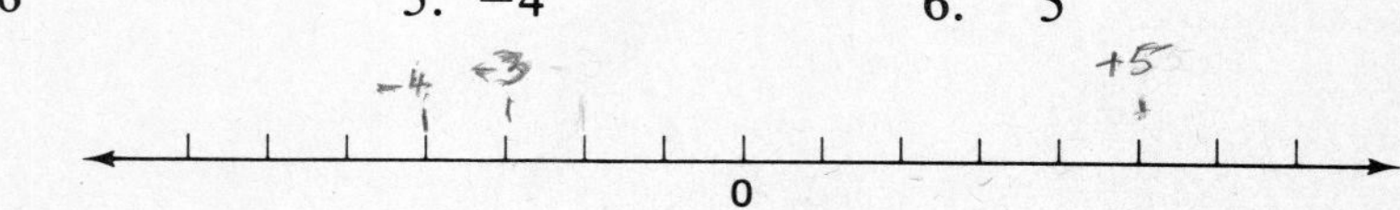

Represent each quantity by a signed number:

7. A bet in which you lose \$5.00.
8. A bill for \$23.50.
9. A mountain 7,288 feet above sea level.
10. Death Valley lies 280 feet below sea level.
11. The year 1990 A.D.
12. Euclid lived around 300 B.C.
13. A football team is penalized 15 yards.
14. You get a bonus of \$500.
15. You are penalized \$25 for a late payment.
16. The boiling point of water is 100° C above zero.
17. A stock goes down $2\frac{1}{4}$ points.

Answers to odd-numbered exercises on page 453.

1.2 ABSOLUTE VALUE

Before we add signed numbers we must first learn a concept called **absolute value.**

The absolute value of a signed number is the number of units that it is away from zero on the number line. For example, +3 is 3 units away from zero on the number line, so the absolute value of +3 is 3, and it is written as $|+3| = 3$.

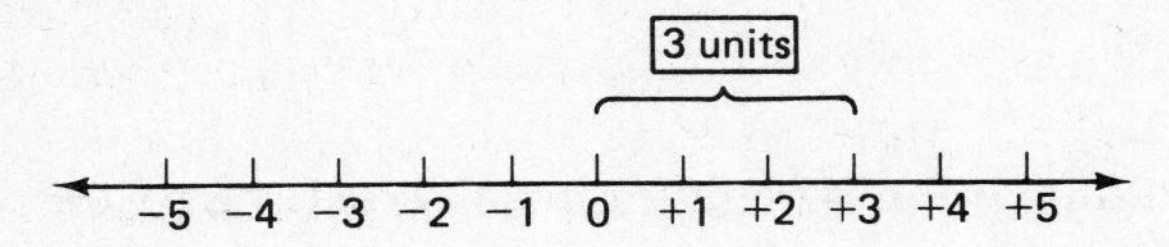

What about the absolute value of −3? Since it is also 3 units away from zero, its absolute value is also 3, which is written as $|-3| = 3$.

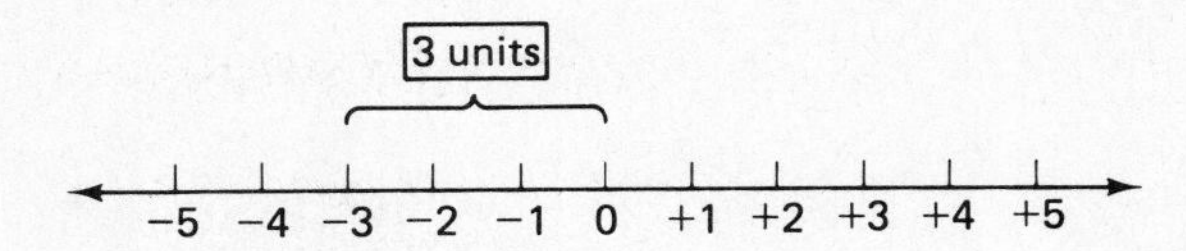

Example 1:

(a) $|+3| = 3$ is read "the absolute value of +3 is 3"
(b) $|-3| = 3$ is read "the absolute value of −3 is 3"
(c) $|-14| = 14$ is read "the absolute value of −14 is 14"
(d) $|0| = 0$ is read "the absolute value of zero is zero"

EXERCISE 1.2

Find each of the following absolute values:

1. $|-8|$ 2. $|+5|$ 3. $|-4|$

4. $|-6|$ 5. $|-1462|$ 6. $|+46|$

7. $|0|$ 8. $|+34|$ 9. $|+6|$

10. $|-62|$

Answers to odd-numbered exercises on page 453.

1.3 GREATER THAN AND LESS THAN

On the number line, as we move to the right the numbers get larger, continuing on and on in such a way that there is no largest integer. Similarly, as we move to the left of zero on the number line the negative numbers have smaller values. Therefore, there is also no smallest negative integer.

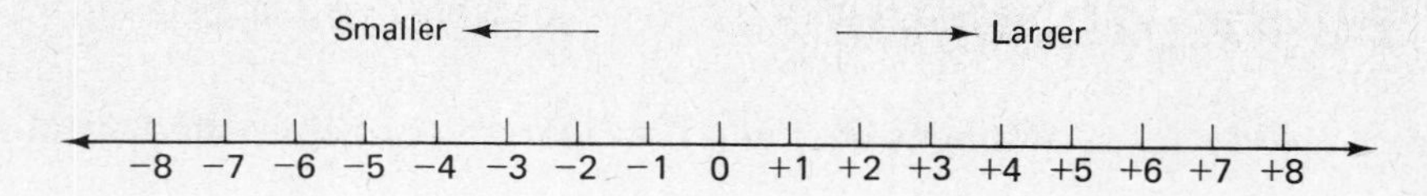

From this we can see that when any two numbers on the number line are compared, the one lying to the right of the other must always be larger. For example, $+6$ is greater than -2 since $+6$ lies to the right of -2 on the number line. This is usually written as

$$+6 > -2$$

and is read "$+6$ is greater than -2."

We can also write

$$-2 < +6$$

which is read "-2 is less than $+6$."

Example 1:

(a) $6 < 18$ is read "6 is less than 18"
(b) $-3 > -8$ is read "-3 is greater than -8"
(c) $4 > -2$ is read "4 is greater than -2"
(d) $-7 < 4 < 6$ is read "-7 is less than 4 and 4 is less than 6"

Note that $4 < 8$ and $8 > 4$ really mean the same thing even though we read them differently.

The symbol

$>$ is read "is greater than"
and
$<$ is read "is less than."

You can remember which way the symbol goes by noticing that the small side of the symbol is on the side of the smaller number and the larger side of the symbol is found on the side of the larger number.

small side smaller number	$<$	large side larger number

EXERCISE 1.3

Indicate which of the following are true and which are false:

1. $-6 < -5$	2. $3 > -4$	3. $7 > 14$
4. $6 < -7$	5. $3 > -4$	6. $-17 > -6$

7. $4 < -8.6$ 8. $3.6 < -1.4$ 9. $-784 < -326$

10. $-\frac{7}{8} < -\frac{1}{4}$ 11. $-7\frac{1}{2} > \frac{2}{3}$ 12. $\frac{5}{16} > -\frac{1}{8}$

Insert > or < between each pair of numbers to make a true statement:

13. 17 14   14. -4 12   15. 23 -5

16. -7 -4 17. 46 49   18. $13\frac{3}{4}$ $-11\frac{1}{2}$

19. 0 14   20. $\frac{5}{8}$ $-2\frac{1}{3}$ 21. 586 -12

22. 0 -4 23. 314 -8.9 24. $-24\frac{2}{3}$ -25

Answers to odd-numbered exercises on page 453.

1.4 ADDING SIGNED NUMBERS

To add signed numbers we again use the number line. To add a positive number move to the right; to add a negative integer, move to the left. Always start at zero.

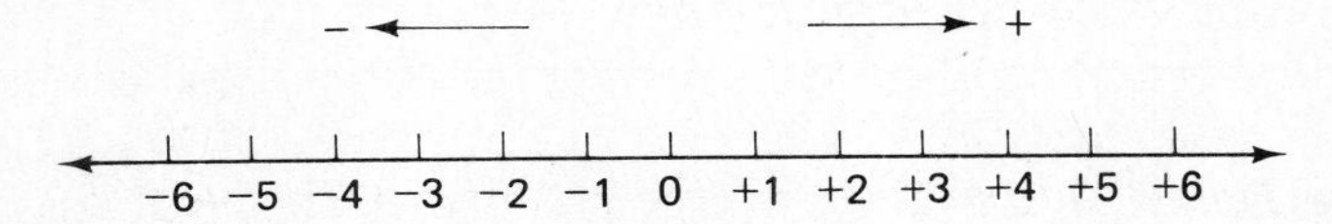

Let's start with an example that we already know the answer to.

Example 1: On the number line 4 + 2 is added by starting at zero, moving 4 spaces to the right, +4, and then moving 2 more spaces to the right, +2. We end at +6, which is the answer. $4 + 2 = 6$

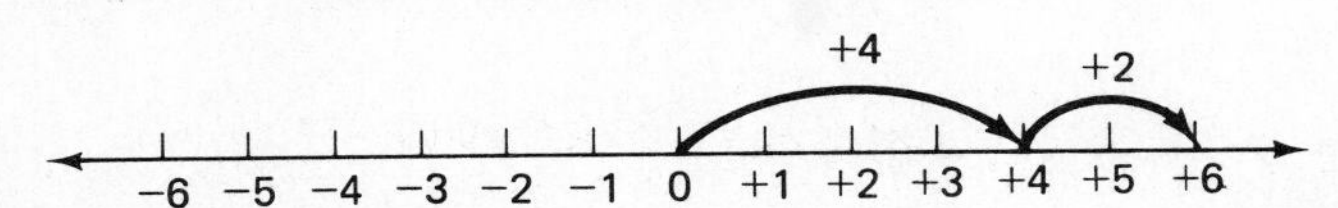

Example 2: $5 + (-2)$ Starting at zero, first move 5 units to the right and then 2 units to the left. The answer is +3 or 3. $5 + (-2) = 3$

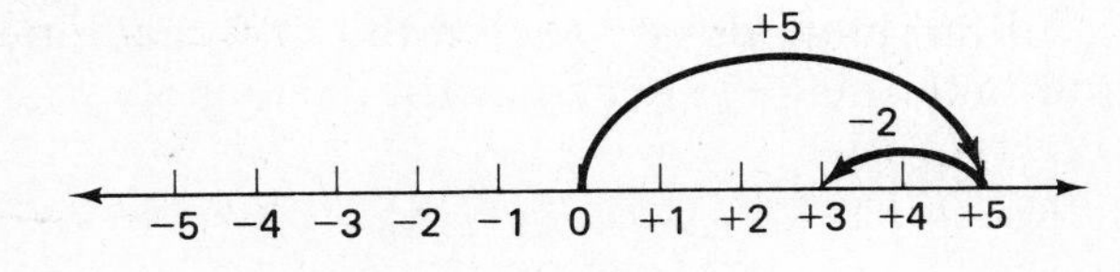

Example 3: $-3 + (-4)$

A move of 3 units to the left plus another move of 4 units to the left leaves us at -7, the answer. $-3 + (-4) = -7$

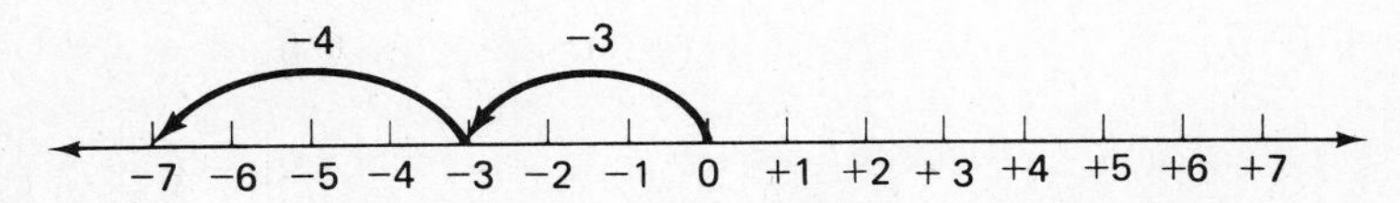

Example 4: $-6 + 2 = -4$

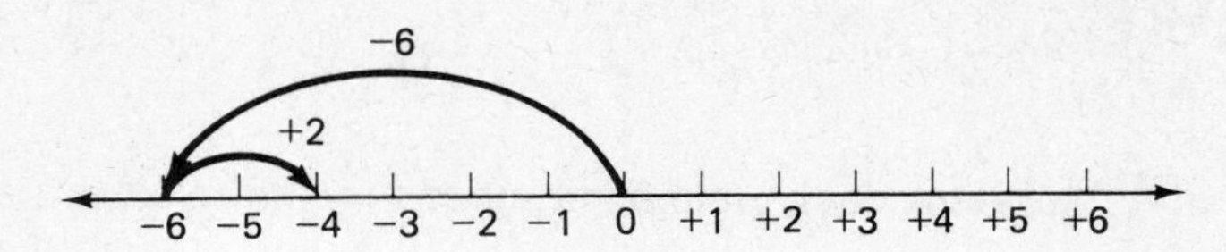

The sign + or − tells us which direction to move in, and the size of the number (its absolute value) tells us how far to move.

The number line provides a good visual explanation of what is taking place when we add signed numbers. However, it is cumbersome to draw a number line everytime we are asked to add two signed numbers, particularly if the numbers are very large. Therefore, we must establish a rule that provides us with the same result as using the number line.

> **To add signed numbers:**
> 1. If both numbers have the same sign, add the absolute values and keep the common sign for the answer.
> 2. If the numbers have different signs, find the difference in their absolute values and take the sign of the number with the larger absolute value for the answer.

Example 5: $3 + 4 = 7$
3 and 4 have the same sign so we add their absolute values and keep the same sign (+).

Example 6: $-4 + 1 = -3$
−4 and +1 have different signs so we find the difference between their absolute values, 4 and 1, giving us 3, and give it the sign of the larger absolute value.

Example 7: $-6 + (-5) = -11$
−6 and −5 have the same sign so add their absolute values ($6 + 5 = 11$) and keep the common sign (−).

Example 8: $2 + (-6) = -4$
2 and −6 have different signs so we find the difference between 2 and 6, giving us 4, and take the (−) sign from the 6 since it is larger than the 2.

EXERCISE 1.4

Add the following:

1. $6 + 8$	2. $4 + 9$	3. $3 + (-6)$
4. $4 + (-7)$	5. $7 + (-5)$	6. $6 + (-8)$
7. $13 + (-6)$	8. $15 + (-3)$	9. $-6 + 5$

10. $-8 + 4$

11. $-13 + 7$

12. $-11 + 6$

13. $-26 + 16$

14. $-28 + 12$

15. $-3 + (-5)$

16. $-6 + (-3)$

17. $-8 + (-9)$

18. $-7 + (-6)$

19. $-12 + (-18)$

20. $-16 + (-24)$

21. $-\frac{3}{8} + \left(-\frac{1}{8}\right)$

22. $-\frac{2}{9} + \left(-\frac{4}{9}\right)$

23. $\frac{1}{7} + \left(-\frac{3}{7}\right)$

24. $\frac{1}{5} + \left(-\frac{2}{5}\right)$

25. In two successive nights of playing poker you lost \$2.58 the first night and won \$1.76 the second night. What were your total winnings or losses?

26. Starting with a balance of zero in his checking account a financier writes checks for \$72.05, \$31.16, \$24.17, and \$12.12. The next day he rushes to the bank and makes a deposit of \$125.00 to cover them. What is the status of his account?

27. On Monday your stock in Hexagon Oil Co. goes up $2\frac{1}{3}$ points. On Tuesday it falls $6\frac{2}{3}$ points. What is the net amount of the change for the two days and in what direction?

28. Death Valley, the lowest point in North America, is 282 feet below sea level. Mount McKinley in Alaska is 20,320 feet above sea level, which is the highest point in North America. What is the difference in altitude between these two places?

29. Place one of the signed numbers $-5, -4, -3, -2, -1, 0, 1, 2,$ and 3 in each circle to make the sum of each side of the triangle equal to -3.

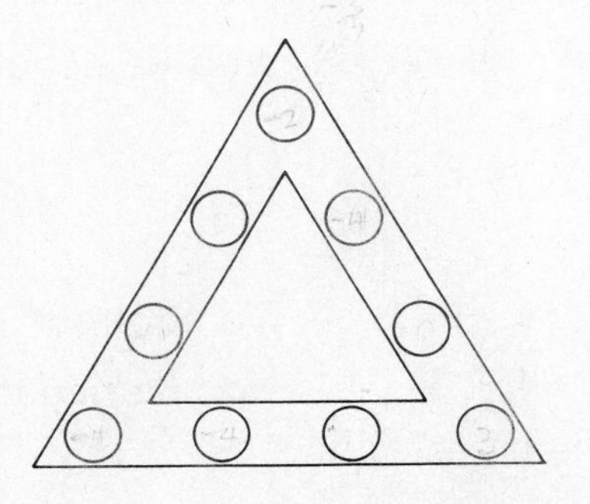

30. Complete the magic squares. The sum of each column, row, and diagonal must be the same.

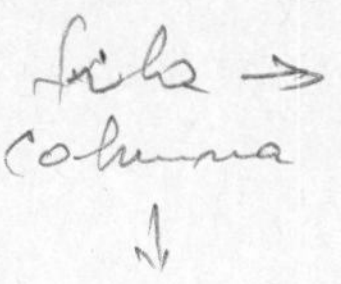

(a)

−1		
4		
3		5

(b)

$-1\frac{1}{2}$		
	−1	
	0	$-\frac{1}{2}$

Answers to odd-numbered exercises on page 454.

1.5 SUBTRACTING SIGNED NUMBERS

The Negative of a Number

The concept of the negative of a number suggests the opposite of that number. The opposite of moving 3 steps to the right on the number line would be moving 3 steps to the left. Since +3 represents a move of 3 steps to the right, the negative of +3 would mean a move of 3 steps to the left, or −3. So the negative of +3 is −3.

Following this same thinking, since −3 represents a move of 3 units to the left, the opposite, or negative, of −3 would be a move of 3 units to the right. So the negative of −3 is +3, and we write this as $-(-3) = +3 = 3$.

Example 1:

(a) The negative of 7 is −7.
(b) The negative of −4 is $-(-4) = 4$.
(c) The negative of −6 is 6.

The rule for subtraction of signed numbers is stated in terms of the rule for adding signed numbers, as follows:

To subtract one signed number from another, add its negative.

Example 2:

$7 - 9$	To subtract 9,
$= 7 + (-9)$	add its negative, −9.
$= -2$	

Example 3:

$-4 - (-3)$	To subtract −3,
$= -4 + 3$	add its negative, +3.
$= -1$	

Example 4:

$6 - (-1)$	To subtract -1,
$= 6 + 1$	add its negative, 1.
$= 7$	

Example 5:

$-8 - 4$	To subtract 4,
$= -8 + (-4)$	add its negative, -4.
$= -12$	

Adding and Subtracting More Than Two Signed Numbers

Example 6:

$4 - 6 - 5$	
$= 4 + (-6) + (-5)$	Change all subtractions to additions.
$= -2 + (-5)$	Add from left to right.
$= -7$	

Example 7:

$6 - (-3) - 7$	
$= 6 + 3 + (-7)$	Change all subtractions to additions.
$= 9 + (-7)$	Add from left to right.
$= 2$	

We frequently encounter problems in algebra that look like the following:

Example 8: $6 + 4 - 7 + 3 - (-8) - 4 - 2$
The easiest way to solve a long problem of this type is to change all subtractions to additions, add all of the positive values, add all of the negative values, and then combine these two results.

$6 + 4 - 7 + 3 - (-8) - 4 - 2$	
$= 6 + 4 + (-7) + 3 + 8 + (-4) + (-2)$	Change subtractions to additions.
$6 + 4 + 3 + 8 = 21$	Add the positive values.
$(-7) + (-4) + (-2) = -13$	Add the negative values.
$21 + (-13) = 8$	Combine the results from the above steps.

Example 9: A man on a diet kept a careful daily record of his weight change. Here is an account.

On Monday he lost 2 pounds.
On Tuesday he lost 4 pounds.
On Wednesday he gained 1 pound.
On Thursday he lost 2 pounds.
On Friday he gained 2 pounds.
On Saturday he lost 2 pounds.
On Sunday he gained 1 pound.

(a) How much weight did he gain?
(b) How much weight did he lose?
(c) What was his net gain or loss for the week?
(d) If he weighed 230 pounds when he started, how much did he weigh at the end of the week?

Solution:

(a) $1 + 2 + 1 = 4$ pounds (gained)
(b) $(-2) + (-4) + (-2) + (-2) = -10$ pounds (lost)
(c) $4 + (-10) = -6$ pounds (net loss)
(d) $230 + (-6) = 224$ pounds at the end of the week

As you go through the exercises, you may notice shortcuts that simplify your work. Go ahead and use any shortcuts you find, providing they always give you correct answers.

EXERCISE 1.5

Subtract:

1. $6 - 8$
2. $9 - 12$
3. $4 - 9$
4. $5 - 7$
5. $16 - 7$
6. $14 - 8$
7. $-6 -5$
8. $-8 -4$
9. $-11 -8$
10. $-14 -6$
11. $-18 -12$
12. $-16 -9$
13. $7 - 7$
14. $5 - 5$
15. $6 - (-4)$
16. $8 - (-5)$
17. $8 - (-9)$
18. $4 - (-8)$
19. $11 - (-5)$
20. $14 - (-3)$
21. $-6 - (-2)$
22. $-8 - (-5)$
23. $-4 - (-8)$
24. $-6 - (-9)$
25. $-11 - (-12)$
26. $-14 - (-11)$
27. $-7\frac{3}{8} - 1\frac{1}{8}$
28. $-4\frac{2}{9} - 3\frac{1}{9}$
29. $-\frac{2}{7} - (-\frac{3}{7})$
30. $\frac{2}{8} - (-\frac{5}{8})$

Combine:

31. $-6 - 3 + 4 - 5 + 1$

$-14 + 5 = -9$ ✓

32. $-6 + 3 - 8 - 4 + 7$

33. $-4 - 6 - 8 - 3 + 1 - 9$

$-30 + 1 = -29$ ✓

34. $8 - 6 + 8 - 6 + 8 - 6$

35. $-3 - (-8) + 2 - 7 - 4$

$= -3 + 8 + 2 - 7 - 4$
$= -14 + 10 = -4$ ✓

36. $-2 - (-4) - (-5) - 5$

37. $-\frac{2}{5} + \frac{3}{10} - \frac{1}{2} - \frac{1}{5} + \frac{7}{10}$

38. $-\frac{3}{8} + \frac{1}{4} - \frac{5}{8} - \frac{1}{2} + \frac{3}{4}$

$-\frac{2}{5} - \frac{1}{2} - \frac{1}{5} + \frac{3}{10} + \frac{7}{10} = -\frac{4}{10} - \frac{5}{10} - \frac{2}{10} + \frac{10}{10} = -\frac{11}{10} + \frac{10}{10} = -\frac{1}{10}$ ✓

4.71
5.32
1.1
11.13
− 4.08
7.05

39. $-2.03 + 4.71 + 5.32 - (-1.1) - 2.05$

$-2.03 - 2.05 + 4.71 + 5.32 + 1.1 =$
$= -4.08 + 11.13 = 7.05$ ✓

40. $3.14 - (-1.41) - 2.24 + 1.73 - 6.09$

41. $7\frac{1}{2} - 2\frac{2}{3} + 5\frac{1}{2} - 7\frac{3}{4} + 6\frac{1}{4} + 1\frac{1}{6}$ $= 7 - 2 + 5 - 7 + 6 + 1 + \frac{1}{2} - \frac{2}{3} + \frac{1}{2} - \frac{3}{4} + \frac{1}{4} + \frac{1}{6} =$

$= 19 - 9 + \frac{2}{2} + \frac{1}{4} + \frac{1}{6} - \frac{2}{3} - \frac{3}{4} = 10 + \frac{5}{4} + \frac{1}{6} - \frac{2}{3} - \frac{3}{4} =$

$= 10 + \frac{15 + 2 - 8 - 9}{12} = 10 + \frac{17 - 17}{12} = 10 + \frac{0}{12} = 10$ ✓

42. $1\frac{1}{4} - \left(-2\frac{5}{8}\right) - \frac{1}{2} + \frac{3}{4} - 7\frac{1}{4}$

43. Big Bang Uranium Co. stock records the following ganancias gains and perdidos losses for the week:

Monday $+\frac{1}{8}$
Tuesday $+\frac{3}{8}$
Wednesday $-\frac{1}{2}$
Thursday $-2\frac{1}{4}$
Friday $+1\frac{1}{2}$

$\frac{1}{8} + \frac{3}{8} - \frac{1}{2} - 2\frac{1}{4} + 1\frac{1}{2} = \frac{4}{8} - \frac{1}{2} - \frac{9}{4} + \frac{3}{2} = \frac{1}{2} - \frac{1}{2} - \frac{9}{4} + \frac{3}{2} = \frac{-9 + 6}{4} = -\frac{3}{4}$

☆ (a) What is the net gain or loss for the week? $= -\frac{3}{4}$

(b) If the stock started the week at a value of $64\frac{1}{2}$, what was its value at the end of the week?

$-2\frac{1}{4} - \frac{1}{8} = -\frac{9}{4} - \frac{1}{8} = \frac{-18 - 1}{8} = \frac{-19}{8}$

☆ $64\frac{1}{2} - \frac{19}{8} = \frac{129}{2} - \frac{19}{8} = \frac{516 - 19}{8} = \frac{497}{8} = 62\frac{1}{8}$

$64\frac{1}{2} - \frac{3}{4} = 64\frac{2}{4} - \frac{3}{4} = 63\frac{6}{4} - \frac{3}{4} = 63\frac{3}{4}$

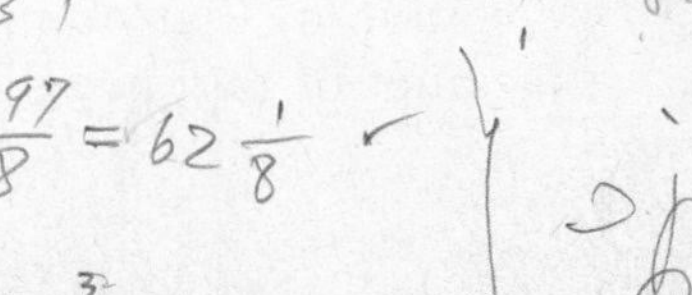

Revisar

44. The Florida Fumblers football team lost 12 yards on their first play and gained 3 yards and 4 yards on each of the next two plays. What was their net total yardage?

Answers to odd-numbered exercises on page 454.

1.6 MULTIPLYING SIGNED NUMBERS

The answer to any multiplication problem is called the **product.** The numbers being multiplied are called **factors.**

$$5 \times 2 = 10$$

factors ↖↗ (5 and 2); product ↖ (10)

There are several different symbols used to indicate multiplication:

1. $5 \times 2 = 10$ — "×" means to multiply.
2. $5 \cdot 2 = 10$ — " · " means to multiply.
3. $(5)(2) = 10$ — Parentheses written right next to each other means to multiply.
4. $5(2) = 10$
5. $(5)2 = 10$ — (4 and 5) Frequently only one set of parentheses is used to mean multiplication.
6. xy — Means x times y. Two letters written right next to each other indicates multiplication.

You should remember that in arithmetic multiplication was defined as repeated addition. For example, $5 \times 2 = 5$ twos $= 2 + 2 + 2 + 2 + 2 = 10$. We can apply this same principle to multiplication of signed numbers as follows:

Example 1: $5 \times (-2) = 5$ negative twos $= (-2)+(-2)+(-2)+(-2)+(-2) = -10$

Example 2: $(-3) \times (2) = 2 \times (-3) = 2$ negative threes $= (-3)+(-3) = -6$

Examples 1 and 2 show that when two numbers with different signs are multiplied, their product is negative.

Example 3: $(-1)(3) = (3)(-1) = 3$ negative ones $= (-1)+(-1)+(-1) = -3$

Example 4: $(-1)(5) = -5 =$ the negative of 5

We can see from examples 3 and 4 that multiplying by -1 gives us the negative of that number. We will use this concept now to show that the product of two negative numbers is positive.

Example 5:

$\boxed{(-2)}(-5)$	$= \boxed{(-1)(2)}(-5)$	since $-2 = (-1)(2)$.
	$= (-1)(-10)$	since $(2)(-5) = -10$.
	$=$ negative of -10	because -1 times a number gives the negative of that number.
	$= 10$	since the negative of a number is found by changing its sign.

Example 6: $(-3)(-4) = (-1)(3)(-4) = (-1)(-12) = 12$

Examples 5 and 6 show us that the product of two negative numbers is a positive number.

These rules for multiplying signed numbers are summarized as follows:

> **To multiply two signed numbers:**
> 1. Multiply their absolute values.
> 2. The product is positive if their signs are alike.
>   The product is negative if their signs are different.

Example 7: Multiply $(-4)(6)$

Solution:

$(-4)(6) = -24$

- The product of the absolute values is $4 \times 6 = 24$.
- The sign is negative because the signs of (-4) and (6) are different.

Example 8: Multiply $(-3)(-6)$

Solution:

$(-3)(-6) = +18$

- $3 \times 6 = 18$
- The sign is positive because the signs of (-3) and (-6) are alike.

Example 9: Multiply $(-3)(5)$

Solution:

$(-3)(5) = -15$

- $3 \times 5 = 15$
- The product is negative since the signs are different.

Example 10: Multiply $\left(-3\frac{1}{3}\right)\left(4\frac{1}{2}\right)$

Solution:

$$\left(-3\frac{1}{3}\right)\left(4\frac{1}{2}\right)$$

$$= \left(-\frac{10}{3}\right)\left(\frac{9}{2}\right)$$

$$= -\left(\frac{\overset{5}{\cancel{10}}}{\underset{1}{\cancel{3}}}\cdot\frac{\overset{3}{\cancel{9}}}{\underset{1}{2}}\right)$$

$$= -\frac{15}{1} = -15$$

Example 11: Multiply $(-1.6)(-3.5)$

Solution:

$$(-1.6)(-3.4)$$
$$= +(1.6 \times 3.4)$$
$$= 5.44$$

Example 12:

(a) $(-7)(-4) = 28$

(b) $(-3)(7) = -21$

(c) $(9)(8) = 72$

(d) $(3.6)(-.002) = -.0072$

(e) $\left(-5\frac{1}{4}\right)\left(-2\frac{2}{7}\right)$

$$= \left(-\frac{21}{4}\right)\left(-\frac{16}{7}\right)$$

$$= +\left(\frac{\overset{3}{\cancel{21}}}{\underset{1}{\cancel{4}}}\cdot\frac{\overset{4}{\cancel{16}}}{\underset{1}{\cancel{7}}}\right)$$

$$= 12$$

Multiplying More Than Two Signed Numbers

Let us extend the notion that we have just learned. Consider a problem in which there are more than two signed numbers to be multiplied. One way to solve the problem is to multiply the first two signed numbers together, multiply this result by the third signed number, multiply that result by the fourth signed number, and so on. To illustrate:

$$(-2)(+1)(-3)(-2)(+4) \qquad \text{multiplying } (-2)(+1) = -2$$
$$= (-2)(-3)(-2)(+4) \qquad \text{multiplying } (-2)(-3) = +6$$
$$= (+6)(-2)(+4) \qquad \text{multiplying } (+6)(-2) = -12$$
$$= (-12)(+4) \qquad \text{multiplying } (-12)(+4) = -48$$
$$= (-48)$$

There is an easier method for solving this type of problem. Each pair of negative signed numbers, when multiplied together, yields a positive answer. Therefore, if there are an even number of negative signs in the problem, the product will be positive. If there are an odd number of negative signs, each pair of negative factors yields a positive answer, but there will be one negative factor remaining, which will make the final product negative. The numerical part of the product will always be the product of the absolute values of all the factors in the problem.

To multiply more than two signed numbers:

1. Multiply the absolute values of the factors.
2. If there are an **even** number of negative signs, the product is **positive.**
   If there are an **odd** number of negative signs, the product is **negative.**

Example 13: Multiply $(-2)(+1)(-3)(-2)(+4)$

Solution: $(2)(1)(3)(2)(4) = 48$ Multiply the absolute values.

Since there are three negative factors, an *odd* number, our answer is *negative*.
Therefore $(-2)(+1)(-3)(-2)(+4) = -48$

Example 14: Multiply $(-2)(-2)(-2)(-2)(-2)$

Solution: $(2)(2)(2)(2)(2) = 32$ Multiply the absolute values.

Five negative signs is an odd number, so our answer is negative.
$(-2)(-2)(-2)(-2)(-2) = -32$

Example 15: Multiply $(-1)(2)(3)(-1)(4)$

Solution: $(1)(2)(3)(1)(4) = 24$

Two negative factors, an even number, tells us our answer is positive.
$(-1)(2)(3)(-1)(4) = +24 = 24$

EXERCISE 1.6

Multiply:

1. $(-4)(-6)$
2. $3(-8)$
3. $(-7)(4)$
4. $(-10)(-10)$
5. $-8 \cdot 2$
6. $(-9)8$
7. $(6)(11)$
8. $(-12)(-8)$
9. $(10)(-6)$
10. $(-13)(-4)$
11. $(-11)(-12)$
12. $(-14)(31)$
13. $33(-16)$
14. $(-30)5$
15. $-14 \cdot 7$
16. $(-7)(-20)$
17. $(-1.2)(3.4)$
18. $(-\frac{2}{3})(\frac{3}{8})$
19. $(4.5)(-12)$
20. $(-400)(-700)$
21. $(-3\frac{1}{5})(1\frac{1}{4})$
22. $(-4\frac{1}{2})(-1\frac{1}{3})$
23. $(-8.6)(4.5)$
24. $(17.2)(-.046)$
25. $(-12)(+4)(-3)(-1)$
26. $(+3)(+1)(+2)(+5)$
27. $(-1)(-2)(-3)(-4)(-5)$
28. $(3)(-2)(1)(1)(1)(2)$
29. $(-1)(1)(-1)(-1)(1)$
30. $(-3)(0)(6)(-2)(-2)(-1)$
31. $(-2.1)(1.4)(2)(-3)$
32. $(-\frac{2}{3})(-\frac{3}{4})(-\frac{1}{8})(+\frac{4}{5})$
33. $(-3)(-3)(-3)(-3)$
34. $(-3)(-3)(-3)(-3)(-3)$
35. $(-1\frac{1}{3})(-\frac{2}{3})(-1\frac{1}{2})(-\frac{3}{4})$
36. $(-\frac{3}{4})(-\frac{5}{8})(4)(\frac{1}{5})(-\frac{8}{15})(1\frac{2}{3})$

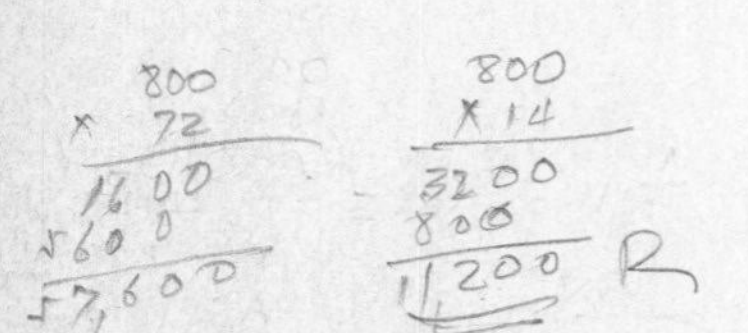

37. If an investor bought 800 shares of stock at $72 per share and then sold them at a loss of $14 per share, find the total amount of the loss and indicate it as a signed number.

Answers to odd-numbered exercises on page 454.

1.7 DIVIDING SIGNED NUMBERS

Division may be shown in three different ways:

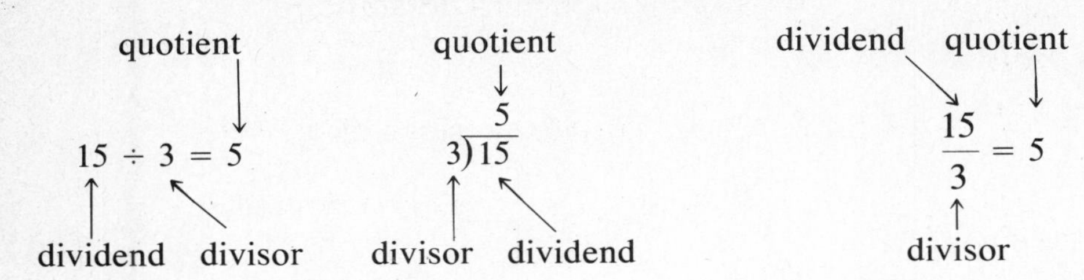

In all cases the number you divide by is the **divisor,** the number being divided is called the **dividend,** and the result is called the **quotient.**

To check a division problem we multiply the divisor by the quotient to obtain the dividend.

$$15 \div 3 = 5 \qquad \text{because } 3 \times 5 = 15$$

Applying this principle to signed numbers gives us rules for dividing signed numbers:

since $(3)(5) = 15$,	$15 \div 3 = 5$
since $(-3)(5) = -15$,	$-15 \div -3 = 5$
since $(3)(-5) = -15$,	$-15 \div 3 = -5$
since $(-3)(-5) = 15$,	$15 \div -3 = -5$

These examples show that division of signed numbers follows the same rule of signs as multiplication of signed numbers. These rules are stated as follows:

To divide one signed number by another:

1. Divide their absolute values.
2. The quotient is positive if their signs are alike.
   The quotient is negative if their signs are different.

Example 1: Divide $-42 \div 6$

Solution: $-42 \div 6 = -7$ $\quad$ $42 \div 6 = 7$, and the answer is negative because the signs are different.

Example 2: Divide $\dfrac{-72}{-9}$

Solution: $\dfrac{-72}{-9} = 8$ $\quad$ $\dfrac{72}{9} = 8$, and the quotient is positive since the signs are alike.

Example 3: Divide $-4\overline{)24}$

Solution: $-4\overline{)24}$ with quotient -6; $4\overline{)24}$ with quotient 6, and the quotient is negative since the signs are different.

Example 4:

(a) $(-28) \div (-4) = 7$ signs alike

(b) $\dfrac{-36}{9} = -4$ signs different

(c) $\dfrac{49}{7} = 7$ signs alike

(d) $-8\overline{)-64}$ with quotient 8 signs alike

(e) $18 \div (-3) = -6$ signs different

Example 5: Divide $-\dfrac{9}{16} \div \dfrac{3}{8}$

Solution:

$-\dfrac{9}{16} \div \dfrac{3}{8}$ Invert the divisor

$= -\dfrac{\overset{3}{\cancel{9}}}{\underset{2}{\cancel{16}}} \times \dfrac{\overset{1}{\cancel{8}}}{\underset{1}{\cancel{3}}}$ and multiply.

$= -\dfrac{3}{2}$ The quotient is negative since the signs are different.

Example 6: Divide $2.4\overline{)-1.44}$

Solution:

$$
\begin{array}{r}
-.6 \\
2{\times}4.\overline{)-1{\times}4.4} \\
1\ 4\ 4 \\
\hline
0
\end{array}
$$

The quotient is negative because the signs are different.

Division Involving Zero

$\dfrac{0}{3} = 0$ because $(0)(3) = 0$

$\dfrac{0}{-3} = 0$ because $(0)(-3) = 0$

Consider $\frac{3}{0} = ?$ What number could replace the question mark so that $? \times 0 = 3$? Since any number multiplied by zero gives zero, there is no possible replacement for the question mark. We conclude then that $\frac{3}{0}$ has no answer.

Next consider $\frac{0}{0} = ?$ What number could replace the question mark so that $? \times 0 = 0$. Since any number times zero is zero, we can replace the question mark by any number we like. This gives us $\frac{0}{0} = 5, -6, 14$, or any other number. This situation is certainly not allowable, so we conclude that division by zero in all cases is not defined.

> Division by zero is undefined.
> Zero may never be used as a divisor.

Example 7:

(a) $\frac{-74}{0} = \text{undefined}$

(b) $\frac{0}{-4} = 0$ because $(0)(-4) = 0$

(c) $\frac{0}{0} = \text{undefined}$

EXERCISE 1.7

Divide:

1. $\frac{-12}{-4}$
2. $\frac{-21}{-7}$
3. $\frac{-12}{-12}$
4. $\frac{-16}{-16}$
5. $\frac{-56}{8}$
6. $\frac{-32}{4}$
7. $\frac{63}{-7}$
8. $\frac{28}{-7}$
9. $\frac{-3}{18}$
10. $\frac{-4}{20}$
11. $\frac{0}{-8}$
12. $\frac{0}{-6}$
13. $\frac{-7}{0}$
14. $\frac{-4}{0}$
15. $-16 \div (-4)$
16. $-24 \div (-8)$
17. $-\frac{3}{4} \div \left(-\frac{1}{2}\right)$
18. $-\frac{3}{8} \div \left(-\frac{5}{6}\right)$
19. $-4\frac{3}{8} \div 1\frac{3}{4}$
20. $-7\frac{1}{7} \div 3\frac{1}{3}$
21. $(8.4) \div (-21)$
22. $(9.6) \div (-16)$
23. $41.8 \div 0$
24. $-54.6 \div 0$

25. A man lost \$24 at the racetrack. If he lost the same amount in each of four races, how much did he lose per race?

26. Eight youngsters playing baseball break a neighbor's window and must share equally the expense of replacing the glass. If the repair costs \$28.80, how much does it cost each youngster?

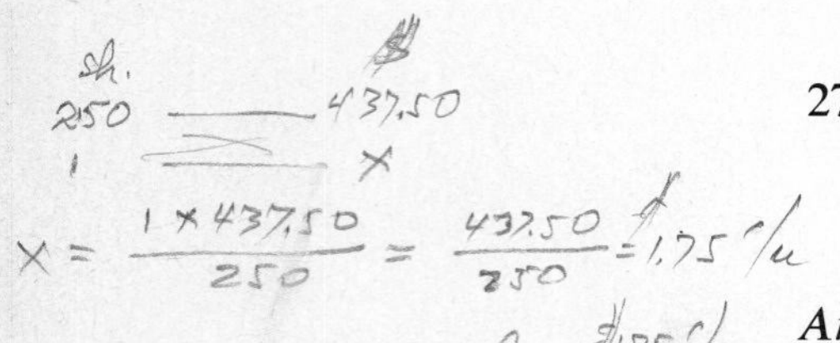

27. An investor sells 250 shares of Heavy Balloon Corporation stock at a total loss of \$437.50. What is his loss per share?

Answers to odd-numbered exercises on page 455.

1.8 PROPERTIES OF SIGNED NUMBERS

Whenever you reverse the order of two signed numbers in an addition problem, the sum does not change.

Example 1:

(a) $6 + 3 = 9$
$3 + 6 = 9$
(b) $(-4) + 5 = 1$
$5 + (-4) = 1$
(c) $(-6) + (-5) = -11$
$(-5) + (-6) = -11$

This property is true in general and is called the **commutative property of addition.**

Commutative property of addition:
If a and b represent any signed numbers, then $a + b = b + a$.

Similarly, if the order of the factors in a multiplication problem is reversed, the product remains the same.

Example 2:

(a) $6 \cdot 3 = 18$
$3 \cdot 6 = 18$
(b) $(-4)(5) = -20$
$(5)(-4) = -20$
(c) $(-6)(-5) = 30$
$(-5)(-6) = 30$

This property is called the **commutative property of multiplication** and is stated formally as follows:

> **Commutative property of multiplication:**
> If a and b represent any signed numbers, then $a \cdot b = b \cdot a$.

Note that in subtraction and division of signed numbers reversing the order *does* make a difference. You get a different result.

Example 3:

(a) $\left.\begin{aligned} 7 - 5 &= 2 \\ 5 - 7 &= -2 \end{aligned}\right\}$ not the same

(b) $\left.\begin{aligned} 20 \div 4 &= 5 \\ 4 \div 20 &= \frac{4}{20} = \frac{1}{5} \end{aligned}\right\}$ not the same

Therefore, neither subtraction nor division of signed numbers is commutative.

When we add three numbers, we have a choice of which two to add first. Does it make any difference? We use parentheses to indicate which two numbers we are adding first.

Example 4: Add $3 + 2 + 6$

(a) $(3 + 2) + 6$ Add 3 + 2 first.
$= 5 + 6$
$= 11$

(b) $3 + (2 + 6)$ Add 2 + 6 first.
$= 3 + 8$
$= 11$

Our example illustrates that when adding three signed numbers it makes no difference which two we add first. This property is always true and is called the **associative property of addition.**

> **Associative property of addition:**
> If a, b, and c are any signed numbers, then $(a + b) + c = a + (b + c)$.

An example suggests that a similar property holds for multiplication of three signed numbers.

Example 5: Multiply $-3 \cdot 4 \cdot 2$

(a) $(-3 \cdot 4) \cdot 2$ Multiply inside the parentheses first.
$= -12 \cdot 2$ $-3 \cdot 4 = -12$
$= -24$

(b) $-3 \cdot (4 \cdot 2)$ Multiply inside the parentheses first.
$= -3 \cdot 8$ $4 \cdot 2 = 8$
$= -24$

As you can see, we obtained the same result no matter which two numbers we multiplied first.

> **Associative property of multiplication**
> If a, b, and c are any signed numbers, then $(a \cdot b) \cdot c = a \cdot (b \cdot c)$.

As is the case with the commutative property, neither subtraction nor division is associative. This is illustrated in the following example.

Example 6:

(a) $(6 - 3) - 7 = 3 - 7 = -4$
$6 - (3 - 7) = 6 - (-4) = 6 + 4 = 10$ Results differ.

(b) $(20 \div -4) \div 2 = -5 \div 2 = -\frac{5}{2}$
$20 \div (-4 \div 2) = 20 \div (-2) = -10$ Results differ.

These properties are quite intuitive, but they are stated here because they are used frequently as we continue on in algebra.

> To summarize, we can change the order or the grouping in any addition or multiplication problem whenever it suits our purpose.

Study the next example carefully to discover still another important property of signed numbers.

Example 7:

(a) Evaluate $4(2 + 7)$
(b) Evaluate $4 \cdot 2 + 4 \cdot 7$

Solution:

(a) $4(2 + 7)$ Add $2 + 7$ inside the parentheses first.
$= 4 \cdot 9$ Multiply.
$= 36$

(b) $4 \cdot 2 + 4 \cdot 7$ Multiply first.

$= 8 + 28$ Add.

$= 36$

Since both results are the same we see that $4(2 + 7) = 4 \cdot 2 + 4 \cdot 7$.

This property is true in general and is called the **distributive property of multiplication over addition.**

> **The distributive property of multiplication over addition:**
> If a, b, and c are signed numbers, then $a \cdot (b + c) = a \cdot b + a \cdot c$.

Example 8: Verify that the distributive property holds in each case.

(a) $3(2 + 6) = 3 \cdot 2 + 3 \cdot 6$

(b) $-7(4 + 2) = (-7)(4) + (-7)(2)$

(c) $12\left(\frac{1}{4} + \frac{1}{6}\right) = 12 \cdot \frac{1}{4} + 12 \cdot \frac{1}{6}$

Solution:

left side	right side
(a) $3(2 + 6)$	$3 \cdot 2 + 3 \cdot 6$
$= 3(8)$	$= 6 + 18$
$= 24$	$= 24$
(b) $-7(4 + 2)$	$(-7)(4) + (-7)(2)$
$= -7(6)$	$= (-28) + (-14)$
$= -42$	$= -42$
(c) $12\left(\frac{1}{4} + \frac{1}{6}\right)$	$= 12 \cdot \frac{1}{4} + 12 \cdot \frac{1}{6}$
$12\left(\frac{3}{12} + \frac{2}{12}\right)$	$= \overset{3}{\cancel{12}} \cdot \frac{1}{\underset{1}{\cancel{4}}} + \overset{2}{\cancel{12}} \cdot \frac{1}{\underset{1}{\cancel{6}}}$
$\overset{1}{\cancel{12}} \cdot \frac{5}{\underset{1}{\cancel{12}}}$	$= 3 + 2$
$= 5$	$= 5$

EXERCISE 1.8

In exercises 1 through 14 answer true or false and give a reason for your answer.

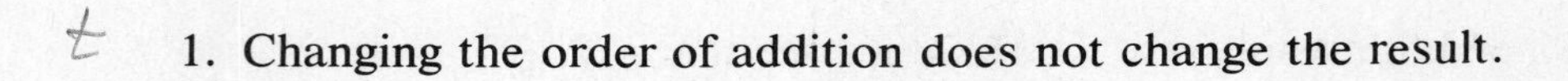

1. Changing the order of addition does not change the result.

2. Changing the order of division does not change the result.

3. Changing the grouping in a subtraction problem does not change the difference.

4. Changing the order of multiplication does change the product.

5. Changing the grouping in an addition problem has no effect on the sum.

6. $6 - 4 = 4 - 6$

7. $r(st) = (rs)t$

8. $\frac{a}{b} = \frac{b}{a}$

9. $(e \div f) \div g = e \div (f \div g)$

10. $(x + 3) + 4 = x + (3 + 4)$

11. $x - 15 = 15 - x$

12. $(7 - 5) - 2 = 7 - (5 - 2)$

13. $6 + m = m + 6$

14. $m \cdot 6 = 6m$

In exercises 15 through 20 verify the distributive property.

15. $6(3 + 5) = 6 \cdot 3 + 6 \cdot 5$

16. $-4(-5 + 7) = (-4)(-5) + (-4)(7)$

17. $-8(4 + 7) = (-8)(4) + (-8)(7)$

18. $\frac{1}{2}(7 + 5) = \frac{1}{2} \cdot 7 + \frac{1}{2} \cdot 5$

19. $4.3(1.2 + .3) = (4.3)(1.2) + (4.3)(.3)$

20. $6.01(-2.1 + .05) = (6.01)(-2.1) + (6.01)(.05)$

Answers to odd-numbered exercises on page 455.

1.9 POWERS OF SIGNED NUMBERS

When the same factor appears many times, a shorthand notation is usually used. For example:

$$2^6 \text{ means } 2 \cdot 2 \cdot 2 \cdot 2 \cdot 2 \cdot 2 = 64$$

In the expression 2^6, 2 is called the **base,** 6 is called the **exponent,** and 2^6 is read "2 to the 6th power."

exponent ↘

$$2^6 = 64$$

base ↗

Example 1:

(a) $2^1 = 2$ — 2 to the 1st power is 2.
(b) $2^2 = 2 \cdot 2 = 4$ — 2 to the 2nd power, or "2 squared," is 4.
(c) $2^3 = 2 \cdot 2 \cdot 2 = 8$ — 2 to the 3rd power, or "2 cubed," is 8.
(d) $2^4 = 2 \cdot 2 \cdot 2 \cdot 2 = 16$ — 2 to the 4th power is 16.

Example 2: Evaluate the following:

(a) $3^2 = 3 \cdot 3 = 9$
(b) $(-3)^2 = (-3)(-3) = 9$

(c) $(-3)^3 = (-3)(-3)(-3) = -27$
(d) $5^1 = 5$
(e) $1^5 = 1 \cdot 1 \cdot 1 \cdot 1 \cdot 1 = 1$
(f) $(-1)^5 = (-1)(-1)(-1)(-1)(-1) = -1$
(g) $\left(\frac{1}{2}\right)^3 = \left(\frac{1}{2}\right)\left(\frac{1}{2}\right)\left(\frac{1}{2}\right) = \frac{1}{8}$
(h) $(10)^3 = (10)(10)(10) = 1000$
(i) $(.1)^3 = (.1)(.1)(.1) = .001$
(j) $\left(-\frac{3}{4}\right)^2 = \left(-\frac{3}{4}\right)\left(-\frac{3}{4}\right) = \frac{9}{16}$
(k) $\left(-\frac{4}{5}\right)^3 = \left(-\frac{4}{5}\right)\left(-\frac{4}{5}\right)\left(-\frac{4}{5}\right) = -\frac{64}{125}$

EXERCISE 1.9

Evaluate the following:

1. 2^4 2. 3^2 3. $(-2)^2$ 4. $(-2)^3$
5. 3^4 6. 5^4 7. $(-2)^6$ 8. $(-3)^4$
9. 2^8 10. 10^2 11. 10^3 12. 10^4
13. 1^4 14. $(-1)^4$ 15. $(-10)^3$ 16. $(\frac{3}{4})^2$
17. $(\frac{1}{2})^2$ 18. $(.4)^2$ 19. $(.03)^2$ 20. $(-.1)^5$
21. $(-\frac{2}{3})^3$ 22. $(-\frac{3}{4})^3$ 23. $(-1)^{50}$ 24. $(-1)^{51}$
25. From exercises 10, 11, and 12, what general rule can be established for powers of 10?

Answers to odd-numbered exercises on page 455.

1.10 ROOTS OF SIGNED NUMBERS

Subtraction is called the *inverse* operation of addition. Division is considered to be the inverse of multiplication. There is also an inverse operation of raising a number to a power, and this is called *finding the root of a number*.

The inverse of squaring is finding the square root. For example, to find the square root of 9, which is written $\sqrt{9}$, we must find a number which when squared gives us 9.

$$\text{Since } 3^2 = 9$$
$$\text{we have } \sqrt{9} = 3$$

In a similar fashion we have

$\sqrt{16} = 4$ because $4^2 = 16$

$\sqrt{4} = 2$ because $2^2 = 4$

$\sqrt{49} = 7$ because $7^2 = 49$

and so on.

Example 1: Find the following square roots:

(a) $\sqrt{36} = 6$ because $6^2 = 36$

(b) $\sqrt{1} = 1$ because $1^2 = 1$

(c) $\sqrt{0} = 0$ because $0^2 = 0$

(d) $\sqrt{121} = 11$ because $11^2 = 121$

You may have noticed that not only does $3^2 = 9$ but also $(-3)^2 = 9$. However, whenever we use the symbol $\sqrt{}$ it is understood to mean the positive square root, which is called the **principal square root.**

$\sqrt{9} = 3$ 3 equals the principal square root of 9

$-\sqrt{9} = -3$ since $\sqrt{9} = 3$, $-\sqrt{9} = -3$

We may be asked to find roots other than square roots. $\sqrt[3]{}$, $\sqrt[4]{}$, $\sqrt[5]{}$ means the third root (cube root), fourth root, fifth root, and so on.

Example 2: Find the following roots:

(a) $\sqrt[3]{8} = 2$ Read "the cube root of 8 is 2" (because $2^3 = 8$).

(b) $\sqrt[4]{81} = 3$ Read "the fourth root of 81 is 3" (because $3^4 = 81$).

(c) $\sqrt[5]{32} = 2$ Read "the fifth root of 32 is 2" (because $2^5 = 32$).

(d) $\sqrt[3]{-8} = -2$ Read "the cube root of -8 is -2" (because $(-2)^3 = -8$).

(e) $\sqrt[5]{-1} = -1$

(f) $-\sqrt[3]{8} = -2$

(g) $-\sqrt[3]{-8} = -(-2) = 2$

EXERCISE 1.10

Find the roots:

1. $\sqrt{25}$
2. $\sqrt{81}$
3. $-\sqrt{36}$
4. $\sqrt{144}$
5. $\sqrt{49}$
6. $\sqrt[3]{27}$
7. $\sqrt[3]{-27}$
8. $-\sqrt[3]{27}$
9. $-\sqrt[3]{-27}$
10. $\sqrt[3]{125}$
11. $\sqrt[6]{1}$
12. $\sqrt[7]{-1}$
13. $\sqrt{400}$
14. $\sqrt[3]{64}$
15. $\sqrt[3]{216}$
16. $\sqrt[3]{-1000}$
17. $\sqrt[6]{64}$
18. $\sqrt[3]{343}$

Answers to odd-numbered exercises on page 456.

1.11 ORDER OF OPERATIONS

If we are given an expression containing more than one operation, we must be careful to perform the operations in the proper order. The following convention is used:

Order of operations:

1. Evaluate the expression inside any parentheses ( ), brackets [ ], or braces { }.
2. Powers and roots are performed.
3. Multiplications and divisions are done as they appear, from left to right.
4. Additions and subtractions are done as they appear, from left to right.

Example 1: Evaluate $27 \div (12 - 3) \cdot 2 + \sqrt{36}$

Solution:

$27 \div (12 - 3) \cdot 2 + \sqrt{36}$	Evaluate inside the parentheses.
$= 27 \div 9 \cdot 2 + \sqrt{36}$	Take the square root.
$= 27 \div 9 \cdot 2 + 6$	Divide and multiply from
$= 3 \cdot 2 + 6$	left to right.
$= 6 + 6$	Add.
$= 12$	

Example 2: Evaluate $\sqrt{49} - 6 + 3^2$

Solution:

$\sqrt{49} - 6 + 3^2$	Do root and power.
$= 7 - 6 + 9$	Subtract and add
$= 1 + 9$	from left to right.
$= 10$	

Example 3: Evaluate $\sqrt{16} - 4(\sqrt{25} - 3) \div 4$

Solution:

$\sqrt{16} - 4(\sqrt{25} - 3) \div 4$	Evaluate the expression inside
$= \sqrt{16} - 4(5 - 3) \div 4$	the parentheses.
$= \sqrt{16} - 4(2) \div 4$	Take the root.
$= 4 - 4(2) \div 4$	Do multiplication and
$= 4 - 8 \div 4$	division from left to right.
$= 4 - 2$	Subtract.
$= 2$	

Example 4: Evaluate $3 \cdot 3^3 - 4\sqrt{64} + 3 \cdot 5$

Solution:

$3 \cdot 3^3 - 4\sqrt{64} + 3 \cdot 5$	Do the powers and roots.
$= \underbrace{3 \cdot 27} - \underbrace{4 \cdot 8} + \underbrace{3 \cdot 5}$	Do multiplications.
$= 81 - 32 + 15$	Subtract.
$= 49 + 15$	Add.
$= 64$	

Example 5: Evalute $9 - (3 \cdot 5^2 + 1) - \sqrt[3]{8}$

Solution:

$9 - (3 \cdot 5^2 + 1) - \sqrt[3]{8}$	Do the power inside the parentheses.
$= 9 - (3 \cdot 25 + 1) - \sqrt[3]{8}$	Multiply.
$= 9 - (75 + 1) - \sqrt[3]{8}$	Add.
$= 9 - 76 - \sqrt[3]{8}$	Take the cube root.
$= 9 - 76 - 2$	Subtract from left to right.
$= -67 - 2$	
$= -69$	

EXERCISE 1.11

Evaluate using the correct order of operations:

1. $6 \div 2 + 4$
2. $9 \div 3 + 6$
3. $6 \div (2 + 4)$
4. $9 \div (3 + 6)$
5. $7 + 3 \cdot 4$
6. $4 - 3^2 + 6 \cdot 5$
7. $\sqrt{36} - 4^2 + 5 - 2$
8. $4 - (\sqrt[3]{8} + 1) \cdot 3$
9. $3\sqrt{9} - 5 + 2^4 \div 8$
10. $3 + 44 \div 11 - 6$
11. $(5^2)\sqrt{4} \div 10$
12. $(6 - 14) - (\sqrt{36} - 5)$
13. $3^3 - 14 \div 2 + 3 \cdot 6$
14. $(3^3 - 14) \div 2 + 3 \cdot 6$
15. $(-2)^3 + \sqrt[3]{-8} \div 2$
16. $-\sqrt{81} \div 9 + 4^2 - (2 + 3^2)$
17. $-\sqrt{16} \cdot 4 - 8 \cdot 0$
18. $(3 \cdot 4^2 - 8) \div (-10)$
19. $(2\sqrt{121} + 3) \cdot 4$
20. $-\sqrt{196} \cdot 0 - 1$

21. $6 - 3^4 + (\sqrt{49} + 9) \div 4$ 22. $\sqrt[3]{125} - 2(4^2 - 6 \div 3)$

$= 6 - 81 + \frac{(7+9)}{4} = -75 + \frac{16}{4} = -75 + 4 = -71$ ✓

Answers to odd-numbered exercises on page 456.

1.12 GROUPING SYMBOLS

In the last section we introduced these grouping symbols:

( ) parentheses

[ ] brackets

{ } braces

These symbols provide a means for changing the normal order of operations. For example, in the expression $6 - 4 \cdot 5$, using the correct order of operations, we multiply first and then subtract.

$$6 - 4 \cdot 5 = 6 - 20 = -14$$

If it were our intention to subtract first and then multiply we would use parentheses to do so:

$$(6 - 4) \cdot 5 = 2 \cdot 5 = 10$$

Grouping symbols can also be used in more complicated expressions. The procedure when grouping symbols occur within other grouping symbols is to evaluate the innermost grouping first. The key phrase to remember is "inside out."

Example 1: Evaluate $14 - [7 - (9 - 4)]$

Solution:

$14 - [7 - (9 - 4)]$	Evaluate the inner parentheses.
$= 14 - [7 - 5]$	Evaluate the brackets.
$= 14 - 2$	Subtract.
$= 12$	

Example 2: Evaluate $7 + 2\{3 - [11 - (3 + 4)]\}$

Solution:

$7 + 2\{3 - [11 - (3 + 4)]\}$	Evaluate $(3 + 4)$.
$= 7 + 2\{3 - [11 - 7]\}$	Evaluate $[11 - 7]$.
$= 7 + 2\{3 - 4\}$	Evaluate $\{3 - 4\}$.
$= 7 + 2(-1)$	Multiply.
$= 7 - 2$	Subtract.
$= 5$	

Example 3: Evaluate $6 - [3 - (4-8)]$

Solution:

$6 - [3 - (4-8)]$	Evaluate (4 − 8).
$= 6 - [3 - (-4)]$	Replace −(−4) by +4.
$= 6 - [3 + 4]$	Evaluate [3 + 4].
$= 6 - 7$	Evaluate 6 − 7.
$= -1$	

Example 4: Compare (a) $3 - 6 \cdot 2$ and (b) $(3 - 6) \cdot 2$

Solution:

(a) $3 - 6 \cdot 2$	Multiply first.
$= 3 - 12$	
$= -9$	
(b) $(3 - 6) \cdot 2$	Evaluate the parentheses first.
$= -3 \cdot 2$	
$= -6$	

EXERCISE 1.12

Simplify:

1. $6(2-7)$

   = 12−42 = −30 ✓

2. $-4(3 - 12)$

3. $19 - [7 + (4-7)]$

   = 19 − [7 + 4 − 7] =
   = 19 − 7 − 4 + 7 =
   = 26 − 11 = 15 ✓

4. $6 + [2 - (5-8)]$

5. $[6 + (2-19)] - 12$

   = [6 + 2 − 19] − 12
   = 8 − 19 − 12 = 8 − 31
   = −23

6. $3 + 5[4 - (6-2)]$

7. $3[4 - 2(3-8)]$

   3[4 − 6 + 16] =
   = 12 − 18 + 48
   = 60 − 18 = 42 ✓

8. $7 - \{4 + [2 + (3-4)]\}$

9. $\{3 - [6 + (8-9)]\} - 5$

   = {3 − [6 + 8 − 9]} − 5
   = {3 − 6 − 8 + 9} − 5
   = 3 − 6 − 8 + 9 − 5
   = 12 − 19 = −7 ✓

10. $2 + [5 - (\sqrt{16} + 3)]$

11. $23 - \{6 - [4 - 3(7-5)]\}$

12. $7 - [4(3)^2 - (7+4)]$

13. $32 - 3\{4 - 7[6-(5-8)]\}$

14. $17 + 4\{5-6[4-3(7-5)]\}$

Answers to odd-numbered exercises on page 456.

1.13 Chapter Summary

Examples

(1.1) Integers or signed numbers are the positive and negative counting numbers.

(1.1) Number line: $-5\ \ -4\ \ -3\ \ -2\ \ -1\ \ 0\ \ +1\ \ +2\ \ +3\ \ +4\ \ +5$

(1.2) The absolute value of a signed number is the number of units that it is away from zero on the number line. It is written $|a|$.

(1.2) $|-6| = 6$
$|+4| = 4$

(1.3) The symbol $>$ is read "is greater than."
The symbol $<$ is read "is less than."

(1.3) $6 > 4$
$-6 < 5$

(1.4) To add signed numbers:
1. If both numbers have the same sign, add the absolute values and keep the common sign for the answer.
2. If the numbers have different signs, find the difference in their absolute values and take the sign of the number with the larger absolute value.

(1.4) $6 + 4 = 10$
$-6 + (-4) = -10$
$-6 + 4 = -2$
$6 + (-4) = 2$

(1.5) To subtract one signed number from another, add its negative.

(1.5) $5 - 9 = 5 + (-9) = -4$
$-4 - (-6) = -4 + 6 = 2$

(1.5) To add or subtract more than two signed numbers:
1. Change all subtractions to additions by adding the negatives.
2. Add all the positive values.
3. Add all the negative values.
4. Add the results from steps 2 and 3.

(1.5) $-4 + 5 - (-7) - 3 + 4 - 1$
$= -4 + 5 + 7 + (-3) + 4 + (-1)$

Add positive values

$5 + 7 + 4 = 16$

Add negative values

$-4 + (-3) + (-1) = -8$

Combine results from above

$16 + (-8) = 8$

(1.6) To multiply two signed numbers:
1. Multiply their absolute values.
2. The product is positive if their signs are alike.
The product is negative if their signs are different.

(1.6) $(-7)(-4) = 28$
$(7)(4) = 28$
$(-7)(4) = -28$
$(7)(-4) = -28$

(1.6) To multiply more than two signed numbers:
1. Multiply the absolute values of the factors.
2. If there are an even number of negative signs, the product is positive. If there are an odd number of negative signs, the product is negative.

(1.6) $(-5)(-1)(3)(1)(2)(-2)$
$= -60$

3 negative signs is an odd number.

$(-2)(-3)(2)(-1)(2)(-3)$
$= 72$

4 negative signs is an even number.

(1.7) To divide one signed number by another:

1. Divide their absolute values.
2. The quotient is positive if their signs are alike. The quotient is negative if their signs are different.

(1.7) $\frac{-36}{-9} = 4$

$\frac{14}{-7} = -2$

$\frac{-18}{6} = -3$

(1.7) Division by zero is undefined.

(1.7) $\frac{0}{-6} = 0$

$\frac{-3}{0}$ is undefined

$\frac{0}{0}$ is undefined

(1.8) Properties of signed numbers:

Commutative property of addition: If a and b represent any signed numbers, then $a + b = b + a$.

Commutative property of multiplication: If a and b represent any signed numbers, then $a \cdot b = b \cdot a$.

Associative property of addition: If a, b, and c are any signed numbers, then $(a+b)+c = a+(b+c)$.

(1.8) $-4 + 7 = 7 + (-4) = 3$

$(-4)(6) = (6)(-4) = -24$

$(-6 + 3) + 2 = -6 + (3 + 2)$
$-3 + 2 = -6 + 5$
$-1 = -1$

Associative property of multiplication: If a, b and c are any signed numbers, then $(a \cdot b) \cdot c = a \cdot (b \cdot c)$.

$(-6 \cdot 3) \cdot 2 = -6 \cdot (3 \cdot 2)$
$-18 \cdot 2 = -6 \cdot 6$
$-36 = -36$

Distributive property of multiplication over addition: If a, b and c are any signed numbers, then $a \cdot (b + c) = a \cdot b + a \cdot c$.

$-2(3+5) = (-2)(3) + (-2)(5)$
$-2 \cdot 8 = -6 + (-10)$
$-16 = -16$

(1.9) Powers of numbers indicate repeated multiplication.

(1.9)

$\underbrace{a \cdot a \cdot a \cdot a \cdot a}_{\text{5 factors}} = a^5$ (5: exponent or power; a: base)

(1.10) Finding a root of a number is the inverse operation of finding the power.
The symbol $\sqrt{\ }$ means the positive square root, called the principal square root.

(1.10) $\sqrt{16} = 4$

$-\sqrt{16} = -4$

$\sqrt[3]{64} = 4$

(1.11) Order of operations:

1. Evaluate the expression inside any parentheses ( ), brackets [ ], or braces { }.
2. Power and roots are performed.

(1.11) $3 + 2(6 + 4) \div (-5)$
$= 3 + 2 \cdot 10 \div (-5)$
$= 3 + 20 \div (-5)$

3. Multiplications and divisions are done as they appear, from left to right.
4. Additions and subtractions are done as they appear, from left to right.

$= 3 + (-4)$

$= -1$

(1.12) When grouping symbols occur within other grouping symbols, evaluate the innermost grouping first. Remember "inside out."

(1.12) $13 - [4 - 2(3 - 6)]$

$= 13 - [4 - 2(-3)]$

$= 13 - [4 + 6]$

$= 13 - 10$

$= 3$

Exercise 1.13 Chapter Review

(1.1) Locate on the number line.

1. -4 2. $+3$ 3. -1

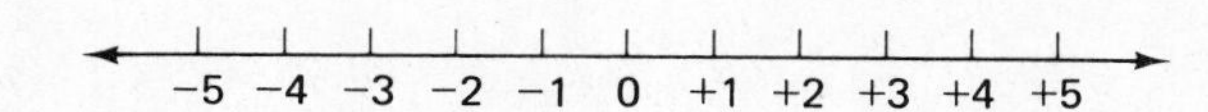

(1.1) Represent each quantity by a signed number.

4. The temperature is six degrees below zero.
5. The stock market goes up eight points.
6. You get a raise of $30 per week.

(1.2) Find the absolute value of each of the following:

7. -4 8. $+5$ 9. -12 10. 0

(1.3) Insert $>$ or $<$ between each pair of numbers to make a true statement.

11. $-3 \quad 5$ 12. $14 \quad -14$ 13. $3 \quad -6$ 14. $-2 \quad -12$

(1.4-1.7) Calculate:

15. $4 - 8$
16. $(-6)(-3)$
17. $(-8) \div (-4)$
18. $3 - (-3)$
19. $-6 - (-5)$
20. $-6 - 8$
21. $-21 \div 7$
22. $(4.6)(-1.12)$
23. $0 - (-9)$
24. $12.8 - 15.6$
25. $-8 \div 0$
26. $-3\frac{5}{8} + 2\frac{3}{8}$
27. $-4 + 6 - 8 - 2 + 1 - 4 + 1$

28. $7 - (-5) + 2 - (-8) + 1 - 2 - 4$

29. $3.81 - (-1.19) - 2.08 - 4.02 - .07 + 1.11$

30. $(-1)(4)(-1)(-1)(2)(1)(-2)(-2)$
31. $(-2)(-3)(1)(-1)(-2)(-2)(1)(-3)$

(1.8) 32. Does changing the grouping in a subtraction problem change the result?

(1.8) 33. Does $(a \div b) \div c = a \div (b \div c)$?

(1.8) 34. Does changing the order in a multiplication problem affect the result?

(1.9-1.10) Evaluate:

35. 3^3 36. 5^2 37. $(-3)^4$

38. 10^3 39. 10^6 40. $(-1)^{30}$

41. $\sqrt{36}$ 42. $\sqrt{49}$ 43. $\sqrt[3]{8}$

44. $-\sqrt{100}$ 45. $\sqrt[3]{1000}$ 46. $\sqrt[3]{-1000}$

(1.11) Evaluate using the correct order of operations:

47. $4 - 3 \times 5$ 48. $6 - 27 \div 3 \cdot 3 + 1$

49. $-\sqrt{25} \cdot 3 + 6 + 3$ 50. $3 - (\sqrt[3]{8} \div 2) - 2 \cdot 1$

(1.12) Evaluate each of the following:

51. $13 - [5 - (6 - 10)]$ 52. $11 + 2(-6 - 1)$

53. $[4 - (2 + 3)] - 8$ 54. $12 - \{4 - [3 + 2(1 - 4)]\}$

Answers to odd-numbered exercises on page 456.

NO!

Chapter 1 Achievement Test

Name ______________________

Class ______________________

This test should be taken before you are tested in class on the material in Chapter 1. Solutions to each problem and the section where the type of problem is found are given on page 457.

Draw a number line and locate the following points.

1. -4
2. 0
3. $+5$

Represent each quantity by a signed number.

4. You overdraw your checking account by \$24. — 4. ____________
5. You receive a bonus of \$50. — 5. ____________

Find the absolute value of each of the following:

6. 0 — 6. ____________
7. -8 — 7. ____________
8. $+13$ — 8. ____________

Insert $>$ or $<$ between each pair of numbers to make a true statement.

9. $-8 \quad -4$ — 9. ____________
10. $6 \quad -5$ — 10. ____________
11. $-4 \quad 0$ — 11. ____________

Evaluate each of the following:

12. $4 - 5$ — 12. ____________
13. $-4 - (-5)$ — 13. ____________
14. $-4 + 5$ — 14. ____________
15. $(-4)(-5)$ — 15. ____________
16. $(-16) \div (-4)$ — 16. ____________

17. $-7 \div 0$

18. $0 \div (-7)$

19. $-3 + 4 - 8 - 2 + 4 + 1 - 5$

20. $-2 + 1 - (-3) - 5 - 2 - (-1)$

21. $(-2)(-1)(2)(-1)(-3)(-2)(2)$

22. $4.19 - 7.35$

23. $-2\frac{1}{5} + 7\frac{3}{5}$

24. 4^3

25. $(-2)^4$

26. $-\sqrt{64}$

27. 10^5

28. $\sqrt[3]{27}$

Evaluate using the correct order of operations:

29. $3 + 4 \cdot 7$

30. $3 - 2^2 + 5 \cdot 2$

31. $2 - 3\,[3-(2+3)]$

32. $4 - \{2\,[4-(3-5)]\}$

17. ____________

18. ____________

19. ____________

20. ____________

21. ____________

22. ____________

23. ____________

24. ____________

25. ____________

26. ____________

27. ____________

28. ____________

29. ____________

30. ____________

31. ____________

32. ____________

2

Algebraic Expressions

2.1 FORMULAS AND THE SUBSTITUTION PRINCIPLE

In almost every field of study, relationships between quantities are expressed using formulas. A formula is a brief, easy-to-use rule for finding the value of an unknown quantity when values of other quantities are known. Formulas are convenient, and they are easier to remember than the English language statement of the rule, as illustrated by the following examples.

Example 1: The area of a triangle is equal to one-half the product of its base and its height. This rather complicated statement translates into the much shorter algebraic formula:

$$A = \frac{1}{2}bh$$

The letters in the formula are usually chosen as the first letter in the word they represent. Here A = area, b = base and h = height.

Example 2: The area of a circle is equal to π (pi) times the square of the radius of the circle. π is a constant approximately equal to 3.14.

The formula that represents this statement is:

$$A = \pi r^2$$

where A = area, π = approximately 3.14, r = radius

Example 3: The amount of interest on a loan is given by the product of the principal, the yearly rate of interest, and the time in years.

The corresponding formula is:

$$I = prt$$

where I = interest, p = principle, r = rate, t = time

In algebraic expressions both **variables** and **constants** are used. Variables are quantities that can assume different values, while constants are quantities that do not change.

In the formula for the area of a circle:

$$A = \pi r^2$$

A and r are variables since their values change from one problem to the next. π is a constant equal to approximately 3.14, and it never changes.

Evaluating Formulas

Formulas are evaluated by substituting numbers for the letters:

To evaluate a formula:

1. Write the formula from the given English statement (if necessary).
2. Replace each letter by its numerical value.
3. Do all arithmetic operations using the correct order of operations.

Example 4: The area (A) of a triangle is given by the formula:

$$A = \frac{1}{2}bh$$

where b = base and h = height.

Find A if b = 16 and h = 6.

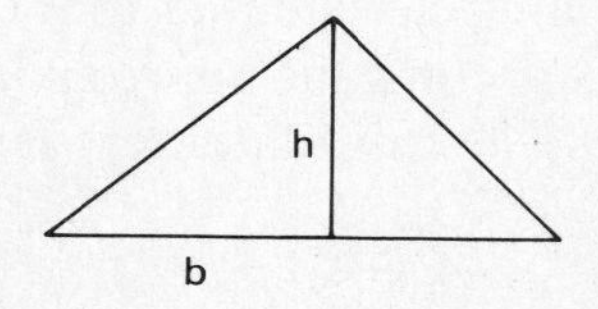

Solution:

$A = \frac{1}{2}bh$ — $b = 16, h = 6$

$A = \frac{1}{2}(\boxed{16})(\boxed{6})$ — Substitute 16 for b and 6 for h.

$= \left(\frac{1}{\cancel{2}_1}\right)(\cancel{16}^{8})(6)$ — Multiply.

$= 48$ square units — (Area is always given in square units.)

Example 5: Celsius (C) and Fahrenheit (F) temperatures are related by the formula:

$$C = \frac{5}{9}(F - 32)$$

If the boiling point of water is 212° Fahrenheit, what is it in Celsius?

Solution:

$C = \frac{5}{9}(F - 32)$ — $F = 212$ — $F = \frac{9}{5}C + 32$ ✓

$= \frac{5}{9}(\boxed{212} - 32)$ — Substitute for F.

$= \frac{5}{9}(180)$ — Work in parentheses first.

$= \frac{5}{\cancel{9}_1}(\cancel{180}^{20})$ — Cancel and multiply.

$= 100$

Therefore 212°F = 100°C. Be sure to use corrrect units.

Example 6: The total resistance (R) of two resistances, (a) and (b), connected in parallel is given by their product divided by their sum. Find the total resistance (R) if $a = 100$ ohms and $b = 400$ ohms (an ohm is a unit of electrical resistance).

Solution:

$R = \frac{a \cdot b}{a + b}$ — Write the formula from the English statement.

$R = \frac{\boxed{100} \cdot \boxed{400}}{\boxed{100} + \boxed{400}}$ — Substitute $a = 100$ and $b = 400$.

$= \frac{40{,}000}{500}$ — Evaluate the product and sum.

$= 80$ ohms — Divide.

Example 7: A bowler's handicap is calculated using the formula $H = .8(200 - A)$ where H is the handicap and A is the bowler's three game average. (a) What is your handicap if your three-game average for the night is 135? (b) Your final score is equal to your actual score added to your handicap. Find your final score if you bowl a game of 142.

Solution:

(a) $H = .8(200 - A)$	$A = 135$
$= .8(200 - (\boxed{135}))$	Substitute for A.
$= .8(65)$	Subtract inside parentheses.
$= 52$	Multiply.
(b) Final score = actual score + handicap	
$= 142 + 52$	Substitute.
$= 194$	Add.

Example 8: You may have heard a weather reporter refer to a number called the *THI* (temperature-humidity index). The *THI* tells you how comfortable or uncomfortable you will be under any specific combination of temperature and humidity conditions. The formula is given as:

$$THI = t - (.55)(1 - h)(t - 58)$$

where t = temperature (in degrees F) and h = relative humidity (expressed as a decimal)

Meteorologists tell us that:

1. At a *THI* of 70 or below, most people are comfortable.
2. At a *THI* of 75, half are comfortable and half are uncomfortable.
3. At a *THI* of 79 or above, almost everyone will be uncomfortable.

Find the *THI* when the temperature is 85°F and the relative humidity is .7:

Solution:

$THI = t - (.55)(1 - h)(t - 58)$	$t = 85, h = .7$
$= \boxed{85} - (.55)(1 - \boxed{.7})(\boxed{85} - 58)$	Substitute for t and h.
$= 85 - (.55)(.3)(27)$	Work inside parentheses.
$= 85 - 4.455$	Multiply.
$THI = 80.545$	Subtract.

Since the *THI* is greater than 79, most people would be uncomfortable.

EXERCISE 2.1

In exercises 1 through 10, evaluate each formula using the values of the letters given. Use a value of $\pi = 3.14$. Some of these formulas should be familiar to you.

1. $A = \ell w$ (area of a rectangle), $\ell = 12$ feet, $w = 9$ feet

   $A = 12\,ft \times 9\,ft = 108\,ft^2$

2. $P = 2\ell + 2w$ (perimeter of a rectangle), $\ell = 12$ feet, $w = 9$ feet

3. $I = prt$ (interest), $p = \$5000$, $r = .15$, $t = 2$

   $I = 5000 \times 0{,}15 \times 2 = 5000 \times 0{,}30 = 1500.0 = 1500$

4. $P = S - C - H$ (net profit), $S = \$40$ (selling price), $C = \$20$ (cost), $H = \$5$ (overhead)

5. $d = rt$ (distance), $r = 55$ miles per hour (rate), $t = 3.2$ hours (time)

   $d = 55\,M/h \times 3.2\,h = 176\,M$

6. $C = \frac{5}{9}(F - 32)$ (temperature), $F = 68$

7. $A = \frac{1}{2}h\,(b_1 + b_2)$ (area of a trapezoid), $h = 6$ (height), $b_1 = 7$ (one base), $b_2 = 9$ (other base).

   $A = \frac{1}{2} \cdot 6(7+9) = 3(7+9) = 3 \times 16 = 48$

8. $F = \frac{g m_1 m_2}{d^2}$ (gravitational force), $g = 32$, $m_1 = 180$, $m_2 = 70$, $d = 10$

9. $A = \pi r^2$ (area of a circle), $r = 4$ meters (radius)

   $A = 3.1416 \times 4^2 = 3.1416 \times 16 = 50.2656$

10. $C = \pi d$ (circumference of a circle), $d = 4.6$ cm (diameter)

In exercises 11 through 25 evaluate each formula using the values of the letters as given. Use $\pi = 3.14$ as an approximate value.

11. $I = \dfrac{E}{R}$ $\qquad E = 114,\ R = 6$

    $I = \dfrac{114}{6} = 19$

12. $S = \frac{1}{2}gt$ $\quad g = 32,\ t = 10$

13. $S = \frac{1}{2}gt$ $\quad g = 32,\ t = 20$

$S = \frac{1}{2} \cdot 32 \cdot 20 = 320$

14. $R = \frac{ab}{a+b}$ $\quad a = 10,\ b = 20$

15. $R = \frac{ab}{a+b}$ $\quad a = 20,\ b = 80$

$R = \frac{20 \cdot 80}{20+80} = \frac{1600}{100} = 16$

16. $V = \frac{4}{3}\pi r^3$ $\quad r = 3$

17. $P = i^2 r$ $\quad i = 3,\ r = 100$

$P = 3^2 \cdot 100 = 9 \times 100 = 900$

18. $A = P(1 + rt)$ $\quad P = 1000,\ r = .2,\ t = 2$

19. $T = \pi\sqrt{\frac{L}{g}}$ $\quad L = 128,\ g = 32$

$T = 3.14 \times \sqrt{\frac{128}{32}} = 3.14 \times \sqrt{4} = 3.14 \times 2 = 6.28$

20. $V = \sqrt{2aS}$ $\quad a = 9,\ S = 18$

21. $d = \sqrt{x^2 + y^2}$ $\quad x = 3,\ y = 4$

$d = \sqrt{3^2 + 4^2} = \sqrt{9+16} = \sqrt{25} = 5$

22. $V = \ell wh$ $\quad \ell = 2.1,\ w = 4.6,\ h = 7$

23. $V = \pi r^2 h$ $\quad r = 5,\ h = 12$

$V = 3.14 \times 5^2 \times 12$ $3.14 \times 25 \times 12 = 942$

24. $S = \frac{a}{1-r}$ $\quad a = -24,\ r = \frac{1}{2}$

25. $S = \frac{a(1-r^n)}{1-r}$ $\quad a = -9,\ r = \frac{1}{3},\ n = 3$

$S = \frac{-9\left[1-\left(\frac{1}{3}\right)^3\right]}{1-\frac{1}{3}} = \frac{-9\left[1-\frac{1}{27}\right]}{\frac{3-1}{3}} = \frac{-9\left[\frac{27-1}{27}\right]}{\frac{2}{3}} = \frac{-9\left[\frac{26}{27}\right]}{\frac{2}{3}} = \frac{-\frac{26}{3}}{\frac{2}{3}} = -\frac{26}{2} = -13$

26. Calculate the temperature-humidity index using the formula $THI = t - (.55)(1 - h)(t - 58)$ when:

    (a) $t = 68°$, $h = .4$

    (b) $t = 90°$, $h = .22$

    (c) $t = 78°$, $h = .95$

27. When a consumer pays off a loan before it is due, part of the interest is refunded using a formula called the *rule of 78*, given by:

$$u = \frac{f \cdot K(K + 1)}{n(n + 1)}$$

How much interest (u) is refunded when:

(a) A 36-payment loan ($n = 36$) is paid off 12 months early ($K = 12$) and the total finance change is \$180 ($f = \$180$)?

(b) The total finance charge is \$158 on a 12-payment loan that is paid off in 10 months? (There are 2 payments left, so $K = 2$.)

(c) A 48-month loan is paid off in 2 years and the total finance charge is \$710? The borrower paid off the loan in half the time—does he get back half of the total finance charge?

28. The volume of a circular post is given by the formula $V = \frac{\pi d^2 L}{4}$ where d = the diameter of the post and L is its length. Using a value $\pi = 3.14$, find the volume of a post with a diameter of 4 inches and a length of 10 feet.

29. The number of stars in each of the triangles at the right is given by the formula $T = \frac{1}{2}n(n + 1)$ where n equals the number of rows in the triangle.

(a) Verify that the formula works for each of the given triangles.

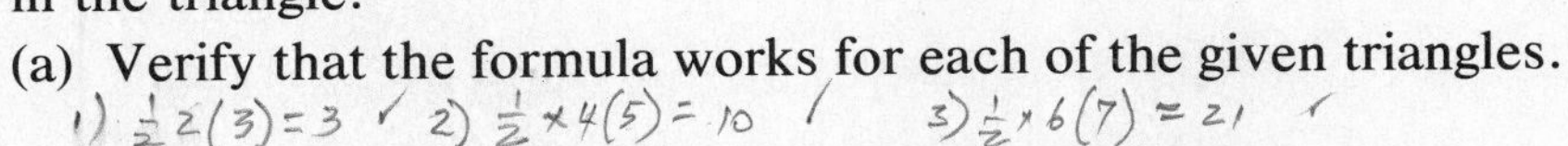

(b) How many stars are there in a triangle with 10 rows?

(c) How many stars are there in a triangle with 100 rows?

Answers to odd-numbered exercises on page 457.

2.2 EVALUATING ALGEBRAIC EXPRESSIONS

Algebraic expressions contain numbers (called constants) and letters (called variables) that represent numbers. In the algebraic expression $-6x^2y + 5z$, -6 and 5 are constants and x, y, and z are variables.

Constants

$$-6xy + 5z$$

Variables

We evaluate algebraic expressions by substituting numerical values for these variables in the same way that we did when we evaluted formulas in the previous section.

> **To evaluate algebraic expressions:**
>
> 1. Replace each variable (letter) by its given numerical value.
> 2. Perform all the arithmetic operations using the correct order of operations.

Example 1: Evaluate $4x^2 - 3y$ if $x = 2$ and $y = 4$.

Solution:

$4x^2 - 3y$	$x = 2$ and $y = 4$.
$= 4 \cdot 2^2 - 3 \cdot 4$	Replace x by 2 and y by 4.
$= 4 \cdot 4 - 3 \cdot 4$	Perform powers first.
$= 16 - 12$	Multiply.
$= 4$	Subtract.

Example 2: Find the value of $-3xy + x^2$ when $x = -2$ and $y = -5$.

Solution:

$-3xy + x^2$	$x = -2, y = -5$
$= -3(-2)(-5) + (-2)^2$	Replace x and y.
$= -3(-2)(-5) + 4$	Do powers first.
$= -30 + 4$	Multiply.
$= -26$	Add.

Example 3: What is the value of $3(x - 2y) - 5z$ if $x = 4$, $y = -7$ and $z = 3$?

Solution:

$3(x - 2y) - 5z$	
$= 3[4 - 2(-7)] - 5(3)$	Replace *x, y* and *z*.
$= 3(4 + 14) - 5(3)$	Multiply inside brackets.
$= 3(18) - 5(3)$	Add inside parentheses.
$= 54 - 15$	Multiply.
$= 39$	Subtract.

Be careful evaluating expressions like -4^2 and $(-4)^2$.

$-4^2 = -16$	The exponent 2 applies only to the 4.
$(-4)^2 = 16$	The exponent applies to -4.

Example 4: Evaluate the following:

(a) x^2 when $x = -3$
(b) $-x^2$ when $x = -3$
(c) $(-x)^2$ when $x = -3$

Solution:

(a) $x^2 = (-3)^2 = 9$
(b) $-x^2 = -(-3)^2 = -9$
(c) $(-x)^2 = [-(-3)]^2 = 3^2 = 9$

EXERCISE 2.2

In exercises 1 through 16 find the value of each expression when $x = 3$, $y = 7$, and $z = -2$.

1. $x + 3y$
$= 3+3(7) =$
$= 3+21 = 24$ ✓

2. $2x - 3y$

3. $x - y$
$3-7=-4$ ✓

4. $x + y + z$

5. $2x - 3y + 2z$
$=2(3)-3(7)+2(-2)$
$= 6-21-4$
$= 6-25 = -19$ ✓

6. $4x - 7y - z$

7. $7xy - 4z + 5y$
$= 7(3)(7)-4(-2)+5(7)$
$= 147+8+35 = 190$ ✓

8. $2(x - y)$

9. $3 - 4z - 2y + x$
$= 3-4(-2)-2(7)+3$
$= 3+8-14+3$
$= 14-14 = 0$ ✓

10. $2(x + z)$

11. $7 - 4(x + y) + 3z$
$=7-4(3+7)+3(-2)$
$= 7-4(10)-6$
$=7-40-6 = 7-46 = -39$ ✓

12. $6x - 4(2x - z)$

13. $7x^2 + 3y + 4z$ 14. $5x^2 + 2y^2 + 2z$ 15. $9x^2 - (y + 2z)^2$

16. $3xy + 2x^2z - 4y$

In exercises 17 through 30 find the value of each expression when $a = -2$, $b = 3$, $x = -4$, $y = 3$, and $z = -1$.

17. x^2 18. $-x^2$ 19. $(-x)^2$

20. $ax^2 - y^2$ 21. $x - y^2 + z$ 22. $2xy + ax$

23. $2abx + xyz$ 24. $\frac{ab}{xy}$ 25. $a - [b - (x - z)]$

26. $2x + 3\{x - 2[y + z]\}$ 27. $a^2 + 2bz - 3x^2y$

28. $\frac{4ax}{b - x}$ 29. $\frac{3x - 2y}{ab - 5z}$

30. $1 - (x - z)^2 + 2abx^2$

Answers to odd-numbered exercises on page 458.

2.3 ADDING AND SUBTRACTING LIKE TERMS

The addition signs (+) and subtraction signs (−) in an algebraic expression divide it up into parts called **terms.**

Expression	*Terms*
$3x^2y + 4xy$	$3x^2y$, $4xz$
$5x^2 + 2x - 1$	$5x^2$, $2x$, -1
$-17xy + 2y^3 + 5$	$-17xy$, $2y^3$, 5

The number in a term is called the **numerical coefficient,** and the variables are called the **literal parts.**

$-5x^3y$

numerical coefficient ↑ ↑ literal part

Two terms are called **like terms** if they differ *only* in their numerical coefficient. That is, like terms must have exactly the same literal parts (the same variables and the same exponents).

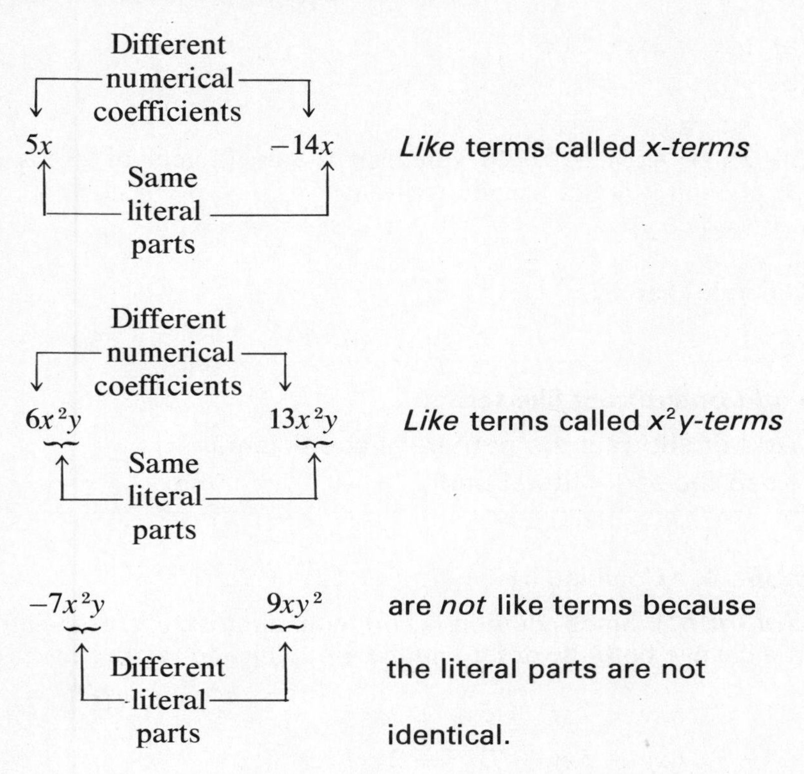

Combining like terms

The distributive property allows us to add or subtract like terms as follows:

Example 1: Combine $6x + 7x$

Solution:

$6x + 7x$

$= (6 + 7)x$ by the distributive property

$= 13x$ adding $6 + 7$

Example 2: Combine $7xy^2 + 9xy^2 + 2xy^2$

Solution:

$7xy^2 + 9xy^2 + 2xy^2$

$= (7 + 9 + 2)xy^2$ distributive property

$= 18xy^2$ adding inside the parentheses

Adding, in both arithmetic and algebra, is really *counting,* and we could say that *6 things + 7 things = 13 things,* where *things* refer to similar objects or quantities. This thinking leads us to the same conclusion:

$$6x + 7x = 13x$$

Example 3:

(a) $14x - 3x = 11x$
(b) $-12x^2y - 11x^2y = -23x^2y$
(c) $4y^2 + 8y^2 + 7y^2 = 19y^2$
(d) $x^2y + 4x^2y = 5x^2y$ (Note that the coefficient of x^2y is 1 even though it is not usually written.)
(e) $4xyz - xyz = 3xyz$

Our rule then is:

To add or subtract like terms:
1. Add or subtract the numerical coefficients.
2. Keep the same literal part.

Example 4: Combine $6xy + 7xy^2 + 5xy + 2xy^2$

Solution: Since addition is both commutative and associative, we can change both the order and the grouping and rewrite the problem.

$$\begin{aligned} & 6xy + 7xy^2 + 5xy + 2xy^2 \\ &= 6xy + 5xy + 7xy^2 + 2xy^2 \\ &= (6xy + 5xy) + (7xy^2 + 2xy^2) \\ &= 11xy + 9xy^2 \end{aligned}$$

Notice that $11xy$ and $9xy^2$ are *not like* terms and cannot be combined further.

Usually the problems are not rewritten and the answers are obtained directly:

Combine these like terms ($3x^3$ and $5x^3$)

$$3x^3 + 7xy + 5x^3 - 9xy = 8x^3 - 2xy$$

Combine these like terms ($7xy$ and $-9xy$)

Example 5:

(a) $-2x^2y + 13xy^2 + 5x^2y - 2xy^2 = 3x^2y + 11xy^2$
(b) $xyz + 4xy^4 - 3xyz - 7xy^4 = -2xyz - 3xy^4$
(c) $-2xy + 7y^3 + 3xy - 5y^3 + 3x^4 = xy + 2y^3 + 3x^4$

(d) $7ab^3 - 4abc - 7ab^3 + 2 = -4abc + 2$

(e) $-2x^2y + 5xy^2 + 3xy - 3$ There are no like terms in this expression so it cannot be simplified any further.

EXERCISE 2.3

In exercises 1 through 10 identify the terms, numerical coefficients, and literal parts of each expression.

1. $2x + 3y$
2. $5x^2y + 3xyz$
3. $-6x^3y + 2x^2y^3$
4. $2x^2 + 3x - 1$
5. $7x^3y - 14xy^3 + 2z - 5$
6. $r^2 + 2s - 3t + 7$
7. $-7xy^4 + 2x^2 - 3xyz$
8. $-4a^2 + 2b^2c + 3b$
9. $-11a^3 - 14b^3 + 17ac^4$
10. $-21x^2 + 6y^2 + 2xz - 3yz + 7$

In exercises 11 through 40 combine where possible.

11. $7x - 4x$
12. $3x + 7x$
13. $10a - 5a$
14. $7x^2 + 2x^2$
15. $6x + 2x + 9x$
16. $3a - 5a + 7a$
17. $2x + x$
18. $6a - a$
19. $3a^2b + 2a^2b - a^2b$
20. $6y + y - (-5y)$
21. $x^2y^2 - 3x^2y^2$
22. $6xyz - 9xyz + 3xyz$
23. $3x^3 - 5x^3 - 7x^3 + x^3$
24. $-16x^2y + 2x^2y - 3x + 5x$
25. $7x^3 + 2x + 3x^3 + 5x$
26. $3a + 2b - 5a + 7b$
27. $7mn - (-3mn) + 4mn$
28. $3a^2 + 5b^2 - 7c^2$
29. $3ab - 7b + 5ab - b + 3$
30. $8w + 5w - 6$

31. $-11cd - 5c^2d + 4cd^2$

32. $-4a - 5b + 7a - 6b + 7$

33. $-7 + 4x^2y - 8 + 5x^2y + 6$

34. $-14rs^2 + 2r^2s^3 - 3r^2s^3 + 2rs^2$

35. $3m + 6 - 4 + 5n -(-n) + 2m$

36. $x^3 - 3x^2 + 2x - x^3 + 3x^2 - 2x$

37. $5a - 6b + 4c + 3a - 5b - 7c$

38. $4c - 3d^2 - 7b - 5c + 3d^2 - 5b$

39. $6x^2y + 5xy^2 - 3xy - 5xyz$

40. $-4x^2y + 7xy - 8x^2y - 6 - 7xy + 5$

Answers to odd-numbered exercises on page 458.

2.4 FIRST RULE OF EXPONENTS

In section 1.9 we introduced the idea of **exponents** or **powers.** We use them whenever a number or variable is to be multiplied by itself.

$$\underbrace{a \cdot a \cdot a \cdot a \cdot a}_{\text{5 factors}} = a^5$$

(5 ← Exponent or power; a ← Base)

In the **exponential expression** a^5, a is the **base** and 5 is the **exponent** or **power.**

Example 1:

(a) $x \cdot x = x^2$ — Read "x squared."

(b) $6 \cdot 6 \cdot 6 = 6^3$ — Read "6 cubed."

(c) $(-2)(-2)(-2)(-2) = (-2)^4$ — Read "negative two to the fourth power."

(d) $7^1 = 7$ — The first power of any number is just the number itself.

$$a^n = \underbrace{a \cdot a \cdot a \cdot a \cdot \ldots \cdot a}_{\text{"}a\text{" used as a factor } n \text{ times}}$$

Example 2: Write without using exponents:

(a) $2^4 = 2 \cdot 2 \cdot 2 \cdot 2 = 16$

(b) $3x^2 = 3 \cdot x \cdot x$ — The exponent, 2, does not apply to the 3.

(c) $(3x)^2 = (3x)(3x)$ — Here the exponent applies to everything in the parentheses.

(d) $6^1 = 6$

(e) $1^6 = 1 \cdot 1 \cdot 1 \cdot 1 \cdot 1 \cdot 1 = 1$ — 1 raised to any power equals 1.

(f) $(-2)^4 = (-2)(-2)(-2)(-2) = 16$

(g) $-2^4 = -(2 \cdot 2 \cdot 2 \cdot 2) = -16$

The exponent applies only to the 2 unless there are parentheses.

(h) $\left(\frac{2}{3}\right)^3 = \frac{2}{3} \cdot \frac{2}{3} \cdot \frac{2}{3} = \frac{8}{27}$

Consider the product $a^3 \cdot a^4$.

Using our definition:

$$a^3 \cdot a^4 = \underbrace{(a \cdot a \cdot a)}_{3 \text{ factors}} + \underbrace{(a \cdot a \cdot a \cdot a)}_{4 \text{ factors}} = \underbrace{a \cdot a \cdot a \cdot a \cdot a \cdot a \cdot a}_{7 \text{ factors}} = a^7$$

So, we have $a^3 \cdot a^4 = a^{3+4} = a^7$. This is true in general, leading us to our first rule of exponents. If you are multiplying two quantities with the same base, keep that base and add the exponents. Written in symbols, we have:

First rule of exponents:

$x^a \cdot x^b = x^{a+b}$

Example 3:

(a) $h^2 \cdot h^3 = h^{2+3} = h^5$

(b) $2^2 \cdot 2^4 = 2^{2+4} = 2^6 = 64$

(c) $x^3 \cdot x = x^3 \cdot x^1 = x^{3+1} = x^4$ — Recall that $x = x^1$.

(d) $10^5 \cdot 10^4 = 10^{5+4} = 10^9 = 1{,}000{,}000{,}000$

(e) $x^4 \cdot y^5$ — These cannot be combined since x and y are different bases.

(f) $x^3 \cdot x^7 \cdot x^4 = x^{14}$ — The rule can be generalized to include more than two factors.

(g) $x^2 \cdot y^4 \cdot x^3 \cdot y^3 = x^5y^7$

EXERCISE 2.4

Simplify using the first rule of exponents:

1. $a^2 \cdot a^3$	2. $h^4 \cdot h^4$	3. $x^5 \cdot x^7$
4. $x^{14} \cdot x^6$	5. $y \cdot y^4$	6. $m^3 \cdot m^7$
7. $10^5 \cdot 10^2$	8. $h^3 \cdot h^9$	9. $3^2 \cdot 3^3$
10. $2^4 \cdot 2^3$	11. $x \cdot x^5$	12. $a^7 \cdot a^{25}$
13. $x^2 \cdot x^3 \cdot x^4$	14. $2^3 \cdot 2^2 \cdot 2$	15. $x^3 \cdot y^5$
16. $10 \cdot 10 \cdot 10$	17. $10^2 \cdot 10^3 \cdot 10$	18. $x \cdot x \cdot x^4$
19. $x^3 \cdot h^4 \cdot m^2$	20. $3 \cdot x^4 \cdot y^3$	21. $x^a \cdot x^b$
22. $a^x \cdot a^y$	23. $a^x \cdot a^2$	24. $x^a \cdot x^b \cdot x^c$
25. $x^a \cdot y^b$	26. $3^2 \cdot 4^2$	27. $x^a \cdot x^a$
28. $x^a \cdot y^a$	29. $x^a \cdot x^b \cdot x^{3a}$	30. $z^2 \cdot w^3 \cdot z^4 \cdot w^5$

Answers to odd-numbered exercises on page 459.

2.5 MULTIPLYING MONOMIALS

Algebraic expressions containing *one term* are called **monomials** (*mono* means one). We apply the commutative and associative properties of multiplication, which allows us to rearrange the factors to suit our purposes. We will multiply the numbers first and then multiply the variables in alphabetical order using the first rule of exponents that we just learned.

To multiply monomials:

1. Multiply the coefficients (numbers).
2. Multiply the variables (letters) in alphabetical order using the first rule of exponents.

Example 1: Multiply $(-3x^2)(4x^5)$

Solution:

$(-3x^2)(4x^5)$	Change the order of factors.
$= (-3 \cdot 4)(x^2 \cdot x^5)$	Multiply the numbers and use
$= -12x^7$	$x^a \cdot x^b = x^{a+b}$ to multiply the variables.

Example 2: Multiply $(-7x^2y)(-4x^3y^3)$

Solution:

$$(-7x^2y)(-4x^3y^3)$$

$= (-7 \cdot -4)(x^2 \cdot x^3)(y \cdot y^3)$ Rearrange the factors.

$= 28x^5y^4$ Multiply the numbers and like variables.

Usually the rearranging of terms is done mentally and the problem is not actually rewritten.

Example 3: Multiply the following:

(a) $(3x^2)(4x) = 12x^3$
(b) $(-5a^2b^3c)(-6a^2bc) = 30a^4b^4c^2$
(c) $(-2xy)(-3xy^2)(-7x^2yz) = -42x^4y^4z$
(d) $(3mn)(6mn)(mn^2) = 18m^3n^4$ (coefficient of mn^2 is 1)
(e) $(6a^2b^3c^2)(a^2bc^4) = 6a^4b^4c^6$

EXERCISE 2.5

Multiply:

1. $(3x)(5x)$
2. $(-2x^2)(3x^4)$
3. $(-7x^4)(-5x^3)$
4. $(-3x^2)(-4x^2)$
5. $(2a^2b^3)(-5a^3b^2)$
6. $(-7m^{10})(-6m^{10}n^2)$
7. $(4x^2yz)(-3xyz^4)$
8. $(2a^2b^3)(-3a^2b^4)(2a^2b)$
9. $(6ab)(3ab)$
10. $(5x^2y)(x^3y)$
11. $(3a)(-a^3)$
12. $(6a^3)(-4a^3)(5a^3b)$
13. $(-3xy)(-xy)(4xy)$
14. $(6a^2b^3)(-a)(-b^2)$
15. $(2st)(3rst)(-r)$
16. $(2x^2y)(-3x^2y^2)(2x^3y^7)$
17. $-3a^2b^3ca^4b^3c^3$

18. $a^7b^3a^4b^8$

19. $(-4x^2y)(-3z)$

20. $(-3a^2b)(-2a)(2b)(4a^2)(3b)$

Answers to odd-numbered exercises on page 459.

2.6 THE DISTRIBUTIVE PROPERTY

In section 1.7 we learned about the distributive property:

$$a(b + c) = ab + ac$$

This rule can be extended to include more than two terms inside the parentheses:

$$a(b + c + d + e + \ldots) = ab + ac + ad + ae + \ldots$$

What this all means is that every term inside the parentheses must be multiplied by the monomial outside the parentheses.

Example 1: Multiply $4(x + y)$

Solution:

$4(x + y)$ — Both x and y must be multiplied by 4.

$= 4x + 4y$

Example 2: $-3x^2(4x^2 + 2x - 6)$

Solution:

$-3x^2(4x^2 + 2x - 6)$

$= (-3x^2)(4x^2) + (-3x^2)(2x) + (-3x^2)(-6)$

$= -12x^4 - 6x^3 + 18x^2$

Each of the 3 terms inside the parentheses is multiplied by $-3x^2$.

Example 3: Multiply the following:

(a) $5a(3a^2b - 2ab + 3b) = 15a^3b - 10a^2b + 15ab$
(b) $-6m^2n(2m^2n - 3mn^2) = -12m^4n^2 + 18m^3n^3$
(c) $-6(3x^3 + 2x^2 - 5x - 1) = -18x^3 - 12x^2 + 30x + 6$

If an expression in parentheses is preceded by only a negative sign, it is understood to be the same as -1. So $-(3x^2 - 2x)$ means $(-1)(3x^2 - 2x) = -3x^2 + 2x$, and $-(2ab^2 + 3abc - 5)$ means $(-1)(2ab^2 + 3abc - 5) = -2ab^2 - 3abc + 5$.

As you can see, the net effect of a single negative sign in front of parentheses is to change the sign of each term inside the parentheses.

Example 4:

(a) $-(6a^2 + 2b) = -6a^2 - 2b$ Change the sign of each term.
(b) $-(3x^2y - 4xy^3 - 7) = -3x^2y + 4xy^3 + 7$
(c) $-(-5x^2 + 1) = 5x^2 - 1$

The monomial factor can also appear on the right side of the parentheses.

Example 5:

(a) $(-3x^2y - 5xy + 2)(-3x^2) = 9x^4y + 15x^3y - 6x^2$
(b) $(9x^2 + 2x - 1)(-4x) = -36x^3 - 8x^2 + 4x$

EXERCISE 2.6

Find the products.

1. $6(x - y)$
2. $-4(x + 2y)$
3. $3(x + 5)$
4. $-7(m - n)$
5. $3(a - 7)$
6. $x(5 - y)$
7. $3x^2(2x + 3)$
8. $(2x + 1)(4)$
9. $(6x + 5)(2x)$
10. $(7x)(4x^2 + 2x - 1)$
11. $(2x - 3y)(3)$
12. $6(a - b + c)$
13. $(-7x^2)(6x^2 + 2x - 5)$
14. $(-6x^2y)(-3x^2 + 7y - 5x^3y)$
15. $(7x^2 + 3x - 1)(-4)$
16. $(y - 6)(-7y)$
17. $xy(-3x + 4y^3 - 7y^7 - 1)$
18. $-(3x^2 - 5)$
19. $-(7x + 5 - 6y^3)$
20. $(-1)(7x + 5 - 6y^3)$
21. $(-3ab)(2a^2 - 6ab - 7ab^2 + 2b^4)$
22. $(7x + 5 - 6y^3)(-1)$
23. $(3x^2yz + 4x^3y^7 - 3x^4y)(-7x^3y)$
24. $-(2x^2 + 3xy + 5y^2)$
25. $10ax^2(3a^2 - 2bc + y^4)$
26. $-(6x^2 - 5x^3 + 7xyz - 8x^4 - 7)$

27. $(4x^4 - 6x^5 + 7x - 1)(-3x^2)$
28. $-(14x^2 - x + 2y - y^2 - 1)$
29. $(-6ab)(-3a^2 + 2b - a + b - 1)$
30. $(-3a^2 + 2b - a + b - 1)(-6ab)$

Answers to odd-numbered exercises on page 459.

2.7 SIMPLIFYING ALGEBRAIC EXPRESSIONS

Algebraic expressions can frequently be very complicated, and it is certainly in our best interest to learn to simplify them as much as possible. To do this we first remove any grouping symbols (parentheses, brackets, and braces) and then combine any like terms. Examine the following examples carefully.

Example 1: Simplify $3x - 6(2x - 3)$

Solution:

$3x - 6(2x - 3)$	Remove the parentheses by mul-
$= 3x - 12x + 18$	tiplying both $2x$ and -3 by -6.
$= -9x + 18$	Combine like terms $3x$ and $-12x$.

Example 2: Simplify $3(2x - 3y) + 5(3x + y)$

Solution:

$3(2x - 3y) + 5(3x + y)$	
$= 6x - 9y + 15x + 5y$	Remove the parentheses.
$= 6x + 15x - 9y + 5y$	Collect and
$= 21x - 4y$	combine like terms.

Example 3: Simplify $7a - 3(a - 4) - (2a + 9)$

Solution:

$7a - 3(a - 4) - (2a + 9)$	The second set of parentheses is preceded by a negative sign which
$= 7a - 3a + 12 - 2a - 9$	changes the sign of each term.
$= 7a - 3a - 2a + 12 - 9$	Collect and
$= 2a + 3$	combine like terms.

Example 4: Simplify $-2\,[4(3x - 2) - 7x] - 4$

Solution:

$-2\,[4(3x - 2) - 7x] - 4$	Remove the innermost parentheses first.
$= -2\,[12x - 8 - 7x] - 4$	Combine $12x$ and $-7x$.
$= -2\,[5x - 8] - 4$	Multiply by -2.
$= -10x + 16 - 4$	Combine 16 and -4.
$= -10x + 12$	

You can readily see that the result is considerably less complicated than the expression with which we started.

Example 5: Simplify $6a - 3b - [5a - 3(2a + b)]$

Solution:

$6a - 3b - [5a - 3(2a + b)]$	Remove the innermost parentheses.
$= 6a - 3b - [5a - 6a - 3b]$	Combine $5a$ and $-6a$.
$= 6a - 3b - [-a - 3b]$	Remove the brackets.
$= 6a - 3b + a + 3b$	Combine like terms.
$= 7a$	

Example 6: Simplify $2x - 3y - 2\{-[x - 2(x - y)] - 3y\}$

Solution:

$2x - 3y - 2\{-[x - 2(x - y)] - 3y\}$	Remove the innermost parentheses.
$= 2x - 3y - 2\{-[x - 2x + 2y] - 3y\}$	Combine x and $-2x$.
$= 2x - 3y - 2\{-[-x + 2y] - 3y\}$	Change the signs in the brackets.
$= 2x - 3y - 2\{x - 2y - 3y\}$	Combine $-2y$ and $-3y$.
$= 2x - 3y - 2\{x - 5y\}$	Remove the braces.
$= 2x - 3y - 2x + 10y$	Combine like terms.
$= 7y$	

EXERCISE 2.7

Simplify as much as possible:

1. $a - 2(a + b)$
2. $y + 3(x - y)$
3. $2m - 5(2m - 3)$
4. $p + 3(p + q)$
5. $2(x - 2y) - 3y$
6. $-3(4x + 2y) - 2x + y$
7. $2x(x - 3) - 2(x^2 - x)$
8. $3(a - b) + 2a - 5b$
9. $-4x(3x - y) + y(4 - x)$
10. $7x - (-3x - 4)$

11. $3a + 4b - (5a + 4b)$

12. $-3(2x + 4y) - 7(-3x - 2y)$

13. $-(m - 2n) + 3(m + 2n)$

14. $4a(a - b) - 3(a^2 - 2ab)$

15. $x(x^2 - 3x + 2) - 4(x^3 + 3x^2 - 2x)$

16. $3x - [2y + 2(4x - 2y)]$

17. $6x + [-(3x - 5) - 8]$

18. $16 - 3[2(x - 4)] - 7$

19. $-7[-4(3x - 2) + 5x] + 4$

20. $3x - [4(2x - y) - 3y] - 1$

21. $7n - 2[3(n - 1) + 4] - 3$

22. $-8 + 3[-(r - s) - 3(r + s)]$

23. $-10[-4(3x - 4) + 16] - 2x$

24. $20 - \{-3a - [4a - (3 - 2a)]\}$

25. $7x - 3\{2x - 2[x - (x - 1)]\}$

26. $6 - \{2y - [3y - (5 - y)]\}$

27. $-\{3 - 5[7(x + 5) - 2x] + 3x\}$

28. $-4\{-2[3(a + 2) - 2a] + 4\}$

29. Try the following:
   (a) Take your age
   (b) Add 4
   (c) Double the result in step b
   (d) Subtract 3 from the result in step c
   (e) Multiply your age by 2 and subtract from the result in step d

The result is 5! Try a different number—the answer is still 5. Let's see why the answer is always 5 by analyzing our problem.

(a) Let x represent your age
(b) Add 4 which gives us $x + 4$
(c) Doubling that result yields $2(x + 4)$
(d) Subtracting 3 gives $2(x + 4) - 3$
(e) Subtracting 2 times your age ($2x$) gives $2(x + 4) - 3 - 2x$
Now let's simplify the expression:

$$\begin{aligned} &2(x + 4) - 3 - 2x \\ &= \cancel{2x} + 8 - 3 - \cancel{2x} \\ &= 5 \end{aligned}$$

The answer is 5 regardless of what value is chosen for x.
Try to make a similar problem of your own.

Answers to odd-numbered exercises on page 460.

2.8 Chapter Summary

(2.1) A **formula** is a mathematical expression that describes the relationship between quantities in the real world.

(2.1) To evaluate a formula:
1. Write the formula from the given English statement (if necessary).
2. Replace each letter by its numerical value.
3. Do all arithmetic operations using the correct order of operations.

(2.2) A **constant** is a quantity that does not change its value. It is usually written as a number.

(2.2) To evaluate an algebraic expression:
1. Replace each variable (letter) by its given numerical value.
2. Perform all the arithmetic operations using the correct order of operations.

(2.3) **Terms.** The + and − signs in a mathematical expression divide it up into terms.
The numerical coefficient is the number part of a term.
The literal part consists of the variables in a term and their exponents.
Like terms are terms that have the same literal parts.
Unlike terms are terms that have different literal parts.

(2.3) To add or subtract like terms:
1. Add or subtract the numerical coefficients.
2. Keep the same literal part.

(2.4) The first rule of exponents is:

$$x^a \cdot x^b = x^{a+b}$$

(2.5) A **monomial** is an algebraic expression containing only one term.

(2.5) To multiply monomials:
1. Multiply the numerical coefficients.
2. Multiply the variables (letters) in alphabetical order using the first rule of exponents.

Examples

(2.1) The area of a trapezoid is given by $A = \frac{1}{2}h\,(b_1 + b_2)$. Find A if $h = 8$, $b_1 = 6$ and $b_2 = 10$.

$$A = \frac{1}{2}h\,(b_1 + b_2)$$
$$= \frac{1}{2} \cdot 8\,(6 + 10)$$
$$= 4 \cdot 16$$
$$= 64 \text{ square units}$$

(2.2) Find the value of $5(2x - y)$ if $x = -3$ and $y = 1$.

$$5(2x - y)$$
$$= 5[2(-3) - 1]$$
$$= 5[-6 - 1]$$
$$= 5(-7)$$
$$= -35$$

(2.3) Given $4x^2y - 3xy + 7$
The terms are $4x^2y$, $-3xy$, and 7.
The numerical coefficients are 4, −3 and 7.
The literal parts are x^2y and xy.

(2.3) $3xy^3 - 4x^2y - 5xy^3 + 7x^2y + 6$

$$= -2xy^3 + 3x^2y + 6$$

(2.4) $x^4 \cdot x^6 = x^{4+6} = x^{10}$

(2.5) -6, $3x^2y$, and $14x^3$ are monomials.

(2.5) $(-4x^3y)(-7x^4y^2) = 28x^7y^3$

(2.6) To multiply an expression enclosed in parentheses by a monomial, multiply each term in the parentheses by the monomial.

(2.6) To remove parentheses preceded by a negative sign, change the sign of each term in the parentheses.

(2.7) To simplify an algebraic expression:

1. Remove the grouping symbols (starting with the innermost grouping symbols first).
2. Combine like terms.

(2.6) $-3x^2y(4x^3y + 2xy - 6)$
$= -12x^5y^2 - 6x^3y^2 + 18x^2y$

(2.6) $-(7x^3y - 4x^2y^4 + 2)$
$= -7x^3y + 4x^2y^4 - 2$

(2.7) $-2(3x - 2y) - (x - 4y)$
$= -6x + 4y - x + 4y$
$= -7x + 8y$

Exercise 2.8 Chapter Review

(2.1) Evaluate each formula using the values of the letters given. Use $\pi = 3.14$ as an approximate value.

1. $I = prt$ $\quad p = 6000, r = .17, t = 2$

2. $V = \pi r^2 h$ $\quad r = 3, h = 5$

3. $A = \frac{1}{2}h(b_1 + b_2)$ $\quad h = 7, b_1 = 4, b_2 = 10$

4. $THI = t - (.55)(1 - h)(t - 58)$ $\quad t = 82, h = .7$

(2.2) Find the value of each expression when $x = 2$, $y = -6$, and $z = -1$.

5. $3x + 2y - 4z$

6. $7x - (x + 2z)$

7. $x - 2xy + yz$

8. $-2xy + 3xz - 4yz$

(2.3) Identify the terms, numerical coefficients, and literal parts of each expression.

9. $3a + 2b$

10. $-6x^2y + 2xy^2 - 5$

11. $-11m^2n + 2mn^2 - 1$

12. $21a^2b - 3ab^2 + 2ab - 5$

(2.3) Combine where possible.

13. $3a - 2b - 5a - 3b$

14. $3m - (-4n) + 2m - 3n$

15. $-3x^2y + 5xy^2 + 2xy - 7x^2y^2$

16. $3ef - (-5ef) - 2$

(2.4) Simplify using the first rule of exponents.

17. $h^4 \cdot h^3$

18. $h^a \cdot h^{2b}$

19. $x^2 \cdot x^4 \cdot x^7$

20. $x^3 \cdot y^4 \cdot z^3$

(2.5) Multiply.

21. $(-2x^2y)(3x^2y^4)$

22. $(-3m^4)(3m^2)(2m)$

23. $(a^2b)(ab^2)(a^2b^2)$

24. $(-a^2)(3a^2b)(-b)$

(2.6) Find the product.

25. $4x^2(2x^2 + 3x)$

26. $(3a + 2b)(-2a)$

27. $-(4a^3 + 3a^2 - 2a - 7)$

28. $(-2h^2)(-5h^3 + 2h^2 - 8h - 8)$

(2.7) Simplify as much as possible.

29. $5a - 3b - 2(2a + b)$

30. $3x - [4(x + 2) - 1]$

31. $-6a(a + b) - 2(a^2 - 2ab)$

32. $-2\{-2[4(x + 2) - 3x] - 2\}$

Answers to odd-numbered exercises on page 460.

Chapter 2 Achievement Test

Name ____________________

Class ____________________

This test should be taken before you are tested in class on the material in Chapter 2. Solutions to each problem and the section where the type of problem is found are given on page 460.

Evaluate each formula using the values of the letters given. Use $\pi = 3.14$ as an approximate value.

1. $V = \frac{4}{3}r^3$ $\quad r = 2$ — 1. ____________

2. $A = \frac{1}{2}bh$ $\quad b = 6, h = 9$ — 2. ____________

Evaluate each expression when $a = -3$, $b = 4$, and $c = -2$.

3. $-3a + 4b - 5c$ — 3. ____________

4. $-2a^2b - 3ac^2 + ab$ — 4. ____________

Identify (a) the terms, (b) the numerical coefficients and (c) the literal parts in the following expression:

5. $-2ab^3 + 6a^2b^2 + 2$ — 5. (a) ____________ (b) ____________ (c) ____________

Combine where possible:

6. $3x^2y - 2xy^2 + 5x^2y^2$ — 6. ____________

7. $2a^2 - 3b^2 + c - 5a^2 - 4b^2 - c$ — 7. ____________

8. $-3rst - (-5rst) - 2r + 1$ — 8. ____________

Simplify where possible:

9. $x^4 \cdot x^9$ — 9. ____________

10. $h^4 \cdot h^7 \cdot h^3$ — 10. ____________

11. $x^3 \cdot y^4 \cdot z^2$ — 11. ____________

12. $x^m \cdot x^n$ — 12. ____________

Multiply.

13. $(7a^2b)(-6a^3b^4)$

14. $(-a^2)(3ab^5)(-b^3)$

15. $4xy(2x + 3y^2)$

16. $(-2x^2)(3x^3 + 2x^2 - 5x - 1)$

17. $-(4x^5 + 3x^3 - 7x - 5)$

18. $(-7x^2 + 2xy - 4)(-3xy^2)$

Simplify as much as possible:

19. $7x - 5(x + 3)$

20. $ab(3a - b) + 2a(b^2 - ab)$

21. $6x - [3(x - 2) + 2x]$

22. $16 - \{-5x - [2x + 2(3x - 1)]\}$

13. ____________________

14. ____________________

15. ____________________

16. ____________________

17. ____________________

18. ____________________

19. ____________________

20. ____________________

21. ____________________

22. ____________________

3
Equations and Inequalities

3.1 THE MEANING OF EQUATIONS AND THEIR SOLUTIONS

Equations are perhaps the most useful and important concept you will learn in algebra. The main reason for studying algebra is to learn to solve problems in the real world, and many of these problems are solved using equations.

An equation contains two algebraic expressions separated by an **equal sign** (=). The expression on the left of the equal sign is called the *left side* or *left member,* and the expression on the right side of the equal sign is called the *right side* or *right member* of the equation.

$$\underbrace{7x - 3}_{\uparrow \text{ Left side or left member}} \overset{\text{Equal sign}}{\overset{\downarrow}{=}} \underbrace{5x - 9}_{\uparrow \text{ Right side or right member}}$$

The symbol = is an important one and means that the left and right sides of an equation are different names for the same thing. This also means that $7x - 3 = 5x - 9$ and $5x - 9 = 7x - 3$ are the same.

Solution to an Equation.

A **solution** to an equation is a value which, when substituted for the variable, makes the equation a true statement.

For example, if we substitute 12 for x in the equation $x + 5 = 17$ we get:

$$x + 5 = 17$$

$$\boxed{12} + 5 = 17 \qquad \text{Substitute 12 for } x.$$

$$17 = 17$$

which is a true statement, and we say that 12 is the *solution* to the equation $x + 5 = 17$, or that $x = 12$ *satisfies* the equation.

The first question to be answered then, is, how do we recognize the solution to an equation if someone gives it to us? This process is called *checking the solution to the equation*.

To check a solution to an equation:

1. Substitute the proposed solution for the variable wherever it appears in the equation.
2. Evaluate each side of the equation.
3. If the left and right sides of the equation are equal, then the proposed solution *satisfies* the equation.
   If the left and right sides are not the same, then the proposed solution does *not satisfy* the equation.

Example 1: Is $y = 5$ a solution to the equation $3y - 5 = 2y$?

Solution:

$$3y - 5 = 2y$$

$$3 \cdot \boxed{5} - 5 \stackrel{?}{=} 2 \cdot \boxed{5} \qquad \text{Substitute 5 for } y \text{ wherever it appears.}$$

$$15 - 5 \stackrel{?}{=} 10 \qquad \text{Evaluate each side of the equation.}$$

$$10 = 10 \qquad \text{Left side = right side.}$$

$y = 5$ is a solution.

Example 2: Is $x = 7$ a solution to $3x + 2 = 4x - 6$?

Solution:

$$3x + 2 = 4x - 6$$

$$3 \cdot \boxed{7} + 2 \stackrel{?}{=} 4 \cdot \boxed{7} - 6 \qquad \text{Substitute 7 for } x.$$

$$21 + 2 \stackrel{?}{=} 28 - 6 \qquad \text{Evaluate each side of the equation.}$$

$$23 \neq 22 \qquad \neq \text{ means is } \textit{not equal to}.$$

$x = 7$ is not a solution since the two sides are not equal.

Example 3:

EQUATION	*SOLUTION*	*BECAUSE*
$y + 4 = 7$	$y = 3$	$\boxed{3} + 4 = 7$
$13 = 18 - x$	$x = 5$	$13 = 18 - \boxed{5}$
$6 - x = 8$	$x = -2$	$6 - \boxed{(-2)} = 8$
$3(z + 2) = 18$	$z = 4$	$3(\boxed{4} + 2) = 18$

EXERCISE 3.1

1. Determine whether $x = 4$ is a solution to each of the following:

   a. $3x - 4 = 8$ b. $x - 2 = 3$ c. $x + 2 = 3x - 6$

   d. $6 - x = 3 + x$ e. $2x - 10 = 2 - x$

2. Does $y = -4$ satisfy the following equations?

   a. $-8 = 2y$ b. $3 + 5y = 17$ c. $-3y + 3 = 15$

   d. $5 = y + 1$ e. $-2y - 5 = y + 7$

3. Is $z = -6$ a solution to the following equations?

   a. $4z = -28$ b. $\frac{12}{z} = -2$ c. $-7 = -1 + z$

   d. $-2(z + 1) = 10$ e. $3(z + 6) = 2z + 12$

Answers to odd-numbered exercises on page 461.

3.2 SOLVING EQUATIONS— THE ADDITION PRINCIPLE

Now that we have learned what a solution to an equation looks like, the next step is to discover how to find one.

In the last section, it was stated that the left and right sides of an equation are different ways of representing the same quantity. In this way, we can think of an equation as a balance, with the left member on one side and the right member on the other. The equation $x - 5 = -1$ could be represented by the following diagram.

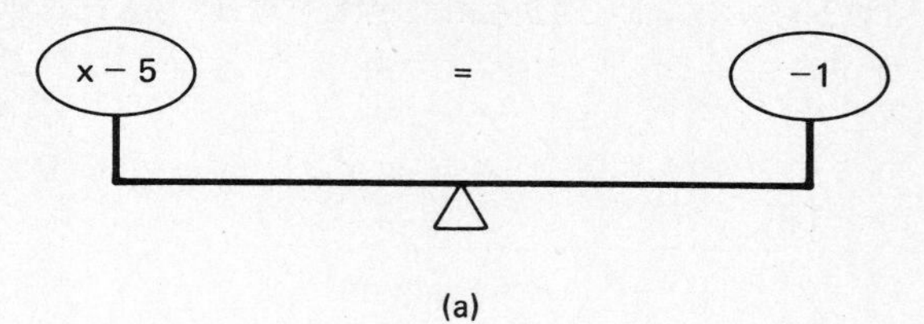

(a)

If we add the same quantity, say 7, to both sides of the balance, it should stay in equilibrium since both sides still carry the same amount.

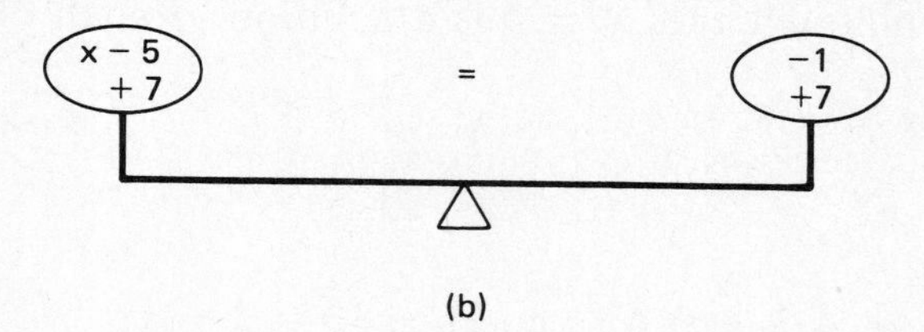

(b)

Algebraically, we are adding 7 to both sides of the original equation:

$$\begin{array}{rcr} x - 5 & = & -1 \\ +7 & & +7 \\ \hline x + 2 & = & 6 \end{array}$$

If you check the equations, $x - 5 = -1$ and $x + 2 = 6$, you will find that $x = 4$ is the solution to both.

$$\begin{array}{rcl} x - 5 = -1 & \quad & x + 2 = 6 \\ \boxed{4} - 5 = -1 & & \boxed{4} + 2 = 6 \\ -1 = -1 & & 6 = 6 \end{array}$$

$x = 4$ is a solution to both equations.

Two equations that have the same solution are called **equivalent equations.** It is apparent from the illustration that we can add the same number to both sides of any equation and obtain a new equation that is *equivalent* to the original one; that is, they both have the same solution.

In a similar way we could show that if we subtract the same quantity from both sides of an equation, the result is an equivalent equation. Actually, this is the same as adding a negative quantity to both sides of the equation. This principle is basic to the solution of equations and is stated as follows:

Addition principle:

The same number may be added to (or subtracted from) each side of any equation and the solution will remain the same.

Let us now apply this principle to finding the solutions to equations. Our purpose is always the same, to isolate the variable all by itself on one side of the equation. This involves adding or subtracting any numbers that are found on the same side of the equation as the variable. Always do the opposite operation—if a number is added, subtract it (i.e., add its negative). Remember, whatever we do to one side of the equation we must do the other side of the equation to keep it in balance.

Example 1: Solve $x + 6 = 14$

Solution:

$x + 6 = 14$	Try to isolate x on the left side.
$x + 6 + \boxed{(-6)} = 14 + \boxed{(-6)}$	Adding (-6) removes the $+ 6$ from the left side.
$x + 0 = 8$	
$x = 8$	The solution is $x = 8$.

We can check by putting $x = 8$ in the original equation.

$$\boxed{8} + 6 = 14$$

Example 2: Solve $x - 7 = -3$

Solution:

$x - 7 = -3$	Isolate x by getting rid of -7.
$x - 7 \boxed{+7} = -3 \boxed{+7}$	Add 7 to both sides.
$x = 4$	The solution is $x = 4$.

Check: $\boxed{4} - 7 = -3$

Example 3: Solve $-9 = y + 3$

Solution:

$-9 = y + 3$	Isolate y by getting rid of 3.
$-9 + \boxed{(-3)} = y + 3 + \boxed{(-3)}$	Add (-3) to both sides.
$-12 = y$	The solution is $y = -12$.

Note that $-12 = y$ and $y = -12$ say the same thing.

Check: $-9 = \boxed{-12} + 3$

Example 4: Solve $3.1 + y = 2.4$

Solution:

$3.1 + y = 2.4$	Isolate y.
$3.1 + \boxed{(-3.1)} + y = 2.4 + \boxed{(-3.1)}$	Add (-3.1) to both sides.
$y = -.7$	The solution is $y = -.7$

Check: $3.1 \boxed{-.7} = 2.4$

In many of the exercises that follow, you will be able to see the solutions immediately. However, you should work through each problem carefully to learn the methods involved. These are the same methods that we will use to solve more difficult equations for which the solutions are not at all obvious.

EXERCISE 3.2

Solve and check the following equations:

1. $x - 4 = 7$
2. $y + 5 = 14$
3. $y - 6 = 5$
4. $-6 + x = 4$
5. $5 = x + 7$
6. $-6 = h - 7$
7. $7 + h = 17$
8. $14 = x - 11$
9. $-13 = -12 + x$
10. $x - 16 = -9$
11. $2 + x = -7$
12. $-6 = x - 4$
13. $4 + y = -11$
14. $12 + h = -14$
15. $-16 = -14 + m$
16. $-71 = s - 26$
17. $3.75 + x = 8$
18. $6.4 + x = -3.6$
19. $x + \frac{3}{4} = 7\frac{3}{4}$
20. $x - 1\frac{2}{3} = 3\frac{2}{3}$
21. $4.62 = x - 3.91$
22. $y + 2\frac{1}{5} = 7\frac{3}{5}$
23. $x - 4.1 = -3.8$
24. $x - 1.75 = 8.5$
25. $4.75 + y = -2\frac{1}{4}$

Answers to odd-numbered exercises on page 461.

3.3 SOLVING EQUATIONS— THE MULTIPLICATION PRINCIPLE

If we return to our idea of the balance representing an equation like $2x = 18$, we have the following diagram:

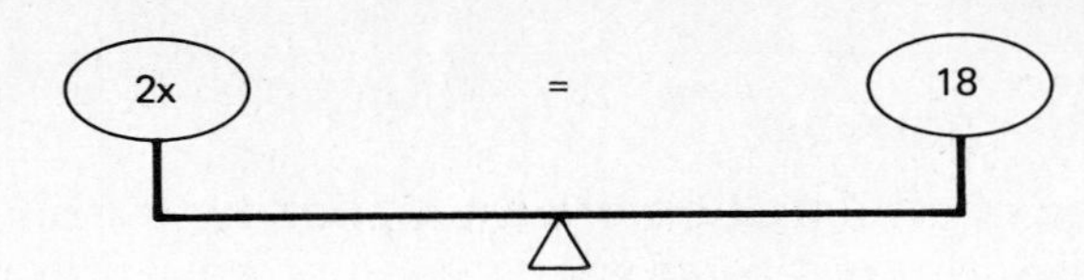

If we decided to multiply each side of our equation by 3, we would have three times as much on each side of the balance and it would remain in equilibrium. Algebraically, it would look like this:

$$2x = 18$$
$$\boxed{3} \cdot 2x = \boxed{3} \cdot 18$$
$$6x = 54$$

If we substitute $x = 9$ in each of our equations, $2x = 18$ and $6x = 54$, we will find it is a solution for each of them.

$$2x = 18 \qquad 6x = 54$$
$$2 \cdot \boxed{9} = 18 \qquad 6 \cdot \boxed{9} = 54$$
$$18 = 18 \qquad 54 = 54$$

$x = 9$ is a solution to both equations.

Dividing by a number such as 3 is the same as multiplying by its reciprocal, $\frac{1}{3}$. The **reciprocal** of a number is 1 divided by the number.

$\frac{1}{5}$ is the reciprocal of 5.

The reciprocal of $\frac{1}{5}$ is:

$$\frac{1}{\frac{1}{5}} = \frac{\frac{1}{1}}{\frac{1}{5}} = \frac{1}{1} \div \frac{1}{5} = \frac{1}{1} \cdot \frac{5}{1} = 5$$

So $\frac{1}{5}$ and 5 are reciprocals of each other.

The reciprocal of $\frac{2}{3}$ is

$$\frac{1}{\frac{2}{3}} = \frac{\frac{1}{1}}{\frac{2}{3}} = \frac{1}{1} \div \frac{2}{3} = \frac{1}{1} \cdot \frac{3}{2} = \frac{3}{2}$$

and this is true in general.

> The reciprocal of $\frac{a}{b}$ is $\frac{b}{a}$.
>
> a and $\frac{1}{a}$ are reciprocals of each other.

It seems reasonable, then, that we can multiply or divide both sides of an equation by the *same* number and produce an equivalent equation. This is the second basic principle that we will use in solving equations.

Multiplication principle:

If both sides of an equation are multiplied by (or divided by) the *same* number (except zero) an equivalent equation results.

To apply this rule, keep in mind that we want to isolate the variable on one side of the equation. If the variable is multiplied by a number, we get rid of it by dividing by the same number. If the variable is divided by a number, we get rid of it by multiplying by that same number. Of course, we must remember that whatever we do to one side of the equation, we must do to the other side of the equation as well.

Example 1: Solve $5y = 30$

Solution:

$5y = 30$ Isolate y on the left side.

$\frac{\cancel{5}y}{\boxed{\cancel{5}}} = \frac{30}{\boxed{5}}$ Divide both sides by 5.

$1 \cdot y = 6$ The 5s cancel.

$y = 6$ $1 \cdot y = y$, which is what we want.

Check: $5 \cdot \boxed{6} = 30 \checkmark$ The solution checks.

We can also obtain the same result by multiplying by the reciprocal of 5, which is $\frac{1}{5}$.

Solution: $5y = 30$

$\boxed{\frac{1}{\cancel{5}}} \cdot \cancel{5}y = \boxed{\frac{1}{5}} \cdot 30$ Multiply both sides by $\frac{1}{5}$.

$1 \cdot y = 6$

$y = 6$

Check: $5 \cdot \boxed{6} = 30 \checkmark$

Example 2: Solve $\frac{1}{4}x = 3$

Solution:

$\frac{1}{4}x = 3$ Get rid of the $\frac{1}{4}$ on the left.

$\boxed{\cancel{4}} \cdot \frac{1}{\cancel{4}}x = \boxed{4} \cdot 3$ Multiply by 4, the reciprocal of $\frac{1}{4}$.

$1 \cdot x = 12$ $1 \cdot x = x$.

$x = 12$

Check: $\frac{1}{4} \cdot \boxed{12} = 3 \checkmark$ The solution checks.

Example 3: Solve $\frac{3}{4}x = 18$

Solution:

$\frac{3}{4}x = 18$ Get rid of the $\frac{3}{4}$.

$\boxed{\frac{4}{3}} \cdot \frac{3}{4}x = \boxed{\frac{4}{3}} \cdot 18$ Multiply by $\frac{4}{3}$, the reciprocal of $\frac{3}{4}$.

$1 \cdot x = 24$

$x = 24$ The solution is $x = 24$.

Check: $\frac{3}{4} \cdot \boxed{24} = \frac{3}{4} \cdot 24 = \frac{18}{1} = 18$ ✓

Example 4: Solve $32 = -8h$

Solution:

$32 = -8h$ Get rid of the -8 on the right.

$\frac{32}{\boxed{-8}} = \frac{-8h}{\boxed{-8}}$ Divide both sides by -8.

$-4 = h$ The solution is $h = -4$.

Check: $32 = (-8)\boxed{(-4)}$ ✓

Example 5: Solve $\frac{x}{7} = -6$

Solution:

$\frac{x}{7} = -6$ We need to get rid of the 7.

$\boxed{7} \cdot \frac{x}{7} = \boxed{(7)}\ (-6)$ Multiply both sides by 7.

$x = -42$ The solution is $x = -42$.

Check: $\frac{\boxed{-42}}{7} = -6$ ✓

Example 6: Solve $-x = -14$

Solution: Since $-x$ is the same as $(-1)x$, we divide both sides of the equation by -1.

$-x = -14$

$-1 \cdot x = -14$ $-x$ is the same as $-1 \cdot x$.

$\frac{-1 \cdot x}{\boxed{-1}} = \frac{-14}{\boxed{-1}}$ Divide by -1 on both sides.

$x = 14$

After you do a few problems of this type, you will probably notice a shortcut. If one side of the equation contains only a variable and a negative sign, to solve for the positive variable just change the sign on both sides of the equation.

Example 7:

(a) Solve $-x = 6$

Solution:

$x = -6$ Change the sign on both sides.

(b) Solve $-y = -8$

Solution:

$y = 8$

EXERCISE 3.3

Solve and check each equation.

1. $6x = 24$
2. $3y = 21$
3. $48 = 6x$
4. $32 = 16y$
5. $-3h = 24$
6. $\frac{y}{7} = 5$
7. $-2 = \frac{y}{-3}$
8. $\frac{1}{4}x = 8$
9. $\frac{1}{5}x = -7$
10. $-72 = -9y$
11. $\frac{t}{-6} = -3$
12. $\frac{1}{-9} = 4$
13. $15t = 6$
14. $\frac{2}{3}x = 12$
15. $\frac{3}{5}y = -15$
16. $39x = -13$
17. $2.4x = -14.4$
18. $4.2 = -3z$
19. $-36 = \frac{-3}{4}x$
20. $-1\frac{5}{16} = \frac{7}{8}y$

Answers to odd-numbered exercises on page 461.

3.4 SOLVING EQUATIONS BY COMBINING RULES

Some equations involve use of both the addition and multiplication principles. Our objective is still the same; try to isolate the variable on one side of the equation. To do this we first get all the terms containing the variable on one side of the equation and the plain numbers on the other side of the equation.

To solve an equation using both the addition and multiplication principles:

1. Put all the terms containing the variable on one side of the equation by using the addition principle.
2. Put all plain numbers on the *other* side of the equation by using the addition principle again.
3. Multiply both sides of the equation by the reciprocal of the coefficient of the variable.

Example 1: Solve $5x + 1 = 3x + 7$

Solution:

$5x + 1 = 3x + 7$	Get all x-terms on the left side.
$5x + \boxed{-3x} + 1 = 3x + \boxed{-3x} + 7$	Add $-3x$ to both sides.
$2x + 1 = 7$	All x-terms are now on the left side.
$2x + 1 \boxed{-1} = 7 \boxed{-1}$	Add -1 to both sides to get all
$2x = 6$	the plain numbers on the right.
$\boxed{\frac{1}{\cancel{2}}} \cdot \cancel{2}x = \boxed{\frac{1}{2}} \cdot 6$	Multiply both sides by $\frac{1}{2}$, the reciprocal of 2.
$x = 3$	The solution is $x = 3$.

Check:

$$5x + 1 = 3x + 7$$
$$5 \cdot \boxed{3} + 1 \stackrel{?}{=} 3 \cdot \boxed{3} + 7$$
$$15 + 1 \stackrel{?}{=} 9 + 7$$
$$16 = 16 \checkmark$$

Did we have to put the x-terms on the *left* and the plain numbers on the *right?* Or could we have done it the other way around? Let's try it and see:

Example 1: (alternate solution): Solve $5x + 1 = 3x + 7$

Solution #2:

$5x + 1 = 3x + 7$	Get all x-terms on the right side.
$5x + \boxed{(-5x)} + 1 = 3x + \boxed{(-5x)} + 7$	Add $(-5x)$ to both sides.
$1 = -2x + 7$	x-terms are all on the right now.
$1 + \boxed{(-7)} = -2x + 7 + \boxed{(-7)}$	Add (-7) to both sides.
$-6 = -2x$	Multiply both sides by $\frac{1}{-2}$ or,
$\frac{-6}{-2} = \frac{-\cancel{2}x}{-\cancel{2}}$	equivalently, divide by -2.
$3 = x$	The solution is the same.

Example 2: Solve $3y - 2 = 13$

Solution:

$3y - 2 = 13$	All y-terms are already on the left.
$3y - 2 \boxed{+2} = 13 \boxed{+2}$	Get rid of -2 on the left.
$3y = 15$	All plain numbers are now on the right.
$\frac{\cancel{3}y}{\boxed{\cancel{3}}} = \frac{15}{\boxed{3}}$	Divide both sides by 3.
$y = 5$	$y = 5$ is the solution.

Check:

$$3 \cdot \boxed{5} - 2 \stackrel{?}{=} 13$$

$$15 - 2 = 13 \ \checkmark$$

Example 3: Solve $-22 - 3x = 5x$

Solution: Since all the plain numbers are already on the left side of the equation, there is some advantage to putting all the variables on the right.

$-22 - 3x = 5x$	
$-22 - 3x \boxed{+3x} = 5x \boxed{+3x}$	Add $3x$ to both sides.
$-22 = 8x$	
$\frac{-22}{\boxed{8}} = \frac{\cancel{8}x}{\boxed{\cancel{8}}}$	Divide both sides by 8.
$\frac{-11}{4} = \frac{-22}{8} = x$	The solution is $x = \frac{-11}{4}$.

Check:

$$-22 - 3\boxed{\left(\frac{-11}{4}\right)} \stackrel{?}{=} 5\boxed{\left(\frac{-11}{4}\right)}$$

$$\frac{-88}{4} + \frac{33}{4} \stackrel{?}{=} \frac{-55}{4}$$

$$\frac{-55}{4} = \frac{-55}{4} \checkmark$$

Example 4: Solve $5 + \frac{y}{6} = 8$

Solution:

$$5 + \frac{y}{6} = 8$$

$$5 + \boxed{(-5)} + \frac{y}{6} = 8 + \boxed{(-5)}$$ Add -5 to both sides.

$$\frac{y}{6} = 3$$

$$\boxed{\cancel{6}} \cdot \frac{y}{\cancel{6}} = \boxed{6} \cdot 3$$ Multiply both sides by 6.

$$y = 18$$

Check:

$$5 + \frac{\boxed{18}}{6} \stackrel{?}{=} 8$$

$$5 + 3 = 8 \checkmark$$

EXERCISE 3.4

Solve each equation and check:

1. $7x = 35 + 2x$
2. $3y - 2 = 13$
3. $4x - 6 = 3x + 2$
4. $6a - 3 = 2a + 5$
5. $6a + 9 = 2a - 7$
6. $6y - 26 = 3y + 1$
7. $-2y + 1 = 13$
8. $5 - 7z = 3z + 15$
9. $-6t - 1 = -5t + 6$
10. $9y + 1 = 4y$
11. $18 - 4y = 5y$
12. $-36 = 24x + 12$

13. $\frac{1}{2}x + 3 = 7$ 14. $m = 5m - 8$ 15. $.6n + .5 = -3.7$

16. $3x - 5.2 = 4.1$ 17. $6 + \frac{P}{5} = 8$ 18. $\frac{h}{3} - 5 = 2$

19. $\frac{x}{4} + 3 = -5$ 20. $-8x + 2 = 3x - 4$ 21. $5x + 6 = 3x + 5$

22. $7y + 4 = -8y + 7$

Answers to odd-numbered exercises on page 462.

3.5 SOLVING EQUATIONS CONTAINING PARENTHESES, FRACTIONS, AND DECIMALS

Equations Containing Parentheses

When an equation contains parentheses, it is necessary to simplify before solving.

Example 1: Solve $4(x + 3) = 8$

Solution: Using the distributive rule we must multiply both x and 3 by 4 on the left side of the equation:

$$4(x + 3) = 8$$

$$4x + 12 = 8 \quad \text{Multiply parentheses by 4.}$$

$$4x + 12 + \boxed{(-12)} = 8 + \boxed{(-12)} \quad \text{Add } (-12) \text{ to both sides.}$$

$$4x = -4$$

$$\frac{\cancel{4}x}{\boxed{\cancel{4}}} = \frac{-4}{\boxed{4}} \quad \text{Divide each side by 4.}$$

$$x = -1$$

Check:

$$4(x + 3) = 8$$

$$4(\boxed{-1} + 3) \stackrel{?}{=} 8$$

$$4(2) \stackrel{?}{=} 8$$

$$8 = 8 \checkmark$$

Example 2: Solve $3x - 2(x + 4) = 6 - (x + 4)$

Solution:

$3x - 2(x + 4) = 6 - (x + 4)$

$3x - 2x - 8 = 6 - x - 4$ Remove parentheses on each side.

$x - 8 = -x + 2$ Combine like terms on each side.

$x \boxed{+x} - 8 = -x \boxed{+x} + 2$ Add x to each side.

$2x - 8 = 2$

$2x - 8 \boxed{+8} = 2 \boxed{+8}$ Add 8 to each side.

$2x = 10$

$\frac{\cancel{2}x}{\cancel{2}} = \frac{10}{2}$ Divide each side by 2.

$x = 5$ 5 is the solution.

Check:

$3x - 2(x + 4) = 6 - (x + 4)$

$3 \cdot \boxed{5} - 2(\boxed{5} + 4) \stackrel{?}{=} 6 - (\boxed{5} + 4)$

$15 - 2(9) \stackrel{?}{=} 6 - 9$

$15 - 18 \stackrel{?}{=} -3$

$-3 = -3$ ✓

Equations Containing Fractions

Equations containing fractions are handled quite easily. Find the least common denominator (LCD) of all the fractions in the equation, multiply each side of the equation by this LCD, and you will find that all the fractions have disappeared. Stated formally:

> **To eliminate fractions from an equation:**
> Multiply each side of the equation by the LCD of all the fractions that appear in the equation.

Example 3: Solve $\frac{1}{2}x + \frac{1}{5} = \frac{2}{5}x - \frac{1}{2}$

Solution:

$\frac{1}{2}x + \frac{1}{5} = \frac{2}{5}x - \frac{1}{2}$

$\boxed{10}\left(\frac{1}{2}x + \frac{1}{5}\right) = \boxed{10}\left(\frac{2}{5}x - \frac{1}{2}\right)$ Multiply each side by 10, the LCD.

$$\boxed{10}\left(\frac{1}{2}x\right)+\boxed{10}\left(\frac{1}{5}\right)=\boxed{10}\left(\frac{2}{5}x\right)-\boxed{10}\left(\frac{1}{2}\right)$$

Apply the distributive rule.

$$\overset{5}{\cancel{10}}\left(\frac{1}{\cancel{2}}x\right)+\overset{2}{\cancel{10}}\left(\frac{1}{\cancel{5}}\right)=\overset{2}{\cancel{10}}\left(\frac{2}{\cancel{5}}x\right)-\overset{5}{\cancel{10}}\left(\frac{1}{\cancel{2}}\right)$$

$$5x + 2 = 4x - 5$$

The fractions are gone!

$$5x + \boxed{(-4x)} + 2 + \boxed{(-2)} = 4x + \boxed{(-4x)} - 5 + \boxed{(-2)}$$

Add $(-4x)$ and (-2) to each side.

$$x = -7$$

$x = -7$ is the solution.

Check:

$$\frac{1}{2}x + \frac{1}{5} = \frac{2}{5}x - \frac{1}{2}$$

$$\frac{1}{2}\cdot\boxed{-7}+\frac{1}{5} \stackrel{?}{=} \frac{2}{5}\cdot\boxed{-7}-\frac{1}{2}$$

$$\frac{-7}{2}+\frac{1}{5} \stackrel{?}{=} \frac{-14}{5}-\frac{1}{2}$$

$$\frac{-35}{10}+\frac{2}{10} \stackrel{?}{=} \frac{-28}{10}-\frac{5}{10}$$

$$\frac{-33}{10} = \frac{-33}{10} \checkmark$$

As you can see, with complicated equations such as these, checking the solution is sometimes more work than finding it in the first place. So the recommendation is to work carefully and in an orderly way in order to eliminate errors.

Example 4: Solve $\frac{2}{3}x + 1 = \frac{1}{2}x + \frac{3}{4}$

Solution:

$$\frac{2}{3}x + 1 = \frac{1}{2}x + \frac{3}{4}$$

$$\boxed{12}\left(\frac{2}{3}x + 1\right) = \boxed{12}\left(\frac{1}{2}x + \frac{3}{4}\right)$$

Multiply each side by 12, the LCD.

$$\boxed{12}\left(\frac{2}{3}x\right) + \boxed{12}\left(1\right) = \boxed{12}\left(\frac{1}{2}x\right) + \boxed{12}\left(\frac{3}{4}\right)$$

Apply the distributive rule.

$$\overset{4}{\cancel{12}}\left(\frac{2}{\underset{1}{\cancel{3}}}x\right) + 12(1) = \overset{6}{\cancel{12}}\left(\frac{1}{\underset{1}{\cancel{2}}}x\right) + \overset{3}{\cancel{12}}\left(\frac{3}{\underset{1}{\cancel{4}}}\right)$$

$$8x + 12 = 6x + 9$$

This is considerably easier to handle than the original equation.

$$8x + \boxed{(-6x)} + 12 + \boxed{(-12)} = 6x + \boxed{(-6x)} + 9 + \boxed{(-12)}$$

Add $(-6x)$ and (-12) to each side.

$$2x = -3$$

$$\frac{\cancel{2}x}{\boxed{\cancel{2}}} = \frac{-3}{\boxed{2}}$$

Divide each side by 2.

$$x = \frac{-3}{2}$$

$x = \dfrac{-3}{2}$ is the solution.

Check:

$$\frac{2}{3}x + 1 = \frac{1}{2}x + \frac{3}{4}$$

$$\frac{\overset{1}{\cancel{2}}}{\underset{1}{\cancel{3}}} \cdot \boxed{\frac{-\overset{-1}{\cancel{3}}}{\underset{1}{\cancel{2}}}} + 1 = \frac{1}{2} \cdot \boxed{\frac{-3}{2}} + \frac{3}{4}$$

$$-1 + 1 = \frac{-3}{4} + \frac{3}{4}$$

$$0 = 0 \ \checkmark$$

Equations Containing Decimals

It is usually more difficult to solve an equation containing decimals than to solve one without decimals. Equations are "cleared" of decimals by using the following rules:

> **To clear an equation of decimals:**
>
> 1. Count the greatest number of decimal places appearing in any one decimal in the equation.
> 2. Multiply each side of the equation by 1 followed by that number of zeros.

Example 5: Solve $.04x + 3.1 = .39x + 3.2$

Solution: The greatest number of decimal places in any one of the decimals is *two* (in both $.04x$ and $.39x$) so we must multiply each side of the equation by 100 (1 followed by *two* zeros). Multiplying by 100 moves the decimal point *two* places to the right.

$$.04x + 3.1 = .39x + 3.2$$

Multiply each side by 100.

$$\boxed{100}(.04x + 3.1) = \boxed{100}(.39x + 3.2)$$

Apply the distributive rule.

$$\boxed{100}(.04x) + \boxed{100}(3.1) = \boxed{100}(.39x) + \boxed{100}(3.2)$$

Multiplying by 100 moves the decimal point two places to the right.

$$.04x + 3.10 = .39x + 3.20$$

$$4x + 310 = 39x + 320$$

$$4x + \boxed{(-39x)} + 310 + \boxed{(-310)} = 39x + \boxed{(-39x)} + 320 + \boxed{(-310)}$$

Add $(-39x)$ and (-310) to both sides.

$$-35x = 10$$

$$\frac{-35x}{\boxed{-35}} = \frac{10}{\boxed{-35}}$$

Divide by -35.

$$x = \frac{10}{-35} = \frac{2}{-7} \text{ or } -.286$$

rounded to the nearest thousandth.

EXERCISE 3.5

Solve and check each equation:

1. $3(x + 1) = 2x + 7$
2. $7(x - 2) = 6x - 5$
3. $7x = 2(x + 20)$
4. $12x = 2(x + 30)$
5. $y + 2(y + 4) = -1$
6. $3(x - 1) + 7 = 2x$
7. $3(x - 4) + 2(x + 6) = 20$
8. $x - 5(x + 2) = 4 - 2(x - 1) + 6$
9. $3(x - 3) + 2(x - 2) = 5(3 - x)$

10. $3(n + 5) = 2(n - 1)$

11. $2(t - 5) = 2(3t - 5)$

12. $.2x = 2.4$

13. $.05x = 7.5$

14. $6.2x = 13.02$

15. $4.1x + 7.8 = 8.62$

16. $7.1x - 4.2 = .1x + .7$

17. $\frac{3}{8}x - \frac{1}{3} = \frac{1}{4}x + \frac{5}{12}$

18. $\frac{1}{15}x + \frac{2}{5} = \frac{1}{3}$

19. $\frac{1}{3}x + \frac{1}{2}x = 5$

20. $\frac{1}{2}y - \frac{1}{5}y = 6$

21. $\frac{1}{3}y - \frac{1}{7}y = 4$

22. $\frac{2}{3}x + 1 = \frac{1}{2}x - \frac{3}{4}$

23. $\frac{2}{3}y + 9 = y$

24. $\frac{1}{3}x + \frac{1}{2} = \frac{3}{2}$

25. $\frac{3}{4}t + \frac{1}{2} = \frac{1}{8}t - 2$

26. $\frac{1}{-2}y + \frac{3}{2} = \frac{2}{3}y - 1$

27. $\frac{1}{3}(x - 4) = \frac{7}{8}(x - 1)$

28. $\frac{5}{16}y - \frac{1}{4} = \frac{1}{2}y - \frac{7}{16}$

29. $.5(w + 1) = 3.5w - 7.5$

30. $5(y - \frac{1}{2}) + \frac{2}{3} = 6 + \frac{1}{5}(2y - \frac{1}{3})$

31. $3(\frac{1}{2}x - 4) = 2 + \frac{1}{2}(x + 1)$

32. $.3x + 2.4 = 1.3x - 4.8$

33. $\frac{2}{3}(x + \frac{1}{3}) = \frac{1}{9}x - 4$

34. $.2x + 3.7 = 4(.8x + 2.1)$

Answers to odd-numbered exercises on page 462.

3.6 CONDITIONAL EQUATIONS, IDENTITIES, AND CONTRADICTIONS

> **Conditional equations** are equations that are true for only certain values of the variable.

For example, when we solve $2x - 1 = 5$ we find that it has only one solution, namely, $x = 3$. The two sides are equal only under the *condition* that $x = 3$. Any other value for x will not make the equation a true statement, so it is a *conditional equation*. All of the equations we have considered so far have been conditional equations.

> An **identity** (or **identical equation**) is an equation that is true for all values of the variable.

For example:

$$3x - 1 = 3x - 1$$

is true no matter what number is substituted for the variable. If you try to get all the x-terms on one side of the equation and all the numbers on the other side, you eventually get $0 = 0$.

If you get $0 = 0$ when trying to solve any equation, then the original equation was an identity.

Example 1: Solve $3(x - 4) + 7 = 3x - 5$

Solution:

$$3(x - 4) + 7 = 3x - 5 \qquad \text{Remove parentheses.}$$

$$3x - 12 + 7 = 3x - 5 \qquad \text{Combine } -12 \text{ and } 7.$$

$$3x - 5 = 3x - 5$$

$$3x \boxed{-3x} - 5 \boxed{+5} = 3x \boxed{-3x} - 5 \boxed{+5}$$

$$0 = 0$$

Since our result is $0 = 0$, we conclude that the given equation is an *identity* and every numerical replacement for the variable is a solution.

> A **contradiction** is an equation that has no solutions.

For example, the equation

$$x = x + 1$$

can't possibly have a solution, since if we replace x by *any* number, the right side of the equation will always be 1 larger than the left. If we try to get all the x-terms on the left, we obtain a strange result:

$$x = x + 1$$

$$x \boxed{-x} = x \boxed{-x} + 1 \qquad \text{Subtract } x \text{ from each side.}$$

$$0 = 1\ ?? \qquad \text{There is no solution.}$$

If the solution of any equation results in the two sides being unequal numbers like $0 = 1$, $5 = 7$, et cetera, the given equation is a *contradiction*, and there is no solution.

Example 2: Solve $6x - 9 = 3(2x - 5)$

Solution:

$$6x - 9 = 3(2x - 5) \qquad \text{Remove parentheses.}$$

$$6x - 9 = 6x - 15$$

$$6x + \boxed{(-6x)} - 9 \boxed{+ 9} = 6x + \boxed{(-6x)} - 15 \boxed{+ 9} \qquad \text{Add } (-6x) \text{ and } 9.$$

$$0 = -6$$

Since $0 \neq -6$ we conclude that there is no solution and the original equation is a *contradiction*.

To summarize:

> **There are three possibilities when attempting to solve an equation:**
>
> 1. Conditional—if a solution results.
> 2. Identity—when $0 = 0$ is the result.
> 3. Contradiction—when the two sides of the *solved* equation are unequal (such as $0 = 6$, etc.)

EXERCISE 3.6

Classify each of the following equations as conditional, an identity, or a contradiction. Solve the conditional equations:

1. $3x = 3(x + 9) - 2$

2. $4x = 5(x - 2) + 7$

3. $3x - 7 = 3(x - 2) - 1$

4. $7x = 4(2 - x) + 5$

5. $6y + 5 = 3(4 + 2y)$

6. $3n = 2 - 3(2 - n)$

7. $8z - 4(3 + 2z) + 5 = -7$

8. $6z - 3(4 - 2z) = 0$

9. $13y + 7(1 - 2y) = -(2 + y)$

10. $4 - 6(t + 2) = 3(2t - 1)$

Answers to odd-numbered exercises on page 462.

3.7 LITERAL EQUATIONS

Literal equations are equations containing more than one letter. They usually contain a variable and constants which are represented by letters instead of numbers. Formulas are good examples of literal equations.

Some examples of literal equations are:

$$A = \pi r^2$$

$$F = \frac{9}{5}C + 32$$

$$A = \ell \cdot w$$

$$P = 2\ell + 2w$$

$$y = mx + b$$

To solve a literal equation for a particular letter means to isolate that letter on one side of the equation. Since letters have the same meaning as numbers in literal equations, we use the same equation-solving techniques applied throughout this chapter.

Example 1: Solve $y = mx + b$ for x.

Solution: If the equation read $7 = 3x + 4$, we would know exactly what to do. To show the similarity let's look at them side by side.

$y = mx + b$	$7 = 3x + 4$
$y + \boxed{(-b)} = mx + b + \boxed{(-b)}$ Add $(-b)$.	$7 + \boxed{(-4)} = 3x + 4 + \boxed{(-4)}$ Add (-4).
$y - b = mx$	$3 = 3x$
$\frac{y-b}{\boxed{m}} = \frac{\cancel{m}x}{\boxed{\cancel{m}}}$ Divide by m.	$\frac{3}{\boxed{3}} = \frac{\cancel{3}x}{\boxed{\cancel{3}}}$ Divide by 3.
$\frac{y-b}{m} = x$	$1 = x$
The solution is $x = \frac{y-b}{m}$	The solution is $x = 1$

As you can see, literal equations are solved using exactly the same methods that we have used all along.

Example 2: Solve $P = 2\ell + 2w$ for w.

Solution: Isolate w on one side of the equation.

$$P = 2\ell + 2w$$

$$P \boxed{-2\ell} = 2\ell \boxed{-2\ell} + 2w \qquad \text{Add } -2\ell \text{ to each side.}$$

$$P - 2\ell = 2w$$

$$\frac{P - 2\ell}{\boxed{2}} = \frac{\cancel{2}w}{\boxed{\cancel{2}}} \qquad \text{Divide each side by 2.}$$

$$w = \frac{P - 2\ell}{2} \qquad \text{is the solution.}$$

Example 3: Solve $d = a + (b - 1)c$ for b.

Solution:

$$d = a + (b - 1)c$$

$$d = a + bc - c \qquad \text{Eliminate parentheses.}$$

$$d \boxed{-a} \boxed{+c} = a \boxed{-a} + bc - c \boxed{+c} \qquad \text{Isolate b-term.}$$

$$d - a + c = bc$$

$$\frac{d - a + c}{c} = \frac{b\cancel{c}}{\cancel{c}} \qquad \text{Divide each side by } c.$$

$$\text{or} \quad b = \frac{d - a + c}{c} \qquad \text{is the solution.}$$

Example 4: Solve $\frac{2}{3}x + \frac{3}{4}y = \frac{1}{2}z$ for y.

Solution:

$$\frac{2}{3}x + \frac{3}{4}y = \frac{1}{2}z \qquad \text{Clear of fractions.}$$

$$\overset{4}{\cancel{12}}\left(\frac{2}{\underset{1}{\cancel{3}}}\right)x + \overset{3}{\cancel{12}}\left(\frac{3}{\underset{1}{\cancel{4}}}\right)y = \overset{6}{\cancel{12}}\left(\frac{1}{\underset{1}{\cancel{2}}}\right)z \qquad \text{Multiply by 12, the LCD.}$$

$$8x + 9y = 6z$$

$$8x + \boxed{(-8x)} + 9y = 6z + \boxed{(-8x)} \qquad \text{Add } (-8x) \text{ to each side.}$$

$$9y = 6z - 8x$$

$$\frac{\cancel{9}y}{\boxed{\cancel{9}}} = \frac{6z - 8x}{\boxed{9}} \qquad \text{Divide each side by 9.}$$

$$y = \frac{6z - 8x}{9}$$

Example 5: Solve $r = \frac{d}{t}$ for t.

Solution:

$$r = \frac{d}{t} \qquad \text{Clear of fractions.}$$

$$r \cdot \boxed{t} = \frac{d}{\cancel{t}} \cdot \boxed{\cancel{t}} \qquad \text{Multiply each side by } t.$$

$$rt = d$$

$$\frac{\cancel{r}t}{\boxed{\cancel{r}}} = \frac{d}{\boxed{r}} \qquad \text{Divide each side by } r.$$

$$t = \frac{d}{r} \qquad \text{is the solution.}$$

EXERCISE 3.7

Solve each literal equation for the letter listed.

1. $a + y = b$ for y
2. $x - y = z$ for x
3. $x - y = z$ for y
4. $ax = b$ for x
5. $x + ay = z$ for y
6. $I = prt$ for r
7. $P = a + b + c$ for b
8. $E = mc^2$ for m

9. $5(x + y) = z$ for y

10. $a(x + y) = z$ for y

11. $A = \frac{1}{2}bh$ for b

12. $h = 3(p + 2k)$ for k

13. $t = 3t(2a + b)$ for b

14. $6s - 4t = 5(s + t)$ for t

15. $7x + 5y = 3(x - y)$ for x

16. $I = \frac{E}{R}$ for R

17. $\frac{PV}{T} = C$ for P

18. $A = P(1 + rt)$ for t

19. $s = \frac{1}{2}gt^2$ for g

20. $A = \frac{h}{2}(b_1 + b_2)$ for b_1

21. $\frac{1}{3}x + \frac{1}{2}y = \frac{2}{3}z$ for y

22. $\frac{1}{2}(x - y) = \frac{1}{4}(x + y)$ for x

23. $C = \frac{5}{9}(F - 32)$ for F

24. $S = \frac{a}{1 - r}$ for r

25. $I = \frac{E}{R + r}$ for R

Answers to odd-numbered exercises on page 463.

3.8 INEQUALITIES

An equation is a statement that two algebraic expressions are equal. An **inequality** is a statement that one algebraic expression is either greater than or less than another one.

$$x - 3 < 6$$
$$2x - 5 > 4x + 7$$
$$5x < 2x - 6$$

are all examples of inequalities.

Fortunately, with only one exception, the rules for solving *inequalities* are the same ones we have learned for solving *equations*.

To solve an inequality, isolate the variable on one side of the inequality sign and put everything else on the other side of the inequality sign.

The following examples will show how the methods are the same or different for equations and inequalities:

Example 1: Starting with the true inequality: $4 < 6$

(a)	$4 < 6$	
	$4 \boxed{+3} < 6 \boxed{+3}$	Add the same number, 3, to both sides of the inequality.
	$7 < 9$	A true inequality results.
(b)	$4 < 6$	
	$4 \boxed{-3} < 6 \boxed{-3}$	Subtract the same number, 3, from both sides of the inequality.
	$1 < 3$	A true inequality results.
(c)	$4 < 6$	
	$\boxed{3}(4) < \boxed{3}(6)$	Multiply both sides of the inequality by the same positive number, 3.
	$12 < 18$	A true inequality results.
(d)	$4 < 6$	
	$\frac{4}{\boxed{2}} < \frac{6}{\boxed{2}}$	Divide both sides of the inequality by the same positive number, 2.
	$2 < 3$	A true inequality results.
(e)	$4 < 6$	
	$\boxed{-3}(4) \; ? \; \boxed{-3}(6)$	Multiply both sides of the inequality by the same *negative* number, −3.
	$-12 \boxed{>} -18$	The *direction* or sense of the inequality sign must be *changed* to obtain a true inequality.
(f)	$4 < 6$	
	$\frac{4}{\boxed{-2}} \; ? \; \frac{6}{\boxed{-2}}$	Divide both sides of the inequality by the same *negative* number, −2.
	$-2 \boxed{>} -3$	The *direction* or sense of the inequality sign must be *changed* to obtain a true inequality.

(a) and (b) illustrate that we can add or subtract the same number on both sides of the inequality.

(c) and (d) illustrate that we can multiply or divide both sides of an inequality by the same *positive* number.

(e) and (f) illustrate that if both sides of an inequality are multiplied or divided by a *negative* number, the direction of the inequality must be reversed.

These procedures for solving an inequality are summarized by the following:

> **To solve an inequality:**
>
> Use all the same rules and procedures for solving equations except, when multiplying or dividing both sides of an inequality by a negative number, the direction of the inequality symbol must be reversed.

The expression $x \leq y$ is read "x is less than or equal to y" and means if either $x < y$ or $x = y$, then $x \leq y$ is true.

$$2 \leq 7 \text{ because } 2 < 7$$

$$2 \leq 2 \text{ because } 2 = 2$$

The expression $x \geq y$ is read "x is greater than or equal to y," and means if either $x > y$ or $x = y$, then $x \geq y$ is true.

$$9 \geq 5 \text{ because } 9 > 5$$

$$6 \geq 6 \text{ because } 6 = 6$$

To solve inequalities with the symbols $\leq$ or $\geq$ we also follow these same rules.

Example 2: Solve $4x - 1 > 2x + 7$

Solution:

$4x - 1 > 2x + 7$	Isolate the x-terms on the left side.
$4x \boxed{-2x} - 1 > 2x \boxed{-2x} + 7$	Subtract $2x$ from each side.
$2x - 1 > 7$	
$2x - 1 \boxed{+1} > 7 \boxed{+1}$	Add 1 to each side.
$2x > 8$	
$\frac{2x}{\boxed{2}} > \frac{8}{\boxed{2}}$	Divide each side by 2.
$x > 4$	is the solution.

To say "$x > 4$ is the solution" means that if x is replaced by any number greater than 4 in the original inequality, a true statement will result.

Example 3: Solve $2x + 3 \leq 5(x - 3)$

Solution:

$2x + 3 \leq 5(x - 3)$	
$2x + 3 \leq 5x - 15$	Clear of parentheses.
$2x + \boxed{(-5x)} + 3 + \boxed{(-3)} \leq 5x + \boxed{(-5x)} - 15 + \boxed{(-3)}$	Add $(-5x)$ and (-3) to each side.

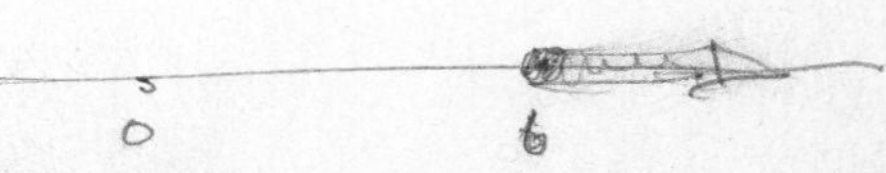

$$-3x \le -18$$

$$\frac{-\cancel{3}x}{\boxed{-\cancel{3}}} \;?\; \frac{-18}{\boxed{-3}}$$

Dividing each side by *negative* 3 changes the direction of the inequality symbol.

$$x \ge 6$$

is the solution.

Example 4: Solve $\frac{2x}{3} < \frac{3x}{4} + 3$

Solution:

$$\frac{2x}{3} < \frac{3x}{4} + 3$$

$$\overset{4}{\cancel{12}}\left(\frac{2x}{\underset{1}{\cancel{3}}}\right) < \overset{3}{\cancel{12}}\left(\frac{3x}{\underset{1}{\cancel{4}}}\right) + 12\,(3)$$

Multiply by the LCD, 12.

$$8x < 9x + 36$$

$$8x \boxed{-9x} < 9x \boxed{-9x} + 36$$

Add -9 to both sides.

$$-x < 36$$

$-x$ is the same as $-1x$.

$$-1x < 36$$

$$\frac{-\cancel{1}x}{\boxed{-\cancel{1}}} \;?\; \frac{36}{\boxed{-1}}$$

Divide each side by -1. Reverse the direction of the inequality.

$$x > -36$$

is the solution.

EXERCISE 3.8

Solve the inequalities:

1. $x + 6 > 3$
2. $x - 5 < 7$
3. $y + 7 \ge -6$
4. $t - 4 \le -5$
5. $3x > 18$
6. $7y \le 28$
7. $-10x < 50$
8. $-4x \ge -28$
9. $6x + 2 > 14$
10. $2y - 5 \ge 15$
11. $3x < 5x + 6$
12. $6x + 3 < 8x - 5$
13. $4(x + 1) > 3x + 7$
14. $6(2 - x) \le 4x - (x - 4)$

15. $5(x + 3) + 3 \geq x$

16. $3(3 + 2y) - (3y + 2) < 12$

17. $4(x - 3) \geq 6(3 - x)$

18. $7(b - 2) \leq 3(b + 2) + 4$

19. $6(y + 2) < 13(y - 2) - 4$

20. $7(t - 3) + 2t < 3(t + 5)$

21. $\frac{a}{3} \leq -5$

22. $\frac{h}{-4} > 7$

23. $\frac{3}{4}x < 12$

24. $\frac{-2}{3}t \geq 6$

25. $\frac{-5}{8}y > -25$

26. $\frac{a}{2} > \frac{a}{5} + 6$

27. $\frac{x}{3} - 1 \geq \frac{x}{2}$

28. $\frac{y}{4} + \frac{y}{3} < \frac{4}{2} - 2$

29. $\frac{2}{3}x + 2 > \frac{1}{2}x - \frac{3}{4}$

30. $\frac{2x}{5} - \frac{1}{4} \leq \frac{3}{4}(x + 2)$

Answers to odd-numbered exercises on page 463.

3.9 Chapter Summary

(3.1) **Solution.** The solution to an equation is a value which when substituted for the variable makes the equation a true statement.

(3.1) To check a solution to an equation

1. Substitute the proposed solution for the variable wherever it appears in the equation.
2. Evaluate each side of the equation
3. If the left and right sides of the equation are equal, then the proposed solution satisfies the equation.
4. If the left and right sides of the equation are not equal, then the proposed solution does not satisfy the equation.

(3.2) **Addition principle:** The same number may be added to (or subtracted from) each side of any equation and the solution will remain the same.

(3.3) **Multiplication principle:** If both sides of an equation are multiplied by (or divided by) the same number (except zero), the new equation is equivalent to the original equation.

(3.3) **Reciprocal.** The reciprocal of $\frac{a}{b}$ is $\frac{b}{a}$.

(3.4) To solve an equation using both the addition and multiplication principles:

1. Put all the terms containing the variable on one side of the equation by using the addition principle.
2. Put all the plain numbers on the other side of the equation by using the addition principle again.
3. Multiply both sides of the equation by the reciprocal of the coefficient of the variable.

(3.5) To eliminate fractions from an equation, multiply each side of the equation by the LCD of all the fractions that appear in the equation.

Examples

(3.1) Is $x = 3$ a solution to $2x + 3 = 5x - 6$? Substitute 3 for x.

$$2 \cdot 3 + 3 \stackrel{?}{=} 5 \cdot 3 - 6$$
$$6 + 3 \stackrel{?}{=} 15 - 6$$
$$9 = 9$$

Yes, $x = 3$ is a solution.

(3.2) Solve $x + 5 = -3$.

$$x + 5 + (-5) = -3 + (-5)$$
$$x = -8$$

(3.3) Solve $\frac{2}{3}x = 12$. Multiply both sides by $\frac{3}{2}$, the reciprocal of $\frac{2}{3}$.

$$\frac{\cancel{3}}{\cancel{2}} \cdot \frac{\cancel{2}}{\cancel{3}}x = \frac{3}{\cancel{2}_{1}} \cdot \cancel{12}^{6}$$
$$x = 18$$

(3.4) Solve $8x + 1 = 5x - 11$.

$$8x + (-5x) + 1 = 5x + (-5x) - 11$$
$$3x + 1 = -11$$
$$3x + 1 + (-1) = -11 + (-1)$$
$$3x = -12$$
$$\frac{3x}{3} = \frac{-12}{3}$$
$$x = -4$$

(3.5) Solve $\frac{1}{2}x + \frac{3}{4} = \frac{1}{3}$. The LCD is 12.

$$(12)\frac{1}{2}x + (12)\frac{3}{4} = (12)\frac{1}{3}$$
$$\cancel{12}^{6} \cdot \frac{1}{\cancel{2}}x + \cancel{12}^{3} \cdot \frac{3}{\cancel{4}} = \cancel{12}^{4} \cdot \frac{1}{\cancel{3}}$$
$$6x + 9 = 4$$
$$6x + 9 + (-9) = 4 + (-9)$$

$$6x = -5$$

$$x = \frac{-5}{6}$$

(3.5) To clear an equation of decimals:

1. Count the greatest number of decimal places appearing in any one decimal in the equation.
2. Multiply each side of the equation by 1 followed by that number of zeros.

(3.5) Solve $.03x + .21 = -.06$.
Multiply by 100.

$$(100).03x + (100).21 = (100)(-.06)$$

$$3x + 21 = -6$$

$$3x = -27$$

$$x = -9$$

(3.6) **Conditional equations** are equations that are true for only certain values of the variable.

(3.6) An **identity** or an **identical equation** is an equation that is true no matter what value is substituted for the variable.

(3.6) A **contradiction** is an equation that has no solutions.

(3.6) When solving an equation, it is:

1. *Conditional* if a solution results.
2. *An identity* if $0 = 0$ results.
3. *A contradiction* if an unequal expression like $0 = 4$ results.

(3.7) **Literal equations** are equations containing more than one letter.

(3.7) Solve $3a + 2x = b$ for x.

$$3a + (-3a) + 2x = b + (-3a)$$

$$2x = b - 3a$$

$$x = \frac{b - 3a}{2}$$

(3.8) An **inequality** is a statement that one algebraic expression is either greater than or less than another algebraic expression.

To solve an inequality use the same rules and procedures for solving equations, except when multiplying or dividing both sides of an inequality by a negative number, the direction of the inequality symbol must be reversed.

(3.8) Solve $3x - 7 < 8$.

$$3x - 7 + 7 < 8 + 7$$

$$3x < 15$$

$$x < 5$$

Exercise 3.9 Chapter Review

(3.1) In exercises 1 through 6 determine whether $x = -3$ is a solution.

1. $-2x = 6$
2. $2x + 1 = 7$
3. $\frac{6}{x} = 2$
4. $3x - 1 = 4x + 4$
5. $2x + 4 = x - 1$
6. $4(x + 1) = -(5 - x)$

(3.2-3.4) Solve each equation and check:

7. $x - 4 = 7$
8. $-3x = -27$
9. $6 + x = 5$
10. $9x = -27$
11. $\frac{h}{3} = -4$
12. $\frac{2}{3}x = -12$
13. $7y + 5 = -6y + 4$
14. $6x - 4 = 7x + 5$
15. $2a + 6 = 6a - 18$
16. $7t - 6 = -4t - 35$

(3.5) Solve each equation and check:

17. $3(x - 2) = 12$
18. $3a - 5(a + 3) = 2 - (a + 4)$
19. $2y - 3(y - 1) = 2(y + 1)$
20. $5(y + 4) + 2 = 3(2 - y) - 4$

21. $.41x + .3 = .39x - .42$

22. $\frac{1}{4}y - 2 = \frac{1}{2}y + \frac{3}{4}$

23. $\frac{1}{2}x + 2 = \frac{1}{3}x - 3$

24. $4(x + .21) = 3(x + .01)$

25. $\frac{2}{3}(x - 3) = \frac{1}{4}(x + 1)$

26. $\frac{1}{2}(t - 2) = \frac{1}{3}(t - 2)$

(3.6) Classify each equation as conditional, an identity, or a contradiction. Solve each conditional equation.

27. $-2(x + 3) = 4x - 6(x + 1)$

28. $6(x - 3) = 2(x + 1) - x$

29. $4(x - 1) = 2x - 2(3 - x)$

30. $3x - (4 + 2x) = 7 - 4x - (-3x + 1)$

31. $4x + 6 = 2(3x - 1)$

32. $1 - 7(x + 1) = -5x + 2(3 - x)$

(3.7) Solve for the indicated variable:

33. $a + x = 2b$ for x

34. $ax + b = 3c$ for x

35. $V = lwh$ for h

36. $x = \dfrac{y}{2z}$ for z

37. $\dfrac{x}{y} = \dfrac{a}{b}$ for y

38. $\dfrac{2a}{b} = c$ for b

39. $\frac{1}{2}(x - 2y) = \frac{1}{3}(x + 2)$ for x

40. $a(b - 2y) = c$ for y

41. $3x + 4y = 5z$ for y

42. $\frac{2}{3}m + \frac{1}{4}n = \frac{1}{6}$ for n

(3.8) Solve the inequalities:

43. $3 + x > 5$

44. $-2x - 1 \geq 7$

45. $-6y \leq 3y - 9$

46. $2 + y > 4y - 6$

47. $3(z - 1) < 5(2 - z)$

48. $6x + 2 \leq 7x - 5$

49. $5y + 6 < 2y - 3$

50. $2z > 3(z + 2)$

51. $\frac{1}{2}x - 5 > \frac{2}{3}x + 1$

52. $\frac{3}{4}(x - 8) \leq \frac{1}{2}(x + 2)$

Answers to odd-numbered exercises on page 463.

Chapter 3 Achievement Test

Name ______________________________

Class ______________________________

This test should be taken before you are tested in class on the material in Chapter 3. Solutions to each problem and the section where the type of problem is found are given on page 464.

Determine whether $y = 4$ is a solution to the following equations:

1. $2y + 3 = 5y - 7$ — 1. ____________

2. $-2(y-2) = 3(3-y) - 1$ — 2. ____________

Solve each equation:

3. $x - 7 = 4$ — 3. ____________

4. $\frac{x}{3} = -2$ — 4. ____________

5. $7 - 2x = -3$ — 5. ____________

6. $3x - 4 = 5x + 10$ — 6. ____________

7. $3(x + 4) = 5(x - 2)$ — 7. ____________

8. $.3x + .4 = .6x - .5$ — 8. ____________

9. $\frac{2}{3}x + \frac{1}{6} = \frac{3}{4}x - \frac{1}{2}$ — 9. ____________

10. $\frac{1}{2}(x - 3) = \frac{2}{3}(2 - x)$

11. Check problem 6.

Classify as conditional, an identity, or a contradiction. Solve the conditional equations:

12. $-2(x + 3) - 4 = 3x - 5(x + 2)$

13. $4(3 + x) = 2x + 2(x - 5)$

14. $2x - (4 - x) = 4(x - 1)$

Solve for the indicated variable.

15. $2x - 3y = -4z$ for y

Solve each inequality.

16. $-3x + 1 < -8$

17. $\frac{1}{4}x - \frac{2}{3} \geq \frac{1}{2}x - 2$

10. ______________________

11. ______________________

12. ______________________

13. ______________________

14. ______________________

15. ______________________

16. ______________________

17. ______________________

4

Applications of Linear Equations

4.1 TRANSLATING FROM ENGLISH TO ALGEBRA

Problems in the real world rarely are stated in algebraic equations. The equations usually come well camouflaged within the English language in the form of word problems. Our first job will be to translate English statements into mathematical expressions and equations.

Certain words and phrases appear over and over again in word problems that tell us which mathematical operations are to be used. Following is a table of such key words and phrases that tell us when we should add, subtract, multiply, or divide and also which words mean "equal to."

ADD	*SUBTRACT*	*MULTIPLY*	*DIVIDE*	*EQUALS*
$a + b$	$a - b$	$a \cdot b$	$a \div b$ or $\frac{a}{b}$	$a = b$
a plus b	a minus b	a times b	a divided by b	a equals b
the sum of a and b	the difference of a and b	the product of a and b	the quotient of a and b	a is the same as b
add a and b	subtract b from a	multiply a and b		a is equivalent to b
b more than a	b less than a	of		a is identical to b
a increased by b	a less b a decreased by b			is, are, was

Now we will consider some English phrases and translate them into mathematical expressions.

Example 1: Write a mathematical expression from each English phrase.

(a) A number increased by 6 equals 9.

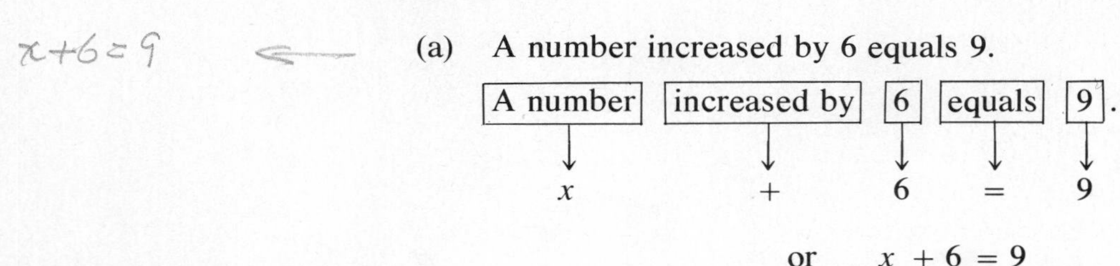

or $x + 6 = 9$

(b) Twice the length is the same as 5 times the width.

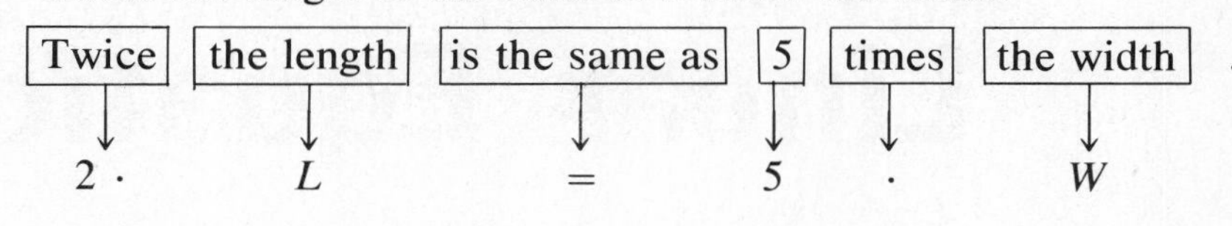

$$2L = 5W$$

(c) When the sum of x and 4 is divided by 3 the result is 6 .

When the [sum of x and 4] [is divided by] [3] the result [is] [6] .

$(x + 4)$ $\div$ 3 $= 6$

$$(x + 4) \div 3 = 6$$

or $$\frac{x + 4}{3} = 6$$

(d) 3 more than Fred's age is 22.

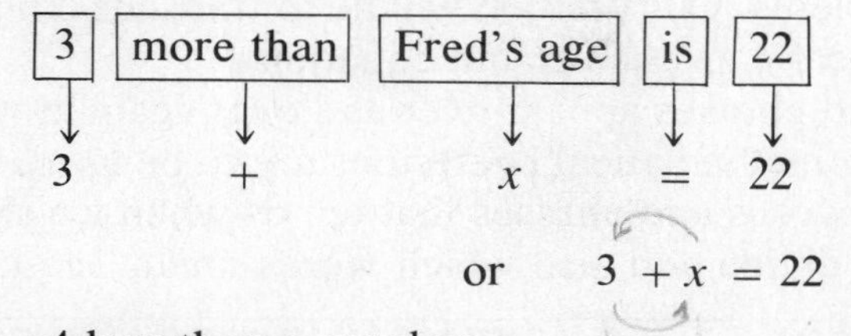

or $3 + x = 22$

(e) 4 less than a number.

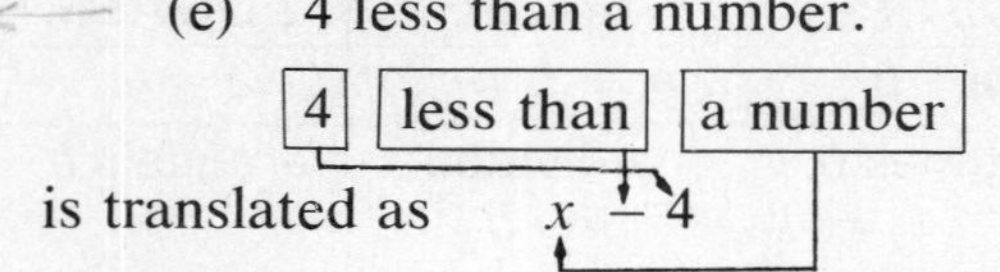

is translated as $x - 4$

(f) $\frac{2}{3}$ of the altitude equals 3 more than twice the base.

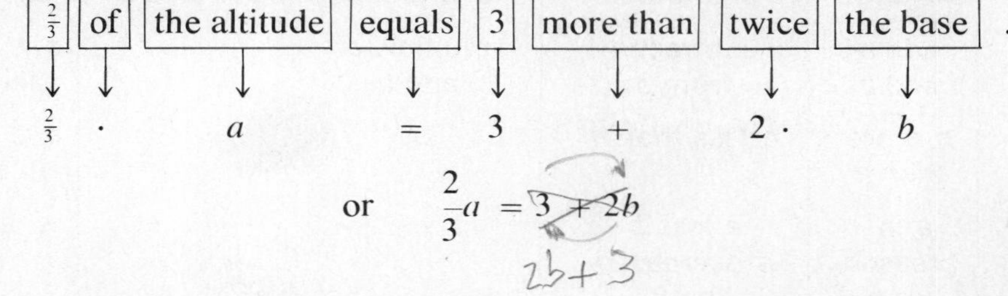

or $$\frac{2}{3}a = 3 + 2b$$

Example 2: Write each phrase as an algebraic expression:

English	*Algebra*
a. A number plus 3 is 14.	$x + 3 = 14$
b. The product of a number and 5 is 35.	$x \cdot 5 = 35$ or $5x = 35$
c. 6 added to a number is fourteen.	$x + 6 = 14$
d. Twice the sum of a number and 6 is 18.	$2(x + 6) = 18$
e. The difference between some number and 4 is 10.	$x - 4 = 10$
f. Two-thirds of an unknown number is 8.	$\frac{2}{3}x = 8$
g. A certain number is 6 less than twice itself.	$x = 2x - 6$
h. A number is 10 more than six times itself.	$x = 6x + 10$
i. The quotient of some number and 6 is 8.	$\frac{x}{6} = 8$

EXERCISE 4.1

pag 464

Write each English sentence as a mathematical equation. Choose any letter to represent an unknown quantity.

1. Six more than a number is twelve. ⇒ $x+6=12$ ✓

2. A number increased by six is fifteen. ⇒ $x+6=15$

3. Seventeen is the same as the sum of four and some number. ⇒  ✓

4. The letter x represents a number that is five less than eleven. ⇒ $x = 11-5$

5. Some number divided by negative four is five. ⇒  ✓

6. Seven times an unknown number is twenty-eight. ⇒ $7x = 28$

7. An unknown number is five greater than eight. ⇒ $x = 8+5$ ✓

8. In six years a boy will be twice as old as he is now.

9. The difference between an unknown number and five is equal to twelve.

10. The product of a number and seven is forty-two.

11. The age of a building in four years will be twice as old as it was last year.

12. Nine is three less than some number.

13. Last year I was half as old as I will be in eighteen years.

14. Twice a number increased by six is twenty-five.

15. An unknown number is six less than twice itself.

16. Six times a number plus four times the same number is thirty.

17. In four years I will be twice as old as I was eight years ago.

18. Fifteen minus five-eighths of a number equals five.

19. If you subtract a number from eight and then add six, the result is ten.

20. Twice the sum of a number and six is twenty-four.

21. Three more than a number is the same as two subtracted from twice the number.

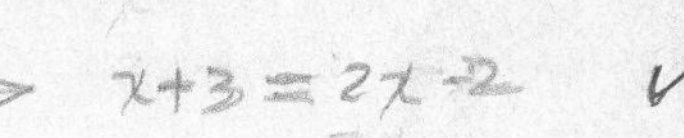

22. When six is added to four times a number, the result is equal to four less than six times the same number.

23. When three times an unknown number is subtracted from sixteen and that result divided by five, the quotient is equal to the unknown number.

24. One number is six times another and their sum is twenty-six.

25. Seven less than twice some number is the same as the quotient of the number and three.

26. Six years ago, a person's age was two-thirds of what it will be six years from now.

Answers to odd-numbered exercises on page 464.

4.2 NUMBER RELATION PROBLEMS

Next we will combine our knowledge of solving equations with our ability to translate from English to mathematics. Solving word problems has traditionally been one of the most difficult parts of algebra for students. Following is a "battle-plan" that has helped many students learn to solve word problems. If you are willing to follow the steps of this strategy carefully through the examples and exercises, your efforts will be rewarded and, indeed, you *will* learn to solve word problems.

Strategy for solving word problems:

1. Read the problem several times, making note of any key words.
2. Write down what you are to find and what is given.
3. Represent one of the unknown quantities by a variable. Express any other unknown quantities in terms of the *same* variable.

4. Find out which expressions are the same and write an equation. Sometimes a simple sketch helps determine which expressions are equal.
5. Solve the equation.
6. Check your solution in the *original word statement* to see whether or not it really meets the conditions of the problem. This will catch an error that might be made in writing the equation.

Example 1: Three times a number increased by seven is equal to thirty-one. Find the number.

Solution: Let x = the unknown number.

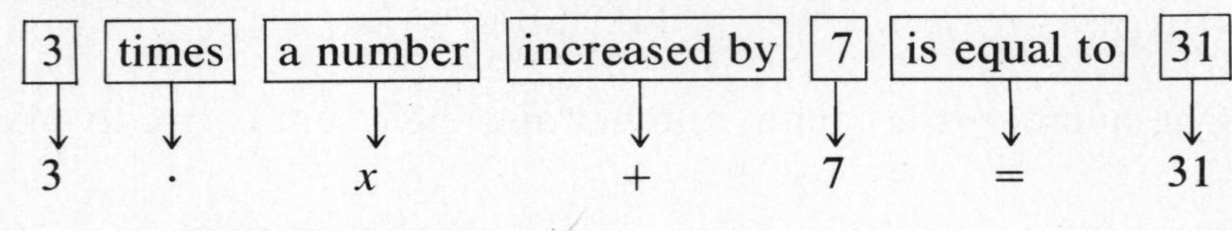

or

$$3x + 7 = 31$$
$$3x + 7 - 7 = 31 - 7$$
$$3x = 24$$
$$x = 8$$

The unknown number is 8.

Now put our solution $x = 8$ back in the original word statement to see if it meets the condition of the problem.

Check:

3	times	a number	increased by	7	is equal to	31
↓	↓	↓	↓	↓	↓	↓
3	·	8	+	7	=	31

$$3 \cdot 8 + 7 \stackrel{?}{=} 31$$
$$24 + 7 \stackrel{?}{=} 31$$
$$31 = 31 \checkmark$$

Example 2: Twice the difference of a number and 6 is 10. Find the number.

Solution: Let x = the desired number.

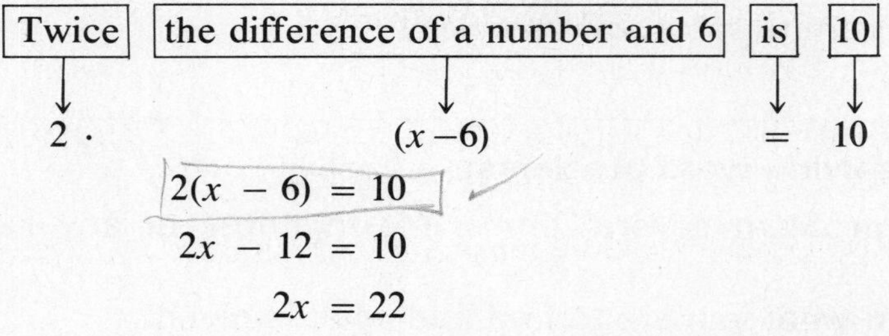

$$2(x - 6) = 10$$
$$2x - 12 = 10$$
$$2x = 22$$
$$x = 11$$

The desired number is 11.

Check: $x = 11$ checks because twice the difference of 11 and 6, $(2 \cdot 5)$, is indeed 10.

Example 3: The sum of two numbers is 84. If one number is one-half the other, find the numbers.

Solution: Let x = one of the numbers, then $\frac{1}{2}x$ = the other number.

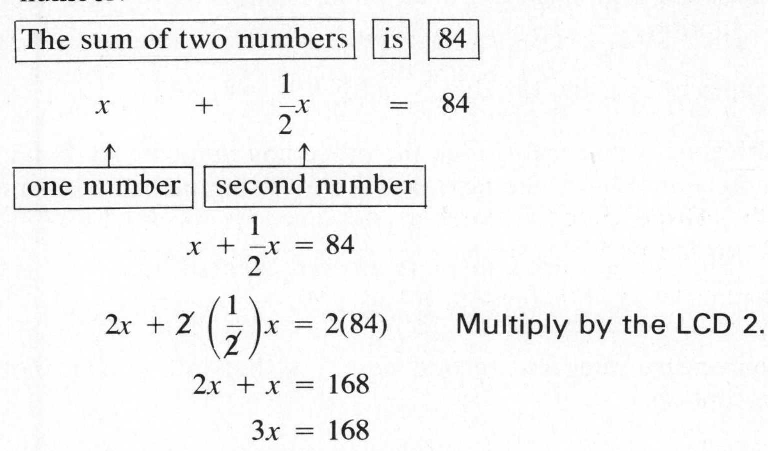

$$x + \frac{1}{2}x = 84$$

$$x + \frac{1}{2}x = 84$$

$$2x + 2\left(\frac{1}{2}\right)x = 2(84) \qquad \text{Multiply by the LCD 2.}$$

$$2x + x = 168$$

$$3x = 168$$

$$x = 56 = \text{one number}$$

$$\frac{1}{2}x = \frac{1}{2}(56) = 28 = \text{second number}$$

So the numbers are 56 and 28.

Check: 28 is one-half of 56 and $28 + 56 = 84$

Example 4: In a math class of 30 students, there are 4 more males than there are females. How many of each sex are in the class?

Solution: Let x = the number of females. Then $x + 4$ = the number of males (since there are 4 more males than females).

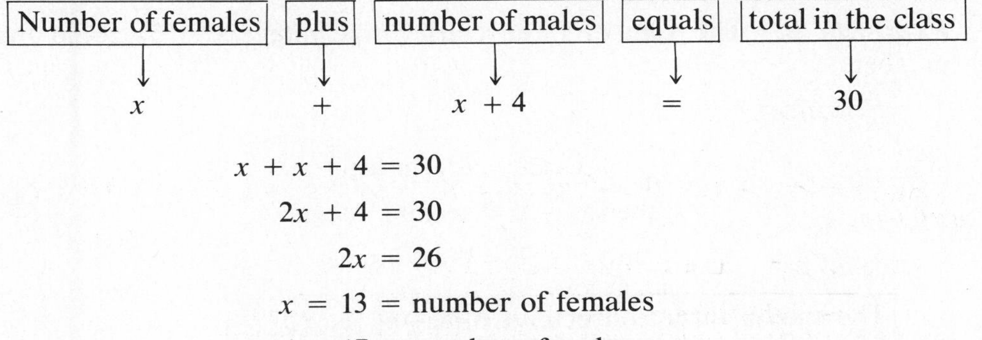

$$x \quad + \quad x + 4 \quad = \quad 30$$

$$x + x + 4 = 30$$

$$2x + 4 = 30$$

$$2x = 26$$

$$x = 13 = \text{number of females}$$

$$x + 4 = 17 = \text{number of males}$$

Check: There are 4 more males than females (17 is 4 greater than 13), and since $13 + 17 = 30$, the answers check.

Consecutive integers are numbers that follow right after one another.

Examples: 5, 6, 7, 8, 9, . . . and −14, −13, −12, −11, ...

If x is some integer, the next consecutive integer is one larger, or $x + 1$. The next one is one larger than that, or $x + 2$, and so on.

> Consecutive integers are represented by:
> $x, x + 1, x + 2, x + 3, \ldots$

Even consecutive integers are even numbers that follow one another without interruption.

Examples: 14, 16, 18, 20, . . . and −10, −8, −6, ...

Each integer is two greater than the preceding number, so if an even integer is represented by x, the next one is $x + 2$, the next $x + 4$, and so on.

> Even consecutive integers are represented by:
> $x, x + 2, x + 4, x + 6, \ldots$

Odd consecutive integers are odd numbers that follow one another without interruption.

Examples: 11, 13, 15, 17, . . . and −5, −3, −1, 1, ...

Each integer is two greater than the last number. So, if x represents some odd integer, the next consecutive odd integer is two greater, or $x + 2$; the next will be $x + 4$, and so on.

> Odd consecutive integers are represented by:
> $x, x + 2, x + 4, x + 6, \ldots$

You will notice that both even and odd consecutive integers are represented by $x, x + 2, x + 4, x + 6, \ldots$. This is because both even and odd consecutive integers differ from the preceding integer by two.

Example 5: The sum of three consecutive integers is 57. Find the numbers.

Solution:

Let x = first integer
then $x + 1$ = second integer
and $x + 2$ = third integer

[The sum of three consecutive integers] [is] [57]

$$x + (x + 1) + (x + 2) = 57$$

$$x + x + 1 + x + 2 = 57$$
$$3x + 3 = 57$$
$$3x = 54$$
$$x = 18 = \text{first integer}$$
$$x + 1 = 19 = \text{second integer}$$

$$x + 2 = 20 = \text{third integer}$$

Check: 18, 19, 20 are consecutive integers and $18 + 19 + 20 = 57$

Example 6: Find three consecutive even integers such that the sum of the first and third integers is equal to twice the middle integer.

Solution:

$$\text{Let } x = \text{first integer}$$
$$x + 2 = \text{second integer}$$
$$x + 4 = \text{third integer}$$

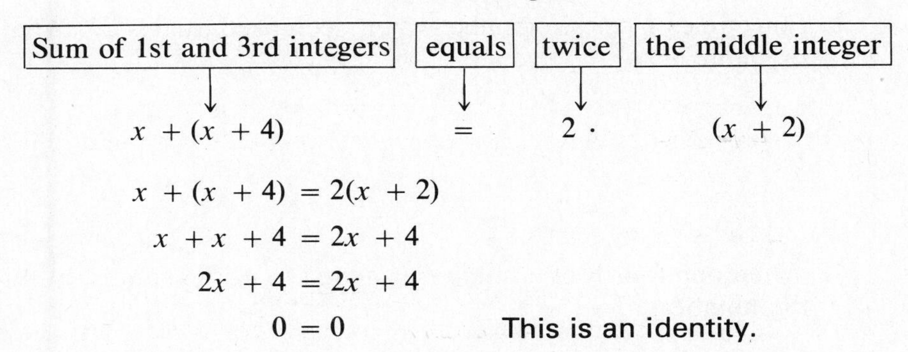

$$x + (x + 4) = 2(x + 2)$$
$$x + x + 4 = 2x + 4$$
$$2x + 4 = 2x + 4$$
$$0 = 0 \qquad \text{This is an identity.}$$

In an identity, any value of the variable will be a solution. This means that if we take any three consecutive even integers, the sum of the first and the third integers will equal twice the middle one. Try a couple of examples and see for yourself!

Example 7: Find three consecutive odd integers such that twice the smallest added to the largest is 67.

Solution:

$$\text{Let } x = \text{first integer}$$
$$x + 2 = \text{second integer}$$
$$x + 4 = \text{third integer}$$

Twice the smallest	added to	the largest	is	67
↓	↓	↓	↓	↓
$2 \cdot x$	$+$	$(x + 4)$	$=$	67

$$2x + x + 4 = 67$$
$$3x + 4 = 67$$
$$3x = 63$$
$$x = 21 = \text{first integer}$$
$$x + 2 = 23 = \text{second integer}$$
$$x + 4 = 25 = \text{third integer}$$

Check: Twice the first is $2(21) = 42$, added to the third, 25, is $42 + 25 = 67$, and it checks.

EXERCISE 4.2

Solve and check the following:

1. When 6 is added to two times a number the result is 22. Find the number.

$2x+6=22$
$2x=22-6$
$2x=16$
$x=\frac{16}{2}=8$ $\boxed{x=8}$ ✓

2. Four times a certain number when decreased by 5 is 27. What is the number?

$4x-5=27$

3. When one-fourth of a number is added to 4, the sum is 10. Find the number.

$4+\frac{1}{4}x=10$
$\frac{16+x}{4}=10$
$16+x=40$
$x=40-16$
$\boxed{x=24}$ ✓

4. Find three consecutive numbers whose sum is 63.

x
$x+1$
$x+2$

$x+x+1+x+2=36$
$3x+3=36$
$3x=36-3$
$3x=33$
$x=\frac{33}{3}$
$\boxed{x=11}$

5. When a number is subtracted from 21 the result is twice the number. Find the number.

$21-x=2x$
$-x-2x=-21$
$-3x=-21$
$x=\frac{-21}{-3}$ $\boxed{x=7}$ ✓

6. Find three consecutive even integers whose sum is 90.

x
$x+2$
$x+4$

$x+x+2+x+4=90$
$3x+6=9$
$3x=90-6$
$3x=84$
$x=\frac{84}{3}$
$\boxed{x=28}$ ✓

7. If 16 is subtracted from two-thirds of a number, the result is 8. Find the number.

$\frac{2}{3}x-16=8$
$\frac{2}{3}x=8+16$
$\frac{2}{3}x=24$
$2x=72$
$x=\frac{72}{2}$
$\boxed{x=36}$ ✓

8. 8 plus an unknown number is the same as twice the difference of the number and 2. What is the number?

9. Find three consecutive odd integers such that the sum of the first two is 3 greater than the third.

10. Find two consecutive even integers such that four times the first is 2 greater than three times the second.

11. The sum of two integers is 124. If the second is 4 more than four times the first, find both integers.

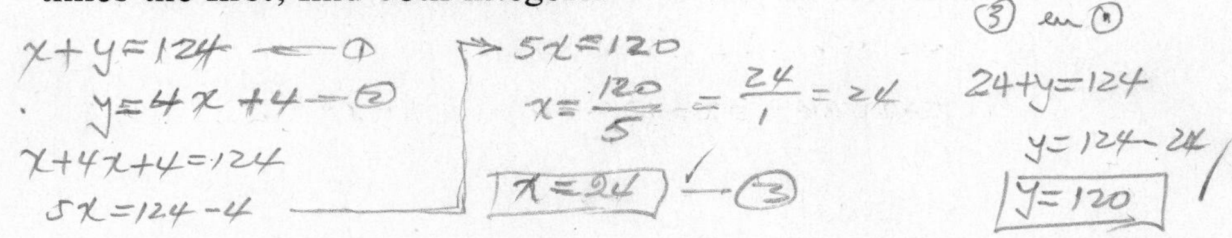

12. A man is three times as old as his son. Find the age of each if the sum of their ages is 48.

13. The total cost of a physics book and a math book is $42. If the math book cost $6 less than the physics book, how much does each book cost?

14. Three-fourths of a number added to twice the number is equal to 44. What is the number?

15. One-third of a number subtracted from one-half of the number is 6. Find the number.

16. When a number is decreased by 18, the result is one-half of the number.

17. Find three consecutive odd integers such that the first plus twice the second plus three times the third is equal to 70.

18. 4 more than a number is 3 less than twice that number. What is the number?

19. Find three consecutive odd integers such that three times the sum of the first and second is 7 more than five times the third.

20. The sum of the first two of three consecutive odd integers added to the sum of the last two is 116. Find the integers.

21. Find four consecutive integers such that twice the sum of the first three is 2 less than five times the fourth.

22. The perimeter of a rectangle is 40. Find the length and width of the rectangle if the length is 4 more than the width.

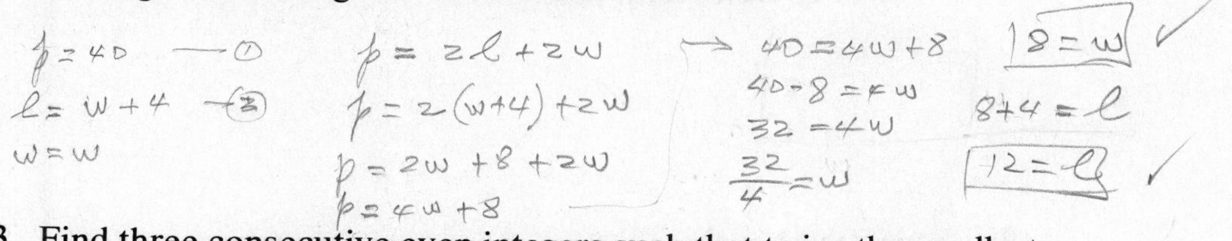

23. Find three consecutive even integers such that twice the smallest is 4 larger than the largest.

24. The sum of two numbers is 175. Three-fourths of the first number is equal to the second number. What are the two numbers?

25. In an election for mayor of a small town, Mrs. Meyers received 6 more than forty times as many votes as her husband, who was running against her. If the total number of votes cast was 498, how many votes did each candidate receive?

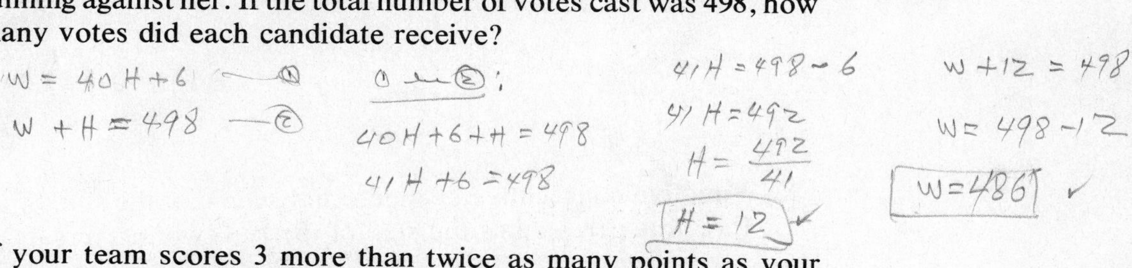

26. If your team scores 3 more than twice as many points as your opponent in a football game, how many points does each team score if the total points scored by both teams is 45?

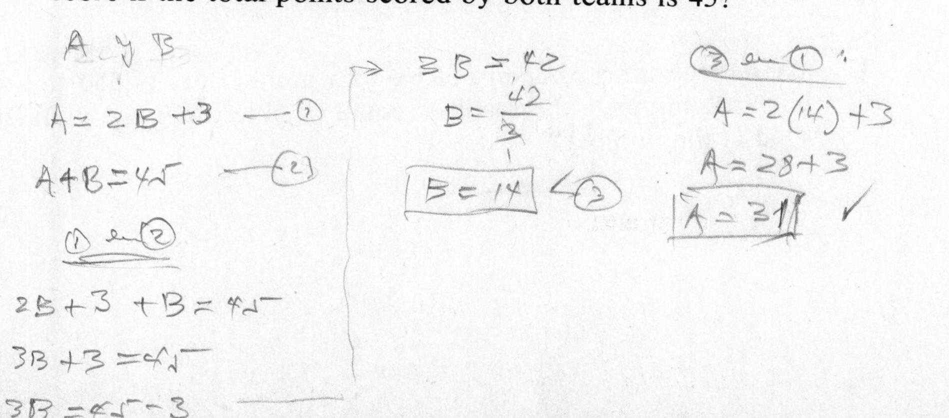

27. Find three consecutive even integers such that four times the first is 12 less than twice the sum of the second and third.

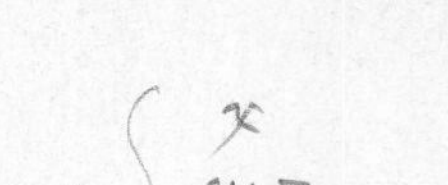

28. Separate 140 into two parts in such a way that the first part is 20 more than the second part.

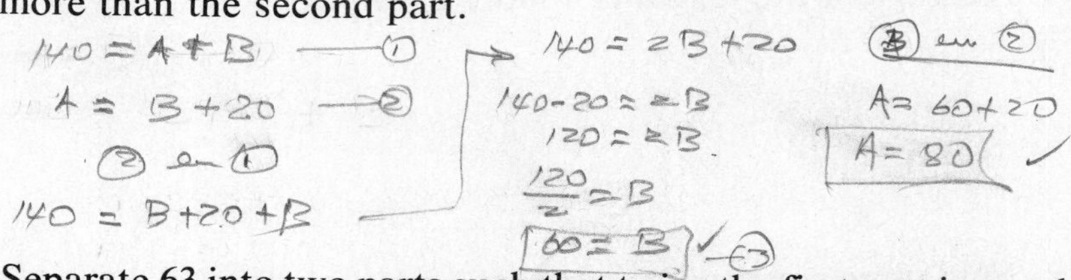

29. Separate 63 into two parts such that twice the first part is equal to the second part.

30. A yardstick is cut into two pieces such that one piece is 7 inches shorter than the other. How long is each piece?

31. A forty-foot rope is cut into two pieces in such a way that one piece is 4 feet longer than twice the other. How long is each piece?

32. Find five consecutive even integers such that the sum of the first three integers equals the sum of the last two integers.

33. When three times the sum of 6 and a number are added to five times the number, the result is equal to nine times the number. What is the number?

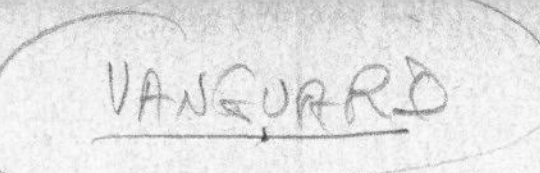

34. Find three consecutive integers whose sum is -105.

35. Find three consecutive odd integers whose sum is -111.

36. A mother is 20 years older than her daughter. Ten years ago the mother was twice as old as the daughter. How old are they now?

37. If you receive a grade of 52 on a history test, what grade must you earn on the second test to bring your average to 70?

38. If you receive grades of 72 and 76 on your first two exams of the semester, what grade would you have to get on the third exam in order to bring your average up to a 90? If the highest grade possible is 100, is this possible?

39. Find four consecutive even integers whose sum is -84.

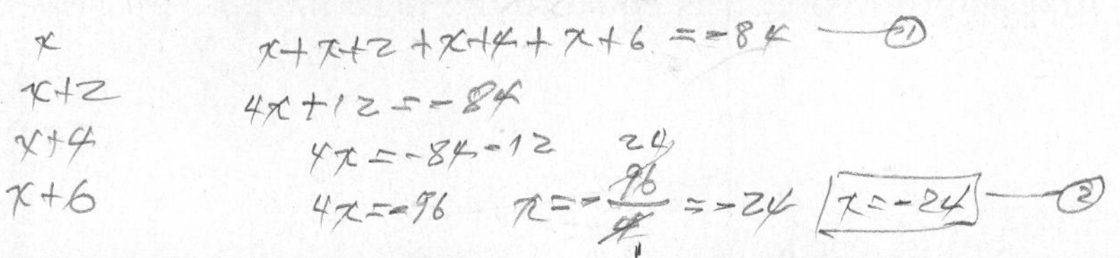

40. Find three consecutive even integers such that the sum of the smallest and the largest is equal to twice the middle integer.

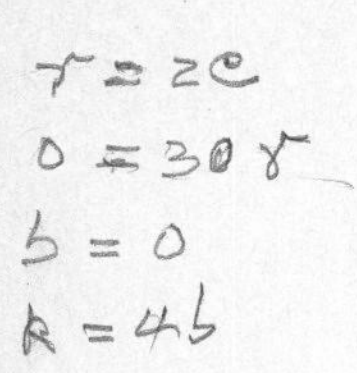

41. A dieter has the following for lunch: a stalk of celery, a radish, an orange, a piece of chocolate layer cake, and a beer. The radish contains twice as many calories as the celery; the orange contains thirty times as many calories as the radish; the beer contains twice as many calories as the orange; and the cake contains four times as many calories as the beer. The total number of calories for the meal is 663. How many calories will the dieter eliminate from the meal if he passes up the celery and the radish?

42. Find three consecutive even integers such that the sum of the smallest and the largest is equal to five times the middle integer.

43. A repairman charges $40 for a service call plus a rate of $25 per hour after the first hour. If his bill was $115, how many hours did he spend on the service call altogether?

44. One-third the sum of a number and 6 is equal to three-fourths of the number. Find the number.

45. Find five consecutive even integers such that the average of the first and fifth integer is equal to the third integer.

46. Find three consecutive even integers such that the sum of the second two is 3 less than twice the first.

47. If you spend half of your money on item A and one-third of your money on item B and you have $30 left, how much money did you have at the beginning?

48. Find three consecutive odd integers such that one-fourth of the sum of the first two is equal to the third.

49. The total cost for a set consisting of a table and six chairs is $727. How much do the table and chairs each cost if eight times the cost of a chair is $1 more than the cost of the table?

50. The length of a rectangle is three less than three times the width. If the perimeter is equal to seven times the width, find the length and width of the rectangle.

Answers to odd-numbered exercises on page 465.

4.3 PERCENT PROBLEMS

A useful application of linear equations is solving word problems involving percent. There are three basic types of percent problems, as illustrated by the following example.

Example 1:

(a) 25% of 80 is what number?
(b) 25% of what number is 50?
(c) 6 is what percent of 24?

Each of the above statements can be represented by an algebraic equation. We always write percent in its decimal form by moving the decimal point two places to the left. For example 25% = .25.

Remembering that the word *of* usually means multiply and *is* means equals,

Solution: (a) [25%] [of] [80] [is] [what number] ?

$$\begin{array}{ccccc} \downarrow & \downarrow & \downarrow & \downarrow & \downarrow \\ .25 & \cdot & 80 & = & x \end{array}$$

$$(.25)(80) = x$$
$$20 = x$$

Solution: (b) [25%] [of] [what number] [is] [50]

$$\begin{array}{ccccc} \downarrow & \downarrow & \downarrow & \downarrow & \downarrow \\ .25 & \cdot & x & = & 50 \end{array}$$

$$.25x = 50$$
$$\boxed{(100)}\ .25x = \boxed{(100)}\ 50 \qquad \text{Multiply by 100 to clear of decimals.}$$
$$25x = 5000$$
$$\frac{\not{25}x}{\not{25}} = \frac{5000}{25}$$
$$x = 200$$

Solution: (c) [6] [is] [what percent] [of] [24] ?

$$\begin{array}{ccccc} \downarrow & \downarrow & \downarrow & \downarrow & \downarrow \\ 6 & = & x & \cdot & 24 \end{array}$$

$$6 = 24x$$
$$\frac{6}{24} = \frac{\not{24}}{\not{24}}x$$
$$.25 = x \qquad \text{Change the decimal to a percent by moving}$$
$$25\% = x \qquad \text{the decimal point two places to the right.}$$

Frequently, word problems involving percent can be rewritten to make them easier to translate into an equation.

Example 2: A house worth \$80,000 is insured for 90% of its value. How much would the owner receive if it were destroyed by fire?

Solution: We are asked to find:

90%	of	80,000	is	what number	?
↓	↓	↓	↓	↓	
.9	·	80,000	=	x	

$$(.9)(80,000) = x$$

$$\$72,000 = x$$

Example 3: The enrollment at Midway Community College was 6400 last year. This year there is an increase of 448 students. What percentage does this increase represent?

Solution: We are being asked to find:

448	is	what percent	of	6400	?
↓	↓	↓	↓	↓	
448	=	x	·	6400	

$$448 = 6400x$$

$$\frac{448}{6400} = \frac{\cancel{6400}}{\cancel{6400}}x$$

$$.07 = x$$

$$7\% = x$$ Write the decimal as a percent.

Example 4: How much money would you have to invest at a rate of 12% to earn \$720?

Solution: We are being asked to find:

12%	of	what number	is	\$720	?
↓	↓	↓	↓	↓	
.12	·	x	=	720	

$$.12x = 720$$

$$(100)\ .12x = (100)\ 720$$ Clear of decimals.

$$12x = 72,000$$

$$\frac{\cancel{12}x}{\cancel{12}} = \frac{72,000}{12}$$

$$x = \$6,000$$

EXERCISE 4.3

Solve each of the following:

1. 40% of 360 is what number?

2. 16 is what percent of 192?

3. 24% of what number is 23.52?

4. Find what percent 12 is of 72.

5. Find 86% of 23.

6. 4% of a number is 72. Find the number.

7. 115% of what number is 80.5?

8. 12 is 36% of what number?

9. 50 is what percent of 275?

10. .4% of 36 is equal to what number?

11. What percent of 45,000 is 150?

12. Find 308% of 41.6.

13. 1.4% of what number is .322?

14. What percent of 90 is 20?

15. 160 is 15% of what number?

16. How much sales tax would you pay on a purchase of $29.00 if the tax rate is 6%?

17. A light bulb manufacturer estimates that 1.2% of the light bulbs that he ships are damaged in transit. How many light bulbs will be damaged in a shipment of 240,000 bulbs?

18. An article which last year sold for $48.00 is increased $6 this year. What percent increase is this?

19. Sales of a small business increased $41,000 over the previous year. If this represented an increase of 26%, what were the previous year's sales?

20. If a student answers 24 out of 30 questions correctly on a test, what percent of the questions were answered correctly?

21. If you got 16 questions wrong out of 80 questions given, what percent did you get wrong? What percent did you get correct?

22. A certain bronze alloy contains 52.3% copper. How many pounds of copper are contained in 600 pounds of the bronze alloy?

23. If a used-car dealer pays $1400 for a car and wants to make a profit of 20%, how much should he sell the car for?

24. How much money must you invest at a rate of 10% per year to earn interest of $30,000 per year?

25. If there are 756 males and 504 females in the freshman class at Smartmore College, what percent are females?

26. If a coat is reduced from $100 to $85 and a suit is reduced from $180 to $165, which was reduced the greater percent?

27. If the interest rate on charge accounts is 1.8% of the unpaid balance per month and your unpaid balance this month is $410, how much interest will you pay this month?

28. The rainfall for the year in Soggybottom County was 46.2 inches. If this represents only 70% of the normal rainfall, what is the normal yearly rainfall?

29. If the dropout rate at a college is 30% in the freshman year, how many students, out of 820 who started, will finish the year?

30. If your salary were to be increased 10% next year and the following year you take a 10% reduction in salary, how would your final salary compare with your present salary? (Hint: Start with some fixed salary and see what happens.)

Answers to odd-numbered exercises on page 465.

4.4 RATIO AND PROPORTION

Ratio

> A ratio is a comparison of two numbers by division.
>
> The ratio of a to b is the fraction $\frac{a}{b}$, $b \neq 0$

Example 1: In a freshman English class, there are 16 females and 9 males. Find the ratio of (a) females to males, (b) males to females, (c) females to total number of students, and (d) males to total number of students.

Solution:

(a) $\dfrac{\text{females}}{\text{males}} = \dfrac{16}{9}$

(b) $\dfrac{\text{males}}{\text{females}} = \dfrac{9}{16}$

(c) $\dfrac{\text{females}}{\text{total \# of students}} = \dfrac{16}{16+9} = \dfrac{16}{25}$

(d) $\dfrac{\text{males}}{\text{total \# of students}} = \dfrac{9}{16+9} = \dfrac{9}{25}$

Example 2: Find the ratio of 2 feet to 4 yards.

Solution: $\dfrac{2\text{ feet}}{4\text{ yards}}$.

You may be tempted to simplify this ratio to $\frac{1}{2}$, but you know that 2 feet is not half of 4 yards. The units in the numerator and denominator must be the same before simplifying. Since 4 yards equals 12 feet, we have

$$\frac{2\text{ feet}}{4\text{ yards}} = \frac{2\text{ feet}}{12\text{ feet}} = \frac{2}{12} = \frac{1}{6}$$

Proportion

> A proportion states that one ratio is equal to another.

Given ratios $\frac{a}{b}$ and $\frac{c}{d}$, the equation

$$\frac{a}{b} = \frac{c}{d}$$

is a proportion. In this proportion a and d are called the **extremes** and b and c are called the **means** of the proportion. There is an important relationship between the means and extremes of any proportion: *The product of the means equals the product of the extremes*. This rule is commonly called the *cross-product rule*.

> **Cross-product rule:**
>
> In any proportion the product of the means equals the product of the extremes.
>
> Symbolically, if $\frac{a}{b} = \frac{c}{d}$ then $b \cdot c = a \cdot d$.

It is easy to see why the cross-product rule works. In the proportion $\frac{a}{b} = \frac{c}{d}$, multiply both sides of the equation by the quantity $b \cdot d$:

$$\frac{a}{b} = \frac{c}{d}$$

$$\not{b} \cdot d \cdot \frac{a}{\not{b}} = b \cdot \not{d} \cdot \frac{c}{\not{d}} \qquad \text{Multiply each side by } b \cdot d.$$

$$d \cdot a = b \cdot c$$

or

$$bc = ad$$

Remember that a proportion is an equation, so if any of the four parts of the proportion is an unknown quantity, we can solve for the unknown quantity by using the cross-product rule.

Example 3: Solve for x if $\frac{3}{4} = \frac{x}{28}$

Solution:

$$\frac{3}{4} = \frac{x}{28}$$

$$4 \cdot x = 3 \cdot 28 \qquad \text{Use the cross-product rule.}$$

$$4x = 84$$

$$\frac{\not{4}x}{\not{4}} = \frac{84}{4}$$

$$x = 21 \qquad \text{is the solution.}$$

Check:

$$\frac{3}{4} \stackrel{?}{=} \frac{21}{28}$$

$$3 \cdot 28 \stackrel{?}{=} 4 \cdot 21$$

$$84 = 84 \ \checkmark$$

Example 4: Solve the proportion $\frac{h}{8} = \frac{21}{6}$

Solution:

$$\frac{h}{8} = \frac{21}{6}$$

$$6h = 8 \cdot 21 \qquad \text{Use the cross-product rule.}$$

$$6h = 168$$

$$\frac{\not{6}h}{\not{6}} = \frac{168}{6}$$

$$h = 28 \qquad \text{is the solution.}$$

Check:

$$\frac{28}{8} \stackrel{?}{=} \frac{21}{6}$$

$$6 \cdot 28 \stackrel{?}{=} 8 \cdot 21$$

$$168 = 168 \ \checkmark$$

Solving Word Problems by Proportion

There is a large category of word problems that are solved quickly and easily by using proportion.

Example 5: If a car requires 7 gallons of gasoline to travel 185 miles, how many gallons of gasoline will be required to make a trip of 550 miles?

Solution: The proportion must be set up in such a way that the comparisons are the same on each side of the equation. Let x = the number of gallons of gasoline required to go 550 miles.

$$\frac{\text{gallons}}{\text{miles}}\Bigg\}\ \frac{7}{185} = \frac{x}{550}\ \Bigg\{\frac{\text{gallons}}{\text{miles}}$$

$185 \cdot x = 7 \cdot 550$ Use the cross-product rule.

$185x = 3850$

$\frac{\cancel{185}x}{\cancel{185}} = \frac{3850}{185}$

$x = 20.8$ gallons rounded off to the nearest tenth of a gallon.

Example 6: If a typist takes 29 minutes to type 3 pages of a manuscript, how long will it take to type 24 pages?

Solution: Let x = the number of minutes to type 24 pages.

$$\frac{\text{minutes}}{\text{pages}}\Bigg\}\ \frac{29}{3} = \frac{x}{24}\ \Bigg\{\frac{\text{minutes}}{\text{pages}}$$

$3x = 29 \cdot 24$

$3x = 696$

$\frac{\cancel{3}x}{\cancel{3}} = \frac{696}{3}$

$x = 232$ minutes

or

$x = 3$ hours 52 minutes

Example 7: A certain drug is administered according to body weight. If 2 cubic centimeters (cc) of the drug are given for each 45 pounds of body weight, how many cc of the drug should be given to a man who weighs 160 pounds?

Solution: Let x = the number of cc given to a 160 pound man.

$$\frac{\text{cc}}{\text{pounds}}\Bigg\}\ \frac{2}{45} = \frac{x}{160}\ \Bigg\{\frac{\text{cc}}{\text{pounds}}$$

$45x = 2 \cdot 160$

$45x = 320$

$$\frac{\cancel{45}x}{\cancel{45}} = \frac{320}{45} = 7.1$$

$x = 7.1$ cc rounded to the nearest tenth of a cc.

In geometry, if the corresponding angles of two triangles have the same measurement, the triangles are *similar*.

The triangles ABC and DEF are similar since the measures of their corresponding angles are equal:

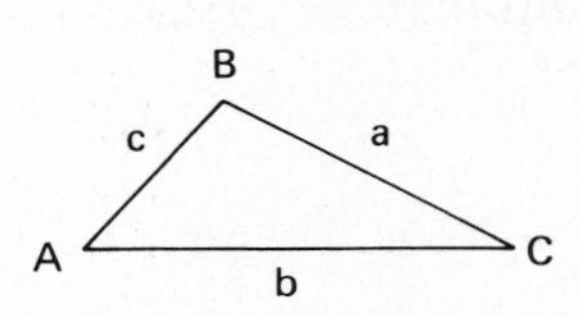

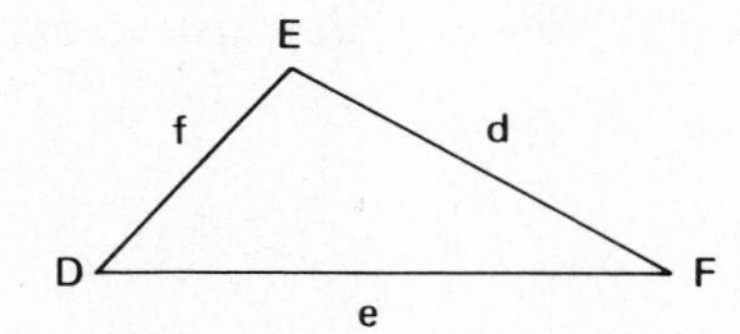

The figure on the right is the *same shape* as the figure on the left, but it is increased in size. The length of the corresponding sides are *proportionately* larger. This is true in general and is stated as follows:

> If one geometric figure is similar to another, then the ratios of the lengths of corresponding sides are equal.

This principal is illustrated in the following example.

Example 8: The measures of the corresponding angles of the two triangles in the figure are equal. Find the length of sides f and e of the triangle on the right if the other dimensions are as given in the figure.

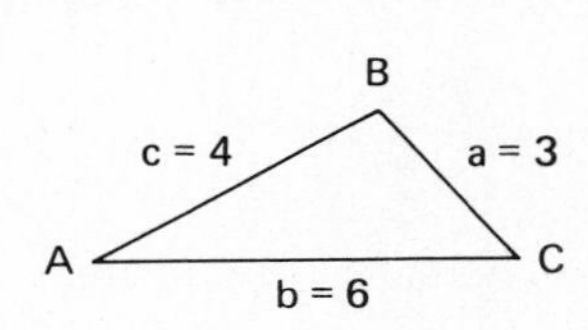

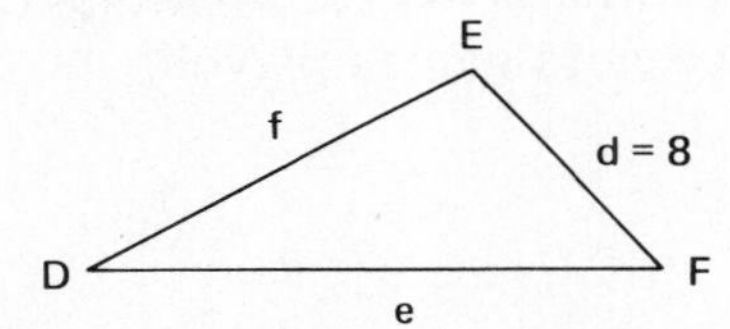

Solution: Since the two triangles are similar, their corresponding sides are in proportion. The side c on the left corresponds to side f on the right and side a on the left corresponds to side d on the right. Three of the four lengths are known and we can solve for the fourth side using a proportion.

$\frac{c}{f} = \frac{a}{d}$ Corresponding sides of similar triangles are in proportion.

$\frac{4}{f} = \frac{3}{8}$ Substitute the known values.

$3f = 32$ Use the cross-product rule.

$$\frac{\cancel{3}f}{\cancel{3}} = \frac{32}{3}$$

$$f = \frac{32}{3} \text{ or } 10\frac{2}{3}$$

To find side e we use a proportion involving side e and its corresponding side b:

$$\frac{b}{e} = \frac{a}{d}$$ Corresponding sides of similar triangles are in proportion.

$$\frac{6}{e} = \frac{3}{8}$$ Substitute the known values.

$$3e = 48$$ Use the cross-multiplication principle.

$$\frac{\cancel{3}e}{\cancel{3}} = \frac{48}{3}$$

$$e = 16$$

EXERCISE 4.4

In exercises 1 through 10 write the required ratios.

1. A baseball team won 18 games and lost 7. Express the ratio of (a) wins to losses, (b) losses to wins, (c) wins to games played, (d) losses to games played.

2. A basketball team won 26 of its first 30 games. Write the ratio of (a) wins to losses, (b) losses to wins, (c) wins to games played, (d) losses to games played.

3. A fruit punch contains 3 parts orange juice, 1 part pineapple juice, and 2 parts ginger ale. Find the ratio of (a) orange juice to ginger ale, (b) orange juice to punch, (c) pineapple juice to ginger ale, (d) ginger ale to punch.

4. A professor teaches 5 classes and the total number of students in all his classes is 135. What is the teacher-student ratio, in its simplest form?

5. What is the ratio of 1 dime to 1 quarter?

6. Find the ratio of $\frac{1}{2}$ hour to $1\frac{1}{4}$ hours.

7. What is the ratio of 2 feet 2 inches to 1 yard?

8. On a trip, a family traveled 242 miles in $5\frac{1}{2}$ hours. How many miles per hour did they average for that trip?

9. The pitch of a roof is defined to be the ratio of the rise to the span. What is the pitch of a roof that has a rise of 8 feet and a span of 24 feet?

10. A tax of $2428.75 is levied on a house that is assessed at a value of $67,000. What is the tax rate per thousand dollars of assessed value?

In exercises 11 through 43 use proportions to solve each problem.

11. $\frac{x}{5} = \frac{20}{4}$

12. $\frac{3}{y} = \frac{15}{45}$

13. $\frac{6}{8} = \frac{7}{x}$

14. $\frac{600}{30000} = \frac{W}{5}$

15. $\frac{3.7}{4.2} = \frac{a}{12.6}$

16. $\frac{h}{8} = \frac{21}{6}$

17. $\frac{m}{8} = \frac{75}{12}$

18. $\frac{x}{6} = \frac{3.5}{10.5}$

19. $\frac{.09}{1} = \frac{18}{z}$

20. $\frac{8}{n} = \frac{20}{250}$

21. $\frac{\frac{2}{3}}{\frac{5}{8}} = \frac{6}{y}$

22. $\frac{\frac{1}{2}}{y} = \frac{\frac{3}{4}}{\frac{7}{8}}$

23. $\frac{2\frac{1}{2}}{n} = \frac{4\frac{1}{4}}{2\frac{2}{3}}$

24. 85 divided by what number is the same as 17 divided by 3?

25. If there are 5 milligrams in a solution of 150 milliliters, how many milligrams of the substance would you find in 500 milliliters of the solution?

26. If a car can travel 297 miles on 9 gallons of gas, how much gas will be required to make a trip of 785 miles?

27. On a map, $\frac{1}{2}$ inch represents a distance of 140 miles. How many miles are represented by $3\frac{3}{4}$ inches?

28. An investor earned $350 on an investment of $2500. How much would he have earned if he had invested $6750?

29. The owner of a house assessed at $52,000 pays taxes of $1142. How much tax does his neighbor pay if his house is assessed at $46,000?

30. If you can average 450 miles of driving per day, how many days will it take you to make a trip of 2415 miles from New York City to Phoenix, Arizona?

31. If a bread recipe calls for 8 cups of flour and the recipe makes 3 loaves, how many cups of flour will be needed if you intend to make 10 loaves?

32. A survey showed that 2 out of every 5 families surveyed in a certain city were hooked up to cable television. If there were 3,580 families in the city, how many of them had cable TV?

33. A time study shows that a worker can assemble 9 items in 30 minutes. How many items can he assemble at the same rate in 7 hours?

34. If a lady loses 5 pounds in the first 2 weeks of her diet, how long will it take her to lose 65 pounds at the same rate?

35. A jogger can run $2\frac{1}{2}$ miles in $22\frac{1}{2}$ minutes. How many minutes will it take her to run 8 miles at the same rate?

36. A drug that is administered according to body weight is given at the rate of 2 milligrams per 15 pounds. How much of the drug should be given to a child who weighs 60 pounds?

37. A basketball player has scored a total of 72 points in his first 5 games. How many points will he score in 33 games if his scoring continues at the same rate?

38. A quality control inspector finds 2 defective parts out of every 175 parts inspected. How many defective parts would you expect to find in a shipment of 2800 parts?

39. A paycheck of $365.00 has $87.60 withheld for taxes. How much tax will you have paid by the time you earn $5000?

40. The scale on a drawing is 1 inch represents 6 feet. What are the dimensions of a house that measures $6\frac{1}{2}$ inches by 5 inches on the drawing?

41. The corresponding angles of the two triangles in the figure have the same measure. Find the lengths of sides a and c.

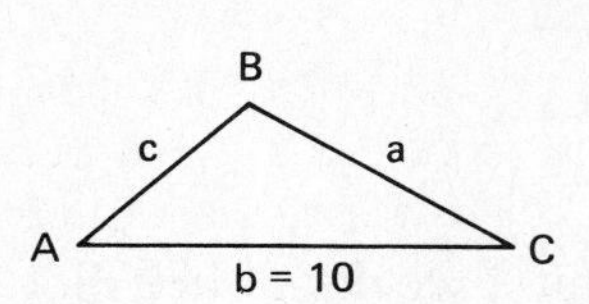

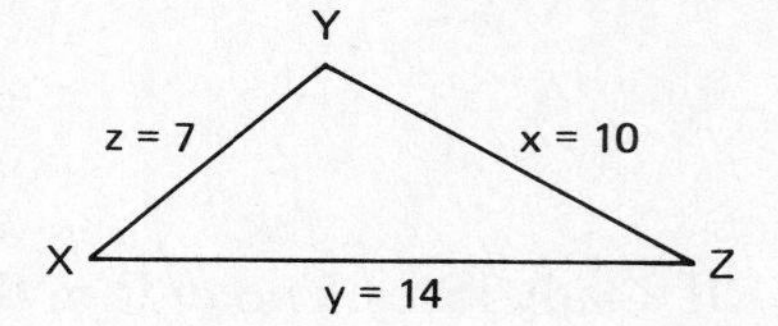

42. A building casts a shadow 90 feet long at the same time that a 4-foot-long stick casts a shadow 6 feet long. How high is the building?

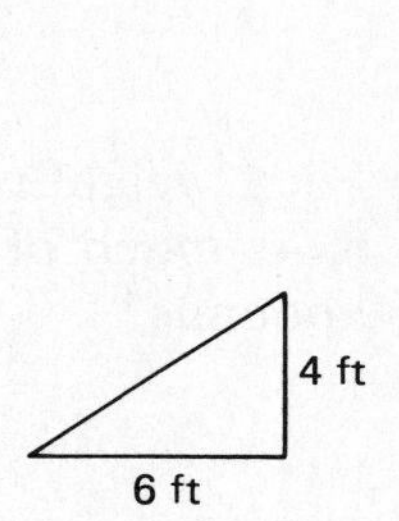

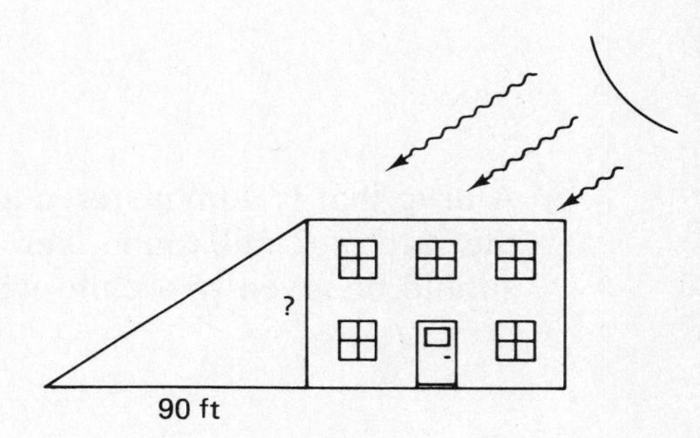

43. A 6-foot man makes a shadow of 8 feet. How high is a tree that casts a shadow 50 feet long?

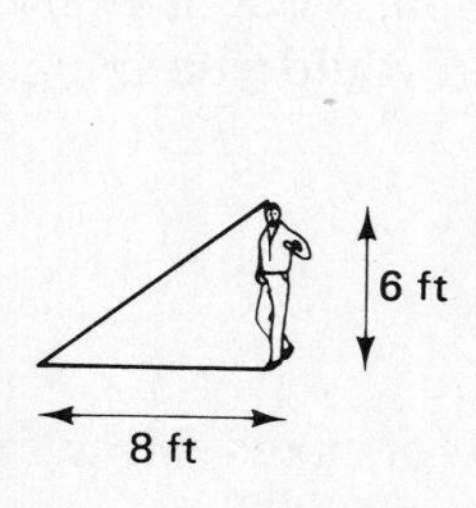

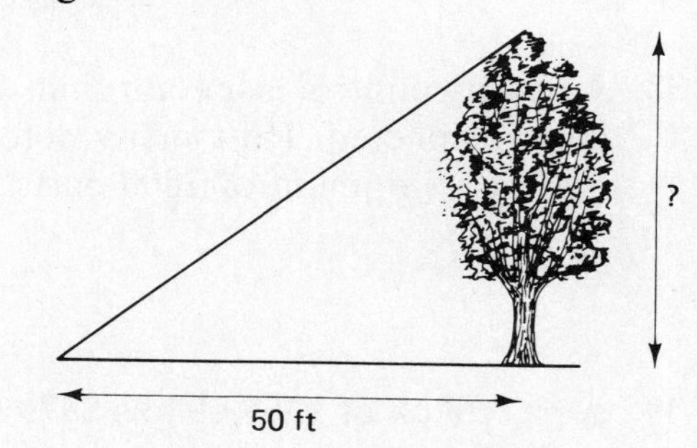

Answers to odd-numbered exercises on page 466.

4.5 UNIT CONVERSION (OPTIONAL)

In the past 35 years the metric system has become the dominant language of measurement in the world. In 1968 Congress formed a committee which concluded that the United States should join the rest of the world in the use of the metric system as the predominant common language of measurement.

However, this changeover is a slow one, and the many people who are more familiar with the English system of units find themselves forced to convert from one system to another. Also, changing from one unit to another within the English system often requires lengthy calculations. We will now investigate a method that uses our knowledge of linear equations to make these unit conversions orderly and error-free.

Following is a unit conversion table that gives common equivalent measurements in the English system, and between the English and metric systems.

LENGTH

ENGLISH	*ENGLISH–METRIC*
12 inches (in) = 1 foot (ft)	1 in = 2.54 centimeters (cm)
3 feet (ft) = 1 yard (yd)	39.37 in = 1 meter (m)
$5\frac{1}{2}$ yd = 1 rod	1 mile = 1.6 kilometers (km)
5280 ft = 1 mile	

WEIGHT

ENGLISH	*ENGLISH–METRIC*
16 ounces (oz) = 1 pound (lb)	1 oz = 28.35 grams (g)
2000 lb = 1 ton	2.2 lb = 1 kilogram (kg)

VOLUME

ENGLISH	*ENGLISH–METRIC*
16 fluid ounces = 1 pint (pt)	1 qt = .946 liter (l)
2 pt = 1 quart (qt)	1.057 qt = 1 liter

TIME (FOR BOTH SYSTEMS)

TIME (FOR BOTH SYSTEMS)
60 seconds = 1 minute
60 minutes = 1 hour
24 hours = 1 day
7 days = 1 week
365 days = 1 year

OTHER MEASURES

1 liter = 1000 cubic centimeters (cc)
7.48 gal = 1 cubic foot (ft^3)
1 ft^3 of water weighs 62.4 lb
1 cc of water weighs 1 g

Our method for unit conversion is based on the premise that a physical quantity remains the same even though the units are changed. For example, 36 inches, 3 feet, and 1 yard all represent the same length even though different units are used to describe it. You will see how this concept is used in the following unit conversion example.

Example 1: Convert 3 meters to feet.

Solution: The distance represented by 3 meters is the same as the distance represented by some unknown number of feet, call it x. This gives us the equation:

$$x \text{ feet} = 3 \text{ meters}$$

Next, look in the conversion table for any combination of unit conversions that will get us from meters to feet. We find:

$$1 \text{ meter} = 39.37 \text{ inches}$$
$$12 \text{ inches} = 1 \text{ foot}$$

In other words, we go from meters to inches and from inches to feet. Since 1 meter equals 39.37 inches, the ratio $\frac{39.37 \text{ inches}}{1 \text{ meter}}$ has the value 1 (since the numerator and denominator represent the same amount). Similarly, $\frac{1 \text{ foot}}{12 \text{ inches}} = 1$ because the numerator and denominator represent the same amount. Now multiply the right side of the equation by these ratios. We are allowed to do this since each of these ratios equals 1.

$$x \text{ feet} = 3 \text{ meters} = 3 \cancel{\text{meters}} \cdot \frac{39.37 \cancel{\text{inches}}}{1 \cancel{\text{meter}}} \cdot \frac{1 \text{ foot}}{12 \cancel{\text{inches}}}$$

Note that meters divided by meters and inches divided by inches are equal to 1 or, said another way, the meters cancel and the inches cancel. This gives us

$$x \text{ feet} = 3 \cdot \frac{39.37}{1} \cdot \frac{1 \text{ foot}}{12} = 9.84 \text{ feet}$$

Our answer, 9.84 feet, is still equal to 3 meters because we obtained it by multiplying 3 meters by ratios that are equal to 1.

Example 2: Change 4 gallons to liters.

Solution: Let x = the number of liters that is the same as 4 gallons.

From the conversion table we have:

$$1 \text{ gallon} = 4 \text{ quarts}$$
$$1 \text{ quart} = .946 \text{ liter}$$

Therefore,

$$x \text{ liters} = 4 \text{ gallons} = 4 \cancel{\text{gallons}} \cdot \frac{4 \cancel{\text{quarts}}}{1 \cancel{\text{gallon}}} \cdot \frac{.946 \text{ liters}}{1 \cancel{\text{quart}}}$$

$$= 4 \cdot 4 \cdot .946 \text{ liters}$$

$$= 15.136 \text{ liters}$$

$$\text{so } 4 \text{ gallons} = 15.136 \text{ liters}$$

We must choose the ratios in such a way that the units cancel. This method actually tells us when we have committed an error. Suppose, for example, that we tried to use the ratio $\frac{1 \text{ quart}}{.946 \text{ liters}}$ instead of $\frac{.946 \text{ liters}}{1 \text{ quart}}$. We would have

$$x \text{ liters} = 4 \text{ gallons} \cdot \frac{4 \text{ quarts}}{1 \text{ gallon}} \cdot \frac{1 \text{ quart}}{.946 \text{ liters}}$$

We find that the units will not cancel. This means we should reverse the ratio and try again.

Example 3: How many seconds are there in 1 week?

Solution: The necessary conversions are:

$$1 \text{ week} = 7 \text{ days}$$

$$1 \text{ day} = 24 \text{ hours}$$

$$1 \text{ hour} = 60 \text{ minutes}$$

$$1 \text{ minute} = 60 \text{ seconds}$$

$$x \text{ seconds} = 1 \cancel{\text{week}} \cdot \frac{7 \cancel{\text{days}}}{1 \cancel{\text{week}}} \cdot \frac{24 \cancel{\text{hours}}}{1 \cancel{\text{day}}} \cdot \frac{60 \cancel{\text{minutes}}}{1 \cancel{\text{hour}}} \cdot \frac{60 \text{ seconds}}{1 \cancel{\text{minute}}}$$

$$= 1 \cdot 7 \cdot 24 \cdot 60 \cdot 60 \text{ seconds}$$

$$= 604{,}800 \text{ seconds}$$

In some cases, measurements are given using more than one unit; for example, miles per hour or feet per second. In these, the word *per* means *divided by*.

Miles *per* hour means miles divided by hours or $\frac{\text{miles}}{\text{hour}}$. Similarly, feet per second is written $\frac{\text{feet}}{\text{second}}$.

Example 4: Change 30 miles per hour to feet per second.

Solution: Since *per* means *divided by* we want to change 30 $\frac{\text{miles}}{\text{hour}}$ to $\frac{\text{feet}}{\text{second}}$. This involves changing the unit of length, miles, to feet and also changing the unit of time, hours, to seconds. From the table, we find

$$1 \text{ mile} = 5280 \text{ feet}$$

and

$$1 \text{ hour} = 60 \text{ minutes}$$

$$1 \text{ minute} = 60 \text{ seconds}$$

Now form ratios from these equalities and multiply in such a way as to produce the desired units, $\frac{\text{feet}}{\text{second}}$.

$$x\ \frac{\text{feet}}{\text{second}} = 30\ \frac{\cancel{\text{miles}}}{\cancel{\text{hour}}} \cdot \frac{5280\text{ ft}}{1\ \cancel{\text{mile}}} \cdot \frac{1\ \cancel{\text{hour}}}{60\ \cancel{\text{minutes}}} \cdot \frac{1\ \cancel{\text{minute}}}{60\text{ seconds}}$$

$$= \frac{30 \cdot 5280\text{ ft}}{60 \cdot 60\text{ seconds}}$$

$$= 44\ \frac{\text{feet}}{\text{second}}$$

Square and Cubic Units

We have been cancelling units just as though they were numbers or algebraic quantities. We continue this mode of thinking now as we consider problems of area and volume. For example, to find the area of a room measuring 12 feet long by 9 feet wide, we use the formula for the area of a rectangle:

$$\text{area} = \text{length} \times \text{width}$$
$$\text{area} = (12\text{ ft})(9\text{ ft}) = 108\text{ ft.}^2$$

We write ft^2 since $\text{ft} \cdot \text{ft} = \text{ft}^2$ just as $5 \cdot 5 = 5^2$ or $x \cdot x = x^2$. We think of 1 ft^2 as having the same meaning as 1 square foot, a unit of area. In a similar fashion, 1 ft^3 is the same as 1 cubic foot, which is a unit of volume.

Changing units involving area or volume involves an extension of the procedure we have already learned.

Example 5: How many square feet are there in 2 square yards?

Solution: Using the notation just discussed, we want to change 2 yd^2 to ft^2. From our conversion table we find 1 yd = 3 ft. Now square both sides of the equation:

$$1\text{ yd} = 3\text{ ft}$$
$$(1\text{ yd})^{\boxed{2}} = (3\text{ ft})^{\boxed{2}}$$ The exponent applies to each quantity in the parentheses.
$$1^2\text{ yd}^2 = 3^2\text{ ft}^2$$
$$1\text{ yd}^2 = 9\text{ ft}^2$$

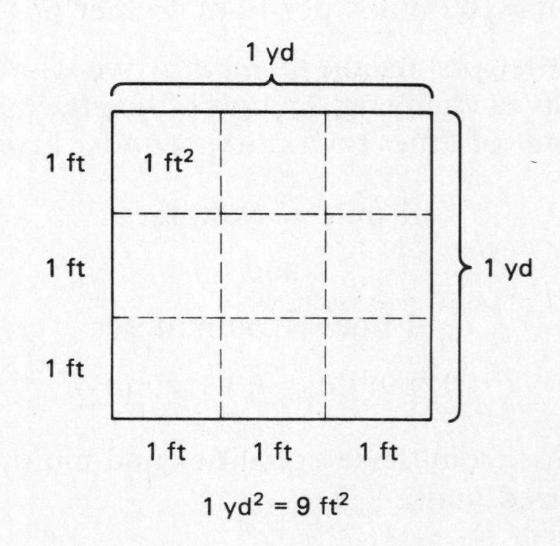

From this equation we obtain the ratio $\frac{9 \text{ ft}^2}{1 \text{ yd}^2}$, which equals 1. Multiplying by this ratio yields

$$x \text{ ft}^2 = 2 \cancel{\text{yd}^2} \cdot \frac{9 \text{ ft}^2}{1 \cancel{\text{yd}^2}} = 18 \text{ ft}^2$$

As before, yd^2 cancels with yd^2 leaving us with the desired units, ft^2.

Example 6: An area of 6 in^2 is equal to how many square centimeters?

Solution: The conversion table indicates that 1 in = 2.54 cm. But we need square units, so square both sides of the equation:

$$1 \text{ in} = 2.54 \text{ cm}$$

$$1 \text{ in}^2 = 6.45 \text{ cm}^2 \text{ (approximately)}$$

Forming the appropriate ratio and multiplying yields:

$$\begin{aligned} x \text{ cm}^2 &= 6 \text{ in}^2 \\ &= 6 \cancel{\text{in}^2} \cdot \frac{6.45 \text{ cm}^2}{1 \cancel{\text{in}^2}} \\ &= 38.7 \text{ cm}^2 \text{ (rounded to the nearest tenth)} \end{aligned}$$

Example 7: How many cubic inches are there in 4 cubic feet?

Solution: 1 ft = 12 in. Since we are dealing with cubic units this time, we must cube both sides of the equation:

$$1 \text{ ft} = 12 \text{ in.}$$

$$1 \text{ ft}^3 = 1728 \text{ in}^3 \quad (12^3 = 12 \cdot 12 \cdot 12 = 1728)$$

$$\begin{aligned} x \text{ in}^3 &= 4 \text{ ft}^3 \\ &= 4 \cancel{\text{ft}^3} \cdot \frac{1728 \text{ in}^3}{1 \cancel{\text{ft}^3}} = 6912 \text{ in}^3 \end{aligned}$$

EXERCISE 4.5

In the following, round the answers to the nearest tenth if necessary.

1. Convert 6 yards to inches.

2. Change 70 miles to kilometers.

3. How many feet are there in 4.3 meters?

4. 6 gallons is equal to how many fluid ounces?

5. 2 days is the same as how many seconds?

6. Mount Everest is 29,028 feet high. How many meters is this?

7. Change 70 kilometers per hour to meters per second.

8. Convert 10 meters per second to miles per hour.

9. In France, a speed sign reads 100, meaning 100 kilometers per hour. How many miles per hour is this?

10. Change 716 square feet to square yards.

11. Change 1 ft^2 to cm^2.

12. How many square meters are there in 20 square yards?

13. How many square rods are there in 1 square mile?

14. 9000 grams is equal to how many pounds?

15. Approximately how many miles high is an airplane that is flying at an altitude of 11,000 feet?

16. How many cubic inches are contained in 1 gallon?

17. If you were building a concrete sidewalk that uses 70 cubic feet of cement, how many cubic yards of cement should be ordered?

18. How many minutes are there in $4\frac{1}{2}$ days?

19. The speed of sound in air is approximately 1100 feet per second. How fast is this in miles per hour?

20. 1 atmosphere of pressure is $14.7\ \frac{\text{lbs}}{\text{in}^2}$. Express this in $\frac{\text{kg}}{\text{cm}^2}$.

21. If you could run 100 meters in 10 seconds (the world record is a little faster than that), how fast would you be running in miles per hour?

22. How many square yards of carpeting are needed to cover a floor 10 feet wide by 14 feet long?

23. How many cubic centimeters (cc) are there in 2 gallons?

24. An automobile engine has a displacement of 4.2 liters. What is this displacement in cubic inches?

25. The weight of a cubic foot of water is approximately 62.4 pounds. From this information, determine how much a cubic inch of water weighs. Round your answer to the nearest thousandth.

26. Which is greater, 10 square meters or 12 square yards?

27. Acceleration due to gravity of a freely falling body is 32 feet per second^2. What is this acceleration in meters per second^2?

28. How many cubic centimeters will a 2.5 liter container hold?

29. A motorcycle engine has a displacement of 400 cubic centimeters. How many cubic inches is this?

30. The flow of natural gas in a pipe is 200 cubic yards per minute. Give this rate of flow in cubic meters per minute.

Answers to odd-numbered exercises on page 466.

4.6 Chapter Summary

(4.2) Strategy for solving word problems

1. Read the problem several times, making note of any key words.
2. Write down what is given and what you are to find.
3. Represent one of the unknown quantities by a variable. Express any other unknown quantities in terms of the same variable.
4. Find out which expressions are the same and write an equation. Sometimes a simple sketch helps determine which expressions are equal.
5. Solve the equation.
6. Check your solution in the original word statement to see if it really meets the conditions of the problem.

(4.2) **Consecutive integers** are integers that follow right after one another, represented by $x, x + 1, x + 2, \ldots$.

(4.2) **Consecutive even integers** are even integers that follow one another, represented by $x, x + 2, x + 4, \ldots$.

(4.2) **Consecutive odd integers** are odd integers that follow one another, represented by $x, x + 2, x + 4, \ldots$.

(4.3) Percent problems can be solved using linear equations. Write percent in its decimal form by moving the decimal point two places to the left.

(4.4) **Ratio** is a comparison of two numbers by division.

(4.4) **Proportion** is a statement that one ratio is equal to another ratio.

(4.4) The **cross-product rule** states that in any proportion the product of the means equals the product of the extremes.

Symbolically, if $\frac{a}{b} = \frac{c}{d}$ then $b \cdot c = a \cdot d$

Examples

(4.2) Find three consecutive odd integers such that the sum of the first two added to twice the third is 110.

x = first integer

$x + 2$ = second integer

$x + 4$ = third integer

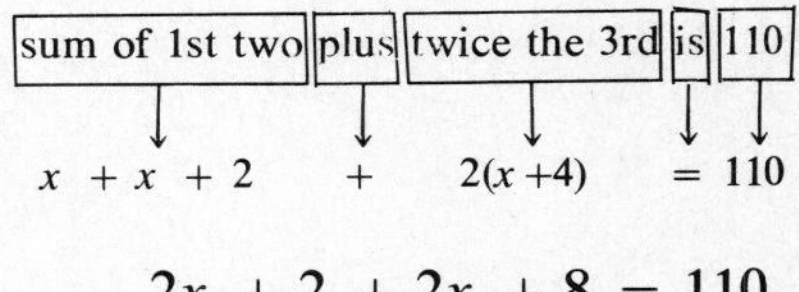

$$2x + 2 + 2x + 8 = 110$$
$$4x + 10 = 110$$
$$4x = 100$$
$$x = 25$$

1st integer = 25

2nd integer = $x + 2$ = 27

3rd integer = $x + 4$ = 29

(4.3)

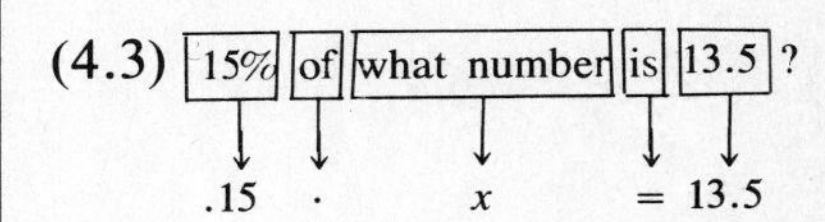

$$.15x = 13.5$$
$$(100)\ .15x = (100)\ 13.5$$
$$15x = 1350$$
$$\frac{\cancel{15}x}{\cancel{15}} = \frac{1350}{15}$$
$$x = 90$$

(4.4) What is the ratio of a nickel to a dollar?

$$\frac{\text{nickel}}{\text{dollar}} = \frac{5 \text{ cents}}{100 \text{ cents}} = \frac{1}{20}$$

(4.4) Solve for n if $\frac{2}{3} = \frac{26}{n}$. Use the cross-product rule

$$2 \cdot n = 3 \cdot 26$$
$$2n = 78$$
$$n = 39$$

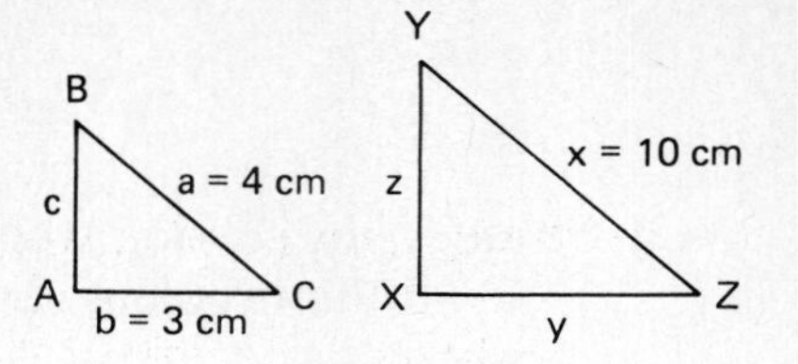

(4.4) **Similar triangles.** Two triangles are similar if the measures of their corresponding angles are equal.

(4.4) If one geometric figure is similar to another, then the ratios of lengths of corresponding sides are equal.

(4.4) Find side y of the triangle on the right if the triangles are similar.

$$\frac{a}{x} = \frac{b}{y}$$

$$\frac{4}{10} = \frac{3}{y}$$

$$4y = 30$$

$$y = \frac{30}{4} = 7\frac{1}{2} \text{ cm}$$

(4.5) Conversions from one unit to another are accomplished by multiplying the quantity to be converted by ratios, each equal to 1, in such a way that all the units cancel except those units you are converting to.

(4.5) Change 25 feet to meters.

1 foot = 12 inches
39.37 inches = 1 meter
} from the table

$$x \text{ meters} = 25 \text{ feet}$$

$$= 25 \not{ft} \cdot \frac{12 \not{in}}{1 \not{ft}} \cdot \frac{1 \not{m}}{39.37 \not{in}}$$

$$= \frac{25 \cdot 12}{39.37}$$

$$= 7.62 \text{ meters}$$

so 25 ft = 7.62 meters

Exercise 4.6 Chapter Review

(4.1) Write each English sentence as a mathematical equation.

1. Six is the same as the sum of some number and four.

2. An unknown number is twelve less than three times itself.

3. In eight years a person will be twice as old as he is now.

4. Six less than three times some number is the same as the quotient of the number and three.

5. Eight years ago, a person's age was three-fourths of what it will be in six years.

(4.2) Solve and check:

6. When 4 is subtracted from three times a number, the result is 17. Find the number.

7. The sum of two integers is 39. If the second is 6 more than twice the first, find both integers.

8. Find three consecutive odd integers such that the sum of the first two is 9 more than the third.

9. Separate 120 units into two parts such that one number is 3 more than twice the other number.

10. When two-thirds of a number is added to one-half of the number, the result is 9 more than the number. Find the number.

11. What grade must you earn on your third exam so that your average will be 90, if you scored 94 and 78 on your first two exams?

(4.3) Solve each percent problem:

12. 24 is what percent of 160?

13. 14 is 28% of what number?

14. 18% of 175 is how much?

15. How much sales tax would you pay on a purchase of $179.00 at a tax rate of 6%?

16. An item increased \$3.60 in price. If this represents a jump in price of 18% what did the item cost before the price increased?

(4.4) 17. A team lost 12 of the first 20 games that it played. Write the ratio of (a) losses to games played, (b) games won to games played, (d) losses to wins.

(4.4) Use proportions to solve the following:

18. $\frac{7}{x} = \frac{4}{56}$

19. $\frac{\frac{1}{2}}{\frac{3}{4}} = \frac{x}{\frac{5}{8}}$

20. If a car can travel 340 miles on 12 gallons of gas, how much gas will be needed to drive 520 miles?

21. The scale on a drawing is $\frac{1}{2}$ inch represents 3 feet. What are the dimensions of a room that measures $2\frac{3}{4}$ inches by $3\frac{1}{2}$ inches on the drawing?

22. The corresponding angles of triangle ABC have the same measure as the angles of triangle RST. Find the length of the two unknown sides *b* and *r*.

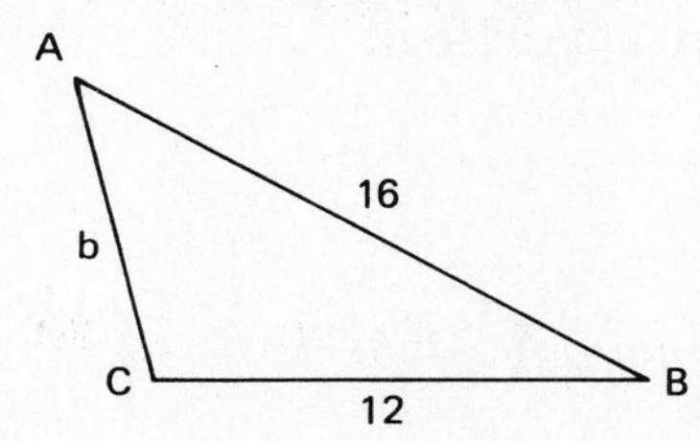

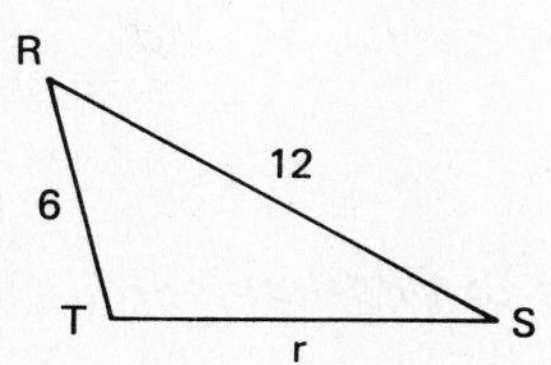

(4.5) 23. Change 72 ounces to pounds.

24. Change 3.2 meters to feet.

25. If it takes 62 liters of gasoline to fill your car's tank, how many gallons is this?

26. An airplane flying at 400 miles per hour is going how many feet per second?

27. In track competition there is a half-mile race and also a 1000-meter race. Which one is longer and by how many feet?

28. Change $\frac{900 \text{ ft}^3}{\text{min}}$ to $\frac{\text{meter}^3}{\text{min}}$

29. If the speed limit is 55 miles per hour, how fast is this in kilometers per hour?

30. Which is greater, 3 square meters or 25 square feet? How much greater? Give your answer in square feet.

Answers to odd-numbered exercises on page 466.

Chapter 4 Achievement Test

Name ______________________________

Class ______________________________

This test should be taken before you are tested in class on the material in Chapter 4. Solutions to each problem and the section where the type of problem is found are given on page 467.

Write each English sentence as a mathematical equation.

1. Twice the sum of some number and five is eighteen. — 1. ____________

2. When four is subtracted from three times some number, the result is the same as twice the sum of the number and five. — 2. ____________

Solve each of the following:

3. When three-fourths of a number is added to 6, the sum is thirty. Find the number. — 3. ____________

4. Find three consecutive even integers such that three times the first integer is equal to 10 more than the sum of the second two integers. — 4. ____________

5. If you receive grades of 75 and 77 on your first two exams, what grade must you receive on the third exam in order to average 80 for the three exams? — 5. ____________

6. 16 is what percent of 20?

6. 

7. 22% of 24 is what number?

7. 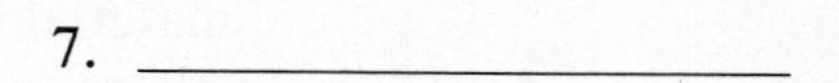

8. How much money would you have to invest at a rate of 14% to earn $10,500?

8. 

9. What is the ratio of 1 yard to 24 feet?

9. ______________

Solve the proportion:

10. $\frac{3}{a} = \frac{15}{6}$

10. ______________

11. A time study shows that a robot assembler can assemble 6 items in 15 minutes. How many items could be assembled in a $7\frac{1}{2}$ hour period at the same rate?

11. ______________

12. How high is a tower that casts a shadow 90 feet long at the same time that a 5-foot stick casts a 3-foot shadow?

12. ______________

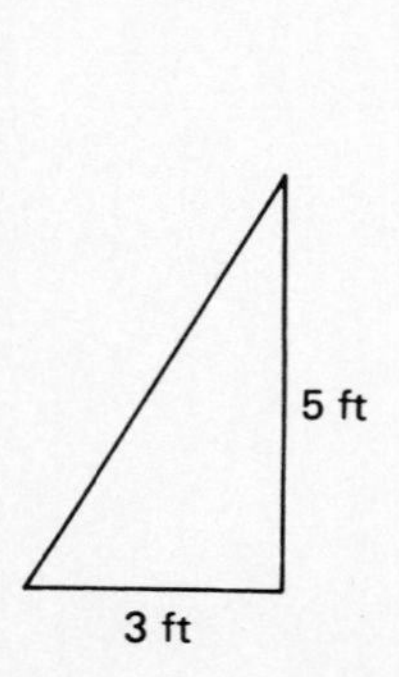

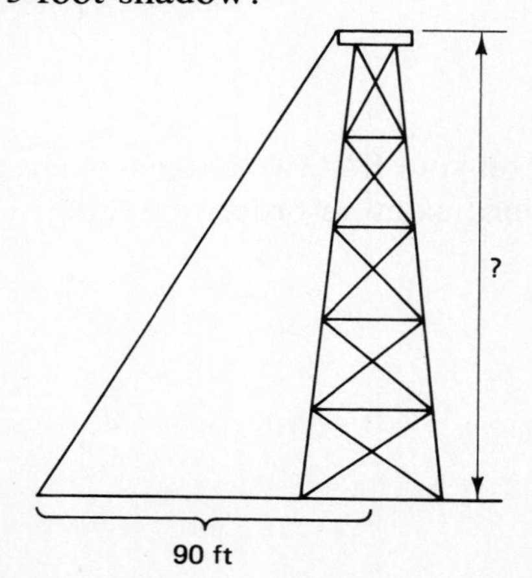

*13. Change 140 feet per second to miles per hour.

13. ________________

*14. How many cubic centimeters will a $3\frac{1}{2}$-liter container hold?

14. ________________

*15. How many square yards of carpeting are necessary to cover a floor measuring 14 feet by 12 feet? Round your answer to the nearest square yard.

15. ________________

5
Exponents

5.1 POSITIVE EXPONENTS

In Section 2.4 the first rule of exponents was given. It says that if you are multiplying two quantities with the same base, keep that base and add the exponents. Symbolically, it is written:

> **Rule I.**
> $x^a \cdot x^b = x^{a+b}$

Example 1:

(a) $x^3 \cdot x^4 = x^{3+4} = x^7$

(b) $2^3 \cdot 2^4 = 2^{3+4} = 2^7 = 128$

(c) $y^4 \cdot y = y^{4+1} = y^5$ — Remember $y = y^1$.

(d) $10^m \cdot 10^n = 10^{m+n}$

(e) $x^2 \cdot y^3$ — cannot be simplified because the bases are different.

(f) $x^2 \cdot x^4 \cdot x^6 = x^{2+4+6} = x^{12}$ — The rule applies to more than two factors.

(g) $x^{3n} \cdot x^{2n} = x^{3n+2n} = x^{5n}$

Now consider expressions of the type where we raise a power to a power. Consider the expression $(x^5)^2$:

$$(x^5)^2 = x^5 \cdot x^5 = x^{5+5} = x^{10}$$

We see that $(x^5)^2 = x^{5 \cdot 2} = x^{10}$. When a quantity raised to a power is again raised to a power, we multiply exponents. This is stated as:

Rule II.

$(x^a)^b = x^{ab}$

Example 2:

(a) $(x^6)^4 = x^{6 \cdot 4} = x^{24}$

(b) $(b^2)^5 = b^{2 \cdot 5} = b^{10}$

(c) $(2^3)^3 = 2^{3 \cdot 3} = 2^9$

(d) $(10^m)^n = 10^{m \cdot n} = 10^{mn}$

(e) $(y^4)^5 = y^{4 \cdot 5} = y^{20}$

(f) $(x^4)^y = x^{4 \cdot y} = x^{4y}$

Next we consider division of two quantities which have the same base raised to powers. Consider the expression $\frac{x^5}{x^2}$.

$$\frac{x^5}{x^2} = \frac{x \cdot x \cdot x \cdot x \cdot x}{x \cdot x} = \boxed{\frac{x \cdot x}{x \cdot x}} \cdot x \cdot x \cdot x = x \cdot x \cdot x = x^3$$

↑ This is equal to 1.

We have $\frac{x^5}{x^2} = x^{5-2} = x^3$.

This is true in general (providing $x \neq 0$) and is stated as follows:

Rule III.

$\frac{x^a}{x^b} = x^{a-b}$

Example 3:

(a) $\frac{x^7}{x^3} = x^{7-3} = x^4$

(b) $\frac{x^5}{x^4} = x^{5-4} = x^1 = x$

(c) $\frac{2^9}{2^7} = 2^{9-7} = 2^2 = 4$

(d) $\dfrac{10^m}{10^n} = 10^{m-n}$

(e) $\dfrac{x^6}{y^2}$ cannot be simplified since the bases are different.

(f) $\dfrac{x^{2n+1}}{x^{2n}} = x^{2n+1-2n} = x^1 = x$

You probably noticed that the examples were chosen so that the exponent in the numerator is greater than the exponent in the denominator. This is not always so, and in the next section we will consider that situation.

If we consider a product raised to a power, for example $(xy)^3$, we find the following:

$$\begin{aligned}(xy)^3 &= xy \cdot xy \cdot xy \\ &= x \cdot x \cdot x \cdot y \cdot y \cdot y \\ &= x^3 \cdot y^3 \\ (xy)^3 &= x^3y^3\end{aligned}$$

What this really means is that each factor in the parentheses is raised to the power. This gives us our next rule.

Rule IV.
$(xy)^a = x^ay^a$

Example 4:

(a) $(mn)^7 = m^7n^7$

(b) $(6x)^3 = 6^3 \cdot x^3 = 216x^3$

(c) $(xyz)^8 = x^8y^8z^8$ The rule applies to more than two factors.

A quotient raised to a power is handled in a similar way.

$$\begin{aligned}\left(\frac{x}{y}\right)^3 &= \frac{x}{y} \cdot \frac{x}{y} \cdot \frac{x}{y} \\ &= \frac{x \cdot x \cdot x}{y \cdot y \cdot y} \\ &= \frac{x^3}{y^3} \\ \left(\frac{x}{y}\right)^3 &= \frac{x^3}{y^3}\end{aligned}$$

As you can see, when a fraction is raised to a power, both the numerator and denominator are raised to the power. This is stated in symbols as:

Rule V.

$$\left(\frac{x}{y}\right)^a = \frac{x^a}{y^a}$$

Example 5:

(a) $\left(\frac{m}{n}\right)^4 = \frac{m^4}{n^4}$

(b) $\left(\frac{6}{x}\right)^2 = \frac{6^2}{x^2} = \frac{36}{x^2}$

(c) $\left(\frac{a}{b}\right)^7 = \frac{a^7}{b^7}$

EXERCISE 5.1

Simplify using the rules of exponents:

1. $x^3 \cdot x^5$
2. $m^2 \cdot m^8$
3. $y^4 \cdot y^4$
4. $(x^3)^5$
5. $(x^5)^3$
6. $(a^2)^7$
7. $\frac{x^5}{x^2}$
8. $h \cdot h^4$
9. $\frac{y^{15}}{y^{13}}$
10. $\frac{h^7}{h}$
11. $\frac{x^4}{y}$
12. $x^7 \cdot y^4$
13. $(xy)^9$
14. $(xyz)^5$
15. $(3x)^3$
16. $\left(\frac{h}{4}\right)^2$
17. $\frac{2^{48}}{2^{46}}$
18. $(2^3)^2$
19. $(2xy)^4$
20. $x^5 \cdot x^h$
21. $x^h \cdot x^5$

22. $\dfrac{x^5}{x^h}$ 23. $\dfrac{x^h}{x^5}$ 24. $(x^5)^h$

25. $(5x)^h$ 26. $\left(\dfrac{x}{5}\right)^h$ 27. $(3^a)^3$

28. $\dfrac{2^a}{2^b}$ 29. $6^2 \cdot 6^x$ 30. $x^3 \cdot y^4 \cdot x^5$

31. $x^{5a} \cdot x^{7a}$ 32. $\dfrac{x^{12n}}{x^{4n}}$ 33. $(6^x)^y$

34. $y^b \cdot y^{2b}$ 35. $(3^x)^x$

Answers to odd-numbered exercises on page 467.

5.2 NEGATIVE AND ZERO EXPONENTS

Negative Exponents

In our discussion of Rule III the exponent in the numerator was always larger than the exponent in the denominator. If that is not the case we have an expression like the following:

$$\frac{x^2}{x^5} = \frac{x \cdot x}{x \cdot x \cdot x \cdot x \cdot x} = \boxed{\frac{x \cdot x}{x \cdot x}} \cdot \frac{1}{x \cdot x \cdot x} = \frac{1}{x \cdot x \cdot x} = \frac{1}{x^3}$$

↑ This is equal to 1.

If we apply our third rule of exponents to the same problem, we will subtract exponents, giving us:

$$\frac{x^2}{x^5} = x^{2-5} = x^{-3}$$

So, we have $\dfrac{x^2}{x^5} = \dfrac{1}{x^3}$

Both are equal to $\dfrac{x^2}{x^5}$, so they are equal to each other.

and $\dfrac{x^2}{x^5} = x^{-3}$

and we conclude that $x^{-3} = \dfrac{1}{x^3}$.

Generalizing gives us our next rule:

Rule VI.

$$x^{-a} = \frac{1}{x^a}.$$

Example 1:

(a) $x^{-8} = \frac{1}{x^8}$

(b) $b^{-5} = \frac{1}{b^5}$

(c) $2^{-5} = \frac{1}{2^5} = \frac{1}{32}$

What would $\frac{1}{x^{-3}}$ be equal to? Applying Rule VI to the denominator, x^{-3}, yields the following:

$$\frac{1}{x^{-3}} = \frac{1}{\frac{1}{x^3}}$$ Rule VI

$$= \frac{\frac{1}{1}}{\frac{1}{x^3}}$$ Now we have one fraction divided by another fraction.

$$= \frac{1}{1} \cdot \frac{x^3}{1}$$ Invert and multiply.

$$= x^3$$

$$\frac{1}{x^{-3}} = x^3$$

This too is true in general:

Rule VII.

$$\frac{1}{x^{-a}} = x^a$$

Example 2:

(a) $\frac{1}{x^{-8}} = x^8$

(b) $\frac{1}{b^{-5}} = b^5$

(c) $\frac{1}{2^{-5}} = 2^5 = 32$

Combining Rules VI and VII gives us an important method for removing negative exponents.

A factor may be moved from numerator to denominator, or vice versa, by changing the sign of the exponent.

Example 3:

(a) $\dfrac{x^3}{\boxed{x^{-4}}} = x^3 \cdot \boxed{x^4} = x^7$

Change $x^{\boxed{-4}}$ to $x^{\boxed{+4}}$ and move from denominator to numerator.

(b) $\dfrac{\boxed{x^{-3}}}{y^2} = \dfrac{1}{\boxed{x^3} \cdot y^2} = \dfrac{1}{x^3y^2}$

Change $x^{\boxed{-3}}$ to $x^{\boxed{+3}}$ and move from denominator to numerator.

(c) $x^{-5}y^3 = \dfrac{x^{\boxed{-5}} \cdot y^3}{1} = \dfrac{y^3}{x^{\boxed{5}}}$

(d) $x^{\boxed{-2}}y^3z^{\boxed{-4}} = \dfrac{y^3}{x^{\boxed{2}}z^{\boxed{4}}}$

(e) $\dfrac{x^{-3}}{y^{-4}} = \dfrac{y^4}{x^3}$

(f) $\dfrac{1}{2^{-4}} = \dfrac{2^4}{1} = 16$

(g) $\dfrac{x^2}{y^{-2}} = x^2y^2$

The next illustration provides a short cut for simplifying a fraction raised to a negative power.

	Rule Used
$\left(\dfrac{x}{y}\right)^{-3} = \dfrac{x^{-3}}{y^{-3}}$	$\left(\dfrac{x}{y}\right)^a = \dfrac{x^a}{y^a}$
$= \dfrac{y^3}{x^3}$	$x^{-a} = \dfrac{1}{x^a}$ and $\dfrac{1}{x^{-a}} = x^a$
$= \left(\dfrac{y}{x}\right)^3$	$\dfrac{x^a}{y^a} = \left(\dfrac{x}{y}\right)^a$

You can see that we simplified the fraction raised to a negative power by inverting the fraction (writing its reciprocal) and changing the sign of the exponent. This occurs frequently enough that it deserves to be stated formally,

Rule VIII.

$$\left(\frac{x}{y}\right)^{-a} = \left(\frac{y}{x}\right)^{a}$$

Example 4:

(a) $\left(\frac{x}{y}\right)^{-5} = \left(\frac{y}{x}\right)^{5}$

(b) $\left(\frac{t}{s}\right)^{-n} = \left(\frac{s}{t}\right)^{n}$

(c) $\left(\frac{2}{3}\right)^{-2} = \left(\frac{3}{2}\right)^{2} = \frac{3^2}{2^2} = \frac{9}{4}$

Zero Exponents

Applying Rule III to the expression $\frac{x^3}{x^3}$ means we subtract exponents, giving us

$$\frac{x^3}{x^3} = x^{3-3} = x^0$$

However, we also know that

$$\frac{x^3}{x^3} = 1$$ since any quantity (except zero) divided by itself equals 1.

So we have $\frac{x^3}{x^3} = x^0$

↖ Both are equal to $\frac{x^3}{x^3}$ so they equal each other. ↙

and $\frac{x^3}{x^3} = 1$

Therefore, we conclude that $x^0 = 1$, which gives us our final rule for exponents.

Rule IX.

$x^0 = 1$

We must make the restriction that $x \neq 0$ here since if we work an expression like $\frac{0^2}{0^2}$ two different ways, we obtain:

$$\frac{0^2}{0^2} = \frac{0}{0}$$

and ↖ Both are equal to $\frac{0^2}{0^2}$ so they are equal to each other. ↙

$$\frac{0^2}{0^2} = 0^{2-2} = 0^0$$

yielding $0^0 = \frac{0}{0}$ and since $\frac{0}{0}$ is undefined, we must also say that 0^0 is undefined. From now on, we will assume that when any variable is raised to the zero power, the variable is not equal to zero.

Example 5:

(a) $7^0 = 1$

(b) $a^0 = 1$ — Remember to exclude $a = 0$.

(c) $3x^0 = 3 \cdot 1 = 3$ — The exponent applies only to x.

(d) $(3x)^0 = 1$ — The exponent applies to the entire expression.

Remember, x^0 is **not** equal to zero, $x^0 = 1$.

EXERCISE 5.2

Simplify leaving answers with only positive exponents.

1. x^{-7}
2. y^{-4}
3. 2^{-3}
4. $\frac{1}{x^{-4}}$
5. $\frac{1}{y^{-1}}$
6. $\frac{1}{2^{-3}}$
7. 10^{-1}
8. $\frac{1}{6^{-2}}$
9. $\frac{x^{-2}}{y^2}$
10. $x^{-2}y^2$
11. $\frac{1}{x^{-2}y^2}$
12. 14^0
13. $\left(\frac{x}{y}\right)^{-5}$
14. $\left(\frac{1}{4}\right)^{-2}$
15. $\left(\frac{a}{b}\right)^{-x}$

16. $\dfrac{1}{4^{-2}}$

17. $\dfrac{m^5}{n^{-3}}$

18. $a^{-5}b^{-3}c^2$

19. h^0

20. $\dfrac{a^0}{b^{-2}}$

21. $\dfrac{5^0}{6^0}$

22. $\left(\dfrac{2}{3}\right)^{-1}$

23. $\left(\dfrac{a}{b}\right)^{0}$

24. $\left(\dfrac{3}{4}\right)^{-2}$

25. $\left(\dfrac{5}{2}\right)^{-3}$

26. $\dfrac{x^{-2}y^0}{x^5y^2}$

27. $m^{-2}n^{-5}q$

Answers to odd-numbered exercises on page 467.

5.3 COMBINING RULES OF EXPONENTS

Expressions containing exponents can be quite complicated and frequently require application of more than one rule to simplify them. In most cases there is more than one order in which the rules may be applied.

Example 1:

	Rule Used
(a) $\dfrac{x^2y^4}{xy^5} = x^{2-1}y^{4-5}$	$\dfrac{x^a}{x^b} = x^{a-b}$
$= x^1 \cdot y^{-1}$	$x^{-a} = \dfrac{1}{x^a}$
$= \dfrac{x}{y}$	
(b) $(3x^2)^3 = 3^3(x^2)^3$	$(xy)^a = x^ab^a$
$= 27x^6$	$(x^a)^b = x^{ab}$
(c) $(3x^2)^{-3} = 3^{-3}(x^2)^{-3}$	$(xy)^a = x^ab^a$
$= 3^{-3}x^{-6}$	$(x^a)^b = x^{ab}$
$= \dfrac{1}{3^3x^6}$	$x^{-a} = \dfrac{1}{x^a}$
$= \dfrac{1}{27x^6}$	

We could have changed the exponent -3 to $+3$ right away by moving the entire quantity $(3x^2)$ into the denominator and changing the sign of the exponent.

Rule used

$$(3x^2)^{-3} = \frac{1}{(3x^2)^3} \qquad x^{-a} = \frac{1}{x^a}$$

$$= \frac{1}{3^3(x^2)^3} \qquad (xy)^a = x^a y^a$$

$$= \frac{1}{27x^6} \qquad (x^a)^b = x^{ab}$$

(d) $\left(\dfrac{x^2y^{-3}}{x^4y^{-8}}\right)^0 = 1$ This simplifies immediately to 1 since it is an expression raised to the zero power.

Rule used

(e) $$\left(\frac{x^{-2}y^4}{x^{-5}y^{-3}}\right)^3 = (x^{-2} \cdot x^5 \cdot y^4 \cdot y^3)^3 \qquad \frac{1}{x^{-a}} = x^a$$

$$= (x^{-2+5} \cdot y^{4+3})^3 \qquad x^a \cdot x^b = x^{a+b}$$

$$= (x^3y^7)^3$$

$$= x^9y^{21} \qquad (x^a)^b = x^{ab}$$

(f) $$-3x^{-2} = \frac{-3}{x^2} \qquad x^{-a} = \frac{1}{x^a}$$

Note that the exponent -2 applies only to the x and not to -3. Also the sign of -3 is not affected by changing the sign of the exponent.

Rule used

(g) $(-3x)^{-2} = \dfrac{1}{(-3x)^2}$ $\qquad x^{-a} = \dfrac{1}{x^a}$, The exponent -2 applies to the entire quantity $(-3x)$.

$$= \frac{1}{(-3)^2x^2} \qquad (xy)^a = x^a y^a$$

$$= \frac{1}{9x^2}$$

Rule used

(h) $$(3x^0y^4)^2 = (3y^4)^2 \qquad x^0 = 1$$

$$= 3^2(y^4)^2 \qquad (xy)^a = x^a y^a$$

$$= 9y^8 \qquad (x^a)^b = x^{ab}$$

Rule used

(i) $\left(\dfrac{2x}{y}\right)^{-3} = \left(\dfrac{y}{2x}\right)^{3}$ $\qquad \left(\dfrac{x}{y}\right)^{-a} = \left(\dfrac{y}{x}\right)^{a}$

$= \dfrac{y^3}{(2x)^3}$ $\qquad \left(\dfrac{x}{y}\right)^{a} = \dfrac{x^a}{y^a}$

$= \dfrac{y^3}{2^3x^3}$ $\qquad (xy)^a = x^a y^a$

$= \dfrac{y^3}{8x^3}$

Rule used

(j) $(2x^{-3})^{-4} = 2^{-4}(x^{-3})^{-4}$ $\qquad (xy)^a = x^a y^a$

$= 2^{-4}x^{12}$ $\qquad (x^a)^b = x^{ab}$

$= \dfrac{x^{12}}{2^4}$ $\qquad x^{-a} = \dfrac{1}{x^a}$

$= \dfrac{x^{12}}{16}$

EXERCISE 5.3

Simplify, leaving your answers with only positive exponents. Try to do some of the problems more than one way by applying the rules of exponents in a different order.

1. $\dfrac{x^2y^3}{x^4y}$
2. $\dfrac{x^6y^3}{xy^5}$
3. $(5x^2)^3$
4. $(-5x^2)^3$
5. $(5x^2)^{-3}$
6. $(-5x^2)^{-3}$
7. $6x^0y^4$
8. $(8x^0y^3)^2$
9. $\dfrac{x^{-2}y^3}{x^{-5}}$
10. $\dfrac{x^{-3}y^{-4}}{x^{-5}y^{-7}}$
11. $\left(\dfrac{x^{-2}y^{-5}}{x^{-3}y^{-6}}\right)^4$
12. $\left(\dfrac{x^3y^{-5}}{x^{-7}y^{-3}}\right)^0$

13. $\left(\dfrac{2x^2}{y}\right)^{-4}$

14. $\left(\dfrac{-4x^3}{y^2}\right)^{-2}$

15. $(6x^2y^{-3})^{-2}$

16. $(-2x^3y^{-5})^{-3}$

17. $\left[(6x)^2\right]^0$

18. $(3x^{-2})^{-4}$

19. $(-2x^{-3})^{-2}$

20. $\left[(-4x^0)^5\right]^0$

21. $\dfrac{5xy^2}{x^{-1}y^{-2}}$

22. $(y^4)^0 \cdot x^{-1}$

23. $\dfrac{x^3y^{-7}}{x^3y^{-4}}$

24. $\left(\dfrac{x^8y^{-5}}{x^8y^{-5}}\right)^{-8}$

Answers to odd-numbered exercises on page 468.

*5.4 SCIENTIFIC NOTATION (OPTIONAL)

In science we often work with very large or very small numbers, like the distance between stars or the size of atoms. To make calculations with extremely large or small numbers, we write them in a form called **scientific notation.** This involves positive and negative powers of 10, which are described in the following chart.

Place	Power	Value
Millions	10^6	= 1,000,000
Hundred-Thousands	10^5	= 100,000
Ten-Thousands	10^4	= 10,000
Thousands	10^3	= 1000
Hundreds	10^2	= 100
Tens	10^1	= 10
Units	10^0	= 1
Decimal Point	.	
Tenths	10^{-1}	= 0.1
Hundredths	10^{-2}	= 0.01
Thousandths	10^{-3}	= 0.001
Ten-Thousandths	10^{-4}	= 0.0001
Hundred-Thousandths	10^{-5}	= 0.00001
Millionths	10^{-6}	= 0.000001

To write a number in scientific notation, we write it as the product of a number between 1 and 10, and a power of 10.

Example 1: Write 6237 in scientific notation.

Solution:

$$6237 = 6.237 \times 1000$$
$$= 6.237 \times 10^3$$

↑ number between 1 and 10 ↑ power of 10

Example 2: Write .006 in scientific notation.

Solution:

$$.006 = 6 \times .001 = 6 \times 10^{-3}$$

These examples lead us to the following procedure for writing ordinary numbers in scientific notation.

To write a number in scientific notation:

1. Start at the right of the first non-zero digit.
2. Count the number of digits from this starting point to the decimal point. (Remember, a decimal point is implied at the end of every whole number, even though it is not written.) This number will be the number part of the exponent, or power, of 10.
3. The sign of the exponent is positive (+) if you count to the right and negative (−) if you count to the left.
4. Write the given number with the decimal point after the first non-zero digit and multiply by the power of 10 found in steps 2 and 3.

Example 3: Write 432.1 in scientific notation.

Solution: Starting at the right of the 4, we count 2 places to the decimal point. The sign is (+) since we moved to the right, so the power is +2. The result is:

$$432.1 = 4.321 \times 10^2$$

Example 4:

605,000 = 6 0 5 0 0 0. ← implied decimal point

5 places to the right →

$$= 6.05 \times 10^5$$

Example 5: Change .000751 to scientific notation.

Start at the right of the 7 (the first non-zero digit) and count 4 places to the left to the decimal point, which gives us an exponent of −4.

$$.000751 = .0\,0\,0\,7\,5\,1$$

4 places to the left

$$= 7.51 \times 10^{-4}$$

Example 6:

$$.0000064 = .0\,0\,0\,0\,0\,6\,4$$

6 places to the left

$$= 6.4 \times 10^{-6}$$

To change numbers from scientific notation to ordinary notation, just reverse the process.

To change from scientific notation to ordinary notation:

1. Move the decimal point as many places as the value of the exponent.
2. Move the decimal point to the right if the exponent is positive (+) and move it to the left if the exponent is negative (−). Add zeros if necessary.

Example 7: Change 3.4×10^5 to ordinary notation.

Solution:

Since the power of 10 is $\boxed{+}$5 move the decimal point 5 places to the *right* (we must add 4 zeros).

$$3.4 \times 10^5 = 3.4\,0\,0\,0\,0 = 340000$$

Example 8: $5.21 \times 10^{-4} = .0\,0\,0\,5.2\,1 = .000521$

Move the decimal point 4 places to the *left* since the exponent is $\boxed{-4}$.

Example 9: $3.201 \times 10^{-8} = .00000003.201 = .00000003201$

8 places to the left

Example 10: $4.58 \times 10^7 = 4.5800000 = 45800000$

7 places to the right

In scientific work many calculations can be simplified by combining the techniques of scientific notation and the rules of exponents.

Example 11: Evaluate $\frac{(640)(.014)}{(.0016)}$

Solution: Change each number to scientific notation and apply the appropriate rules of exponents.

$$\frac{(640)(.014)}{(.0016)} = \frac{(6.4 \times 10^{2})(1.4 \times 10^{-2})}{(1.6 \times 10^{-3})}$$

$$= \frac{\overset{4}{\cancel{(6.4)}}(1.4)}{\underset{1}{\cancel{1.6}}} \times \frac{10^{2} \cdot 10^{-2}}{10^{-3}}$$

$$= 5.6 \times 10^{2} \cdot 10^{-2} \cdot 10^{3}$$

$$= 5.6 \times 10^{3} \text{ in scientific notation}$$

or 5600 in ordinary notation

Example 12: Evaluate $120{,}000 \times 8{,}000{,}000$

Solution:

$$120{,}000 \times 8{,}000{,}000 = 1.2 \times 10^{5} \times 8.0 \times 10^{6}$$

$$= (1.2)(8) \times 10^{5} \times 10^{6}$$

$$= 9.6 \times 10^{11}$$

If you performed this calculation on most calculators, the answer would look like this:

$\boxed{9.6 \quad 11}$ which means 9.6×10^{11}

Example 13: Evaluate $\frac{684{,}000}{40{,}000{,}000}$

Solution:

$$\frac{684{,}000}{40{,}000{,}000} = \frac{6.84 \times 10^{5}}{4.0 \times 10^{7}}$$

$$= \frac{6.84}{4.0} \times \frac{10^{5}}{10^{7}}$$

$$= 1.71 \times 10^{-2} \text{ in scientific notation}$$

or .0171 in ordinary notation.

Example 14: Evaluate $(5{,}400{,}000)(700{,}000)$

Solution:

$$(5{,}400{,}000)(700{,}000) = (5.4 \times 10^{6})(7.0 \times 10^{5})$$

$$= (5.4)(7) \times 10^{6} \cdot 10^{5}$$

$$= 37.8 \times 10^{11}$$

Notice that our answer is not in scientific notation since 37.8 is not a number between 1 and 10. To remedy this, write 37.8 in scientific notation as follows:

$$\boxed{37.8} \times 10^{11} = \boxed{3.78 \times 10^{1}} \times 10^{11} = 3.78 \times 10^{12}$$

Now our answer is in scientific notation.

EXERCISE 5.4

Complete the table

	ORDINARY NOTATION	*SCIENTIFIC NOTATION*
1.		3.48×10^{3}
2.	56700	
3.	.0000061	
4.	34	
5.		3.1×10^{-6}
6.		4.01×10^{2}
7.	84.26	
8.		2.63×10^{-1}
9.	467,000,000	
10.	.00000008112	
11.		4.66×10^{-10}
12.	.6	
13.		7.42×10^{1}
14.	3.62	
15.	.014	

16. The earth moves about the sun at the rate of 107,700 kilometers per hour. Express this number in scientific notation.

17. There is a number used in chemistry, known as Avogadro's Number, which is equal to 6.02×10^{23}. Write this number in ordinary notation.

18. The closest star is called alpha Centauri, and it is 2.58×10^{13} miles away. Write this distance in ordinary notation.

19. The deepest part of the Pacific Ocean, near the Mariana Islands, is 36,000 feet deep. Write this number in scientific notation.

20. In chemistry and physics, the weight of atoms is measured in atomic mass units (amu). One atomic mass unit is equal to 1.66×10^{-24} grams. Express this number in ordinary notation and you will see why scientists use scientific notation.

In exercises 21 through 30, evaluate each expression using scientific notation and the rules of exponents. Express your answer in scientific notation.

21. $(120000)(40000000)$

22. $(41000000)(.000002)(.0002)$

23. $(.00000006)(.0000002)(.00000004)$

24. $\dfrac{(260000)(.00015)}{(.00013)}$

25. $\dfrac{65000000}{130}$

26. $\dfrac{14.4}{9000}$

27. $\dfrac{(.006)(240000)}{.00016}$

28. $\dfrac{(.015)(.0005)}{(2500000)}$

29. $\dfrac{(65000)(900)}{(.0009)(1300)}$

30. $\dfrac{(72000)(9600000)(.00035)}{(480000)(.084)(60000)}$

Answers to odd-numbered exercises on page 468.

5.5 Chapter Summary

Examples

The Rules of Exponents

(5.1) I. $x^a \cdot x^b = x^{a+b}$

(5.1) II. $(x^a)^b = x^{ab}$

(5.1) III. $\dfrac{x^a}{x^b} = x^{a-b}$
(5.2)

(5.1) IV. $(xy)^a = x^a y^a$

(5.1) V. $\left(\dfrac{x}{y}\right)^a = \dfrac{x^a}{y^a}$

(5.2) VI. $x^{-a} = \dfrac{1}{x^a}$

(5.2) VII. $\dfrac{1}{x^{-a}} = x^a$

(5.2) VIII. $\left(\dfrac{x}{y}\right)^{-a} = \left(\dfrac{y}{x}\right)^a$

(5.2) IX. $x^0 = 1$

*(5.4) To write a number in scientific notation
1. Start at the right of the first non-zero digit.
2. Count the number of digits from the starting point to the decimal point. This number is the number part of the exponent, or power, of 10.
3. The sign of the exponent is positive if you count to the right and negative if you count to the left.
4. Write the given number with the decimal point after the first non-zero digit and multiply by the power of 10 found in steps 2 and 3.

*(5.4) To change from scientific notation to ordinary notation:
1. Move the decimal point as many places as the value of the exponent.
2. Move the decimal point to the right if the exponent is positive and to the left if it is negative. Add zeros if necessary.

Examples

(5.1) $x^3 \cdot x^4 = x^{3+4} = x^7$

(5.1) $(x^2)^4 = x^{2 \cdot 4} = x^8$

(5.1) $\dfrac{x^6}{x^4} = x^{6-4} = x^2$

(5.2) $\dfrac{x^3}{x^8} = x^{3-8} = x^{-5}$

(5.1) $(xy)^3 = x^3 y^3$

(5.1) $\left(\dfrac{x}{y}\right)^3 = \dfrac{x^3}{y^3}$

(5.2) $x^{-3} = \dfrac{1}{x^3}$

(5.2) $\dfrac{1}{x^{-3}} = x^3$

(5.2) $\left(\dfrac{x}{y}\right)^{-3} = \left(\dfrac{y}{x}\right)^3$

(5.2) $(-7)^0 = 1$

(5.4) 78,400,000 = 7.8400000.
7 places→
= 7.84×10^7

(5.4) .00000642 = .000006.42
←6 places
= 6.42×10^{-6}

(5.4) $2.04 \times 10^3 = 2040$

(5.4) $4.166 \times 10^{-5} = .00004166$

Exercise 5.5 Chapter Review

(5.1) Simplify using rules of exponents:

1. $a^6 \cdot a^5$
2. $n^5 \cdot n^7$
3. $(x^3)^7$
4. $\dfrac{x^8}{x^3}$
5. $(x^5)^4$
6. $h \cdot h^6$
7. $\dfrac{y^7}{y}$
8. $(xy)^6$
9. $\left(\dfrac{x}{y}\right)^8$
10. $(3mn)^2$
11. $x^4 \cdot y^7$
12. $\dfrac{x^{3a}}{x^{2a}}$

(5.2) Simplify leaving your answers with only positive exponents:

13. x^{-5}
14. y^{-3}
15. $\dfrac{1}{x^{-4}}$
16. $\dfrac{1}{h^{-1}}$
17. 7^{-1}
18. 8^0
19. $\left(\dfrac{m}{n}\right)^{-5}$
20. $\left(\dfrac{a}{b}\right)^{-c}$
21. $x^{-3}y^5$
22. $\dfrac{m^0}{n^{-1}}$
23. $\left(\dfrac{2}{3}\right)^{-2}$
24. $\dfrac{a^{-5}b^4}{c^{-2}}$

(5.3) Simplify leaving your answers with only positive exponents:

25. $\dfrac{x^5y^7}{x^3y}$
26. $\dfrac{(xy)^4}{x^3y^5}$
27. $(-3x^2)^2$
28. $(-3x^2)^{-2}$
29. $(7x^0y^5)^2$
30. $\dfrac{a^{-2}y}{a^{-5}y}$

31. $\dfrac{a^4b^3}{x^3y^7}$

32. $(8a^2)^{-3}$

33. $\left(\dfrac{2x^2}{5x^{-3}}\right)^{-2}$

34. $\left(\dfrac{7x^{-2}y^{-8}}{52x^7y^{-4}}\right)^0$

35. $\left(\dfrac{-5ab^2}{3ab^3}\right)^{-2}$

36. $\left(\dfrac{4x^2y^3z^{-3}}{12x^{-3}yz^{-2}}\right)^{-3}$

*(5.4) Complete the table:

	ORDINARY NOTATION	SCIENTIFIC NOTATION
37.	760000	
38.		3.2×10^4
39.	.000015	
40.	235000000	
41.		2.8×10^{-8}
42.		4.6×10^1
43.	.015	
44.		5.77×10^{-10}
45.	8.88	
46.		5.62×10^6

*(5.4) Evaluate each expression using scientific notation and the rules of exponents. Express your answer in scientific notation.

47. (380000)(7000000)

48. $\dfrac{68000}{340000000}$

49. (.00018)(90000)(.004)

50. $\dfrac{(7600)(.0004)}{(.00038)}$

51. $\dfrac{.00078}{26000}$

52. $\dfrac{(.42)(.003)(.02)}{(21)(.06)(400)}$

Answers to odd-numbered exercises on page 468.

Chapter 5 Achievement Test

Name ______________________

Class ______________________

This test should be taken before you are tested in class on the material in Chapter 5. Solutions to each problem and the section where the type of problem is found are given on page 469.

In exercises 1 through 12, simplify, leaving answers with only positive exponents.

1. $a^4 \cdot a^{-6}$

2. $\dfrac{x^4}{x^9}$

3. $(2xy)^3$

4. $(-2x^3y^2)^2$

5. $x \cdot x^7$

6. $\left(\dfrac{a}{b}\right)^{-5}$

7. $\dfrac{x^{-2}y}{x^4y^3}$

8. $\dfrac{-6x^0y^4}{y^4}$

9. $\dfrac{1}{h^{-7}}$

10. $\left(\dfrac{-3x^4y^{-2}}{-2x^5y^{-5}}\right)^0$

11. $\left(\dfrac{2x^3}{3x^{-4}}\right)^{-2}$

12. $\left(\dfrac{3}{4}\right)^{-3}$

1. ______________

2. ______________

3. ______________

4. ______________

5. ______________

6. ______________

7. ______________

8. ______________

9. ______________

10. ______________

11. ______________

12. ______________

Write in scientific notation:

13. .00000074

14. 5,630,000,000

Write in ordinary notation:

15. 3.14×10^{6}

16. 7.52×10^{-5}

*Calculate using scientific notation and rules of exponents:

17. (460,000) (.00004)

18. $\dfrac{(220)(.005)(.00003)}{(.0002)(1500)}$

13. ____________________

14. ____________________

15. ____________________

16. ____________________

17. ____________________

18. ____________________

6
Polynomials

6.1 BASIC CONCEPTS

In Chapter 2 we learned some definitions which need to be recalled now:

Terms: The + and − signs in a mathematical expression divide it up into *terms*.

Numerical Coefficient: The number part of a term.

Literal Part: The letter or variable part of a term.

Like Terms: Terms with exactly the same literal parts.

Unlike Terms: Terms with different literal parts.

Now we will define a new expression which is the subject of this chapter.

Polynomial: An algebraic expression consisting of the sum or difference of one or more terms. Polynomials will *never* have a variable in any denominators.

Examples of polynomials are:

$$6x^2$$
$$7x^3 - 5$$
$$x^2 + 2x - 7$$
$$-6x^5 + 2x^4 - 3x^3 - 2x + 1$$

The expression $4x^2 + \frac{3}{x} - 2$ is not a polynomial since it contains a variable x in the denominator of the second term.

Polynomials are named by the number of terms they contain:

POLYNOMIAL	*NAME*	*NUMBER OF TERMS*
$-3x^2$	**Monomial**	1
$3x^2 + 1$	**Binomial**	2
$-2x^2 + 6x - 5$	**Trinomial**	3

When a polynomial contains more than three terms it is simply called a polynomial.

Degree of a Polynomial: The highest power of the variable occurring in any term of the polynomial.

Example 1:

(a) $6x^{\boxed{3}} + 5$ of degree 3

(b) $2x^{\boxed{2}} + 7x - 2$ of degree 2

(c) $7x$ of degree 1 ($7x = 7x^{\boxed{1}}$)

(d) -12 of degree 0 ($-12 = -12x^{\boxed{0}}$)

Descending Order: When the polynomial is written with the term containing the highest power of the variable first, and then the next highest power of the variable, and so on.

Example 2:

$7x^{\boxed{4}} - 3x^{\boxed{2}} + 4$ The powers get smaller as we go from left to right.

The usual way to write a polynomial is in *descending order*.

Example 3: Write the given polynomials in descending order; give the number of terms in each, its name, and its degree:

(a) $3 + 4x^3 - 5x$

(b) $7x^3 - 4x^4 - 5x$

(c) -3

(d) $4y^5$

(e) $7x^2 - 1$

(f) $7 + 3x - 2x^2 + 6x^6$

(g) $6x^{-2} + 5$

	POLYNOMIAL IN DESCENDING ORDER	NUMBER OF TERMS	NAME	DEGREE
(a)	$4x^3 - 5x + 3$	3	Trinomial	3
(b)	$-4x^4 + 7x^3 - 5x$	3	Trinomial	4
(c)	-3	1	Monomial	0 ($-3 = -3x^0$)
(d)	$4y^5$	1	Monomial	5
(e)	$7x^2 - 1$	2	Binomial	2
(f)	$6x^6 - 2x^2 + 3x + 7$	4	Polynomial	6
(g)	$6x^{-2} + 5 = \frac{6}{x^2} + 5$	contains a variable in a denominator so it is *not* a polynomial.		

EXERCISE 6.1

Fill in the following table.

	Polynomial	*Descending Order*	*No. of Terms*	*Name*	*Degree*
1.	$2x - 7 + 3x^2$				
2.	$3x^3 - 7x + 2 - 5x^2$				
3.	$-124x + 6$				
4.	$-7y^{18} + 1$				
5.	$x^3 - 2x + \frac{1}{x}$				
6.	$14x^3 - 18x^2 + 2x - 1$				
7.	-4				
8.	$-16x + 6x^3$				
9.	$6x$				
10.	$13y^4 - 18y^5$				
11.	$6x^4 - 7x^{-3} + 2$				
12.	14				
13.	$-23x^5$				
14.	$3x^2 + 5x^4 - 6x + 7x^3 - 5$				
15.	$-6 + 14z - 52z^7$				

Answers to odd-numbered exercises on page 469.

6.2 ADDING AND SUBTRACTING POLYNOMIALS

In Section 2.3 we learned to combine like terms by adding their numerical coefficients and keeping the same literal part.

Example 1:

(a) $\boxed{3}x^2y + \boxed{5}x^2y = \boxed{8}x^2y$
(b) $\boxed{-7}x^4 + \boxed{5}x^4 = \boxed{-2}x^4$

To add polynomials horizontally:

1. Write the polynomials in descending order.
2. Combine all the like terms occurring in the polynomials being added.

Example 2: Add $3x^3 + 5x^2 - 2x + 1$ and $2x^3 + 6x^2 + 5x - 8$

Solution:

$3x^3 + 5x^2 - 2x + 1 \boxed{+} 2x^3 + 6x^2 + 5x - 8$

$= \underbrace{3x^3 + 2x^3} + \underbrace{5x^2 + 6x^2} - \underbrace{2x + 5x} + \underbrace{1 - 8}$ Group like terms.

$= \quad 5x^3 \quad + \quad 11x^2 \quad + 3x \quad -7$ Combine like terms.

$= 5x^3 + 11x^2 + 3x - 7$

Example 3: Add $-6 + 3x^3 - 7x^2$ and $3x^4 - 6x^2 - 5x^3 + 2x$

Solution:

$3x^3 - 7x^2 - 6 \boxed{+} 3x^4 - 5x^3 - 6x^2 + 2x$ Write in descending order.

$= 3x^4 + \underbrace{3x^3 - 5x^3} - \underbrace{7x^2 - 6x^2} + 2x - 6$ Group like terms.

$= 3x^4 \quad -2x^3 \quad -13x^2 \quad + 2x - 6$ Combine like terms.

$= 3x^4 - 2x^3 - 13x^2 + 2x - 6$

After a little practice you will probably find it unnecessary to group like terms before combining them. Study the next example carefully:

Example 4: Add $-4x^3 - 6x^4 + 2x^2 - 11$ and $-x^2 + 6 + 2x^4$

Solution:

$-6x^4 - 4x^3 + 2x^2 - 11 \boxed{+} 2x^4 - x^2 + 6$ Write in descending order.

$= -4x^4 - 4x^3 + x^2 - 5$ Combine like terms.

Polynomials can also be added *vertically:*

To add polynomials vertically:

1. Write the polynomials in descending order with like terms under one another.
2. Combine like terms.

Example 5: Add $6x^3 - 9x^2 + 2x - 1$ and $4x^3 + 6x^2 - 5x - 6$

Solution:

$$\begin{array}{rrrr} 6x^3 & - 9x^2 & + 2x & - 1 \\ 4x^3 & + 6x^2 & - 5x & - 6 \\ \hline 10x^3 & - 3x^2 & - 3x & - 7 \end{array}$$

Write like terms under one another.
Combine like terms.

Example 6: Add $4x^3 + 6x^4 - 3x^2 + 1$ and $-4 + 3x^3 - 2x^2 + 3x$

Solution:

$$\begin{array}{rrrrr} 6x^4 & + 4x^3 & - 3x^2 & & +1 \\ & 3x^3 & - 2x^2 & + 3x & -4 \\ \hline 6x^4 & + 7x^3 & - 5x^2 & + 3x & -3 \end{array}$$

Write like terms under one another, leaving space for missing terms.
Combine like terms.

Subtracting Polynomials

In Section 2.3 we removed parentheses preceded by a negative sign by changing the sign of each term in the parentheses. We will make use of this procedure in subtracting polynomials.

To subtract polynomials:

1. Write each polynomial in descending order.
2. Change the sign of each term of the polynomial being subtracted.
3. Combine like terms.

Example 7: Subtract $6x^2 - 3x^3 + 2x - 1$ from $7x^3 - 3x + 1 - 4x^2$

Solution:

$(7x^3 - 4x^2 - 3x + 1) - (-3x^3 + 6x^2 + 2x - 1)$ Write in descending order.

$= 7x^3 - 4x^2 - 3x + 1 \boxed{+} 3x^3 \boxed{-} 6x^2 \boxed{-} 2x \boxed{+} 1$ Change the sign of each term in the polynomial being subtracted.

$= 10x^3 - 10x^2 - 5x + 2$ Combine like terms.

To do the same subtraction problem vertically, change the sign of each term in the polynomial being subtracted, place the like terms under one another, and combine.

Example 8: Subtract $6x^2 - 3x^3 + 2x - 1$ from $7x^3 - 3x + 1 - 4x^2$

Solution:

$$\begin{array}{rrrr} 7x^3 & - 4x^2 & - 3x & + 1 \\ \boxed{+} 3x^3 & \boxed{-} 6x^2 & \boxed{-} 2x & \boxed{+} 1 \\ \hline 10x^3 & - 10x^2 & - 5x & + 2 \end{array}$$

Write in descending order.
Change the sign of each term.
Combine like terms.

Example 9: Subtract $3x + 4x^3 - 2x^2 + 6$ from $7 - 5x + 5x^3$ (a) horizontally and (b) vertically.

Solution: (a)

$5x^3 - 5x + 7 - (4x^3 - 2x^2 + 3x + 6)$	Write in descending order.
$= 5x^3 - 5x + 7 \boxed{-} 4x^3 \boxed{+} 2x^2 \boxed{-} 3x \boxed{-} 6$	Change the sign of each term being subtracted.
$= x^3 + 2x^2 - 8x + 1$	Combine like terms.

Solution: (b)

Leave space for missing term.

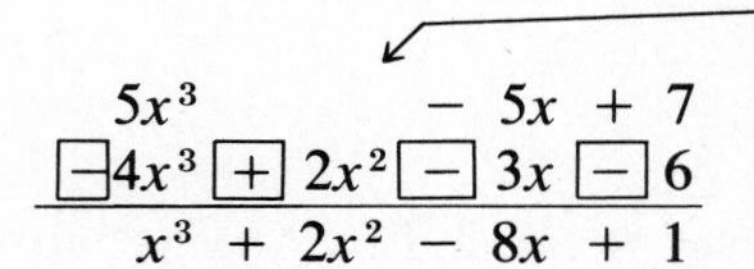

$$\begin{array}{rrrr} 5x^3 & & -\ 5x & +\ 7 \\ \boxed{-}4x^3 & \boxed{+}\ 2x^2 & \boxed{-}\ 3x & \boxed{-}\ 6 \\ \hline x^3 & +\ 2x^2 & -\ 8x & +\ 1 \end{array}$$

Write in descending order.
Change the sign of each term.
Combine like terms.

EXERCISE 6.2

Add the polynomials:

1. $6x^2 + 3x - 1$ and $4x^2 - 6x - 5$

2. $-3x + 4x^3 - 9 + x^2$ and $5x - 3x^2 + 2x^3 + 7$

3. $6y + 7y^2 - 5$ and $5y + 6y^2 + 10$

4. $7n^5 + 2n^3 - 3n$ and $7n - 5n^2 + 2n^3 + 14n^4$

5. $-3a^3 - 4a^4 + 17$ and $6a^4 - 24a - 15$

6. $-7t^3 + 8t^2 - 6 + 6t$, $14t + 5t^3 + 8$ and $8t^3 + 12t - 4t^2$

7. $5y + 14y^5 + 12y^3$, $3y^2 + 6y - 14$, and $-7y^5 - 3y^2 + 7$

8. $-14w^3 + 3w^7 - 8w$, $-w + w^3$, and $7w^4 + 3w - 8w^3$

9.
$$\begin{array}{rrrr} 6x^3 & + 2x^2 & - 8x & - 4 \\ 7x^3 & + 5x^2 & & + 8 \\ \hline \end{array}$$

10.
$$\begin{array}{rrrr} 7y^3 & - 8y^2 & & + 2 \\ -7y^3 & & +8y & - 9 \\ \hline \end{array}$$

11.
$$\begin{array}{rrrrr} -24a^4 & + 7a^3 & & - a & - 4 \\ & 4a^3 & + 2a^2 & + a & + 5 \\ a^4 & - 6a^3 & & & - 7 \\ \hline \end{array}$$

12.
$$\begin{array}{rrrrr} 7t^4 & - 8t^3 & - 6t^2 & & \\ & 8t^3 & - 6t^2 & + 2t & + 6 \\ 5t^4 & & & & - 6 \\ \hline \end{array}$$

Subtract the polynomials:

13. Subtract $3x^2 + 4x - 6$ from $9x^2 - 7x + 5$

14. Subtract $8a + 7a^2 + 9$ from $4a + 6 - 8a^2$

15. Subtract $y^3 - 12y^2 - 8 + 4y$ from $6y + 9y^3$

16. Subtract $-8x - 14x^2 + 2x^3 - 9$ from $8x - 2x^3 + 5$

17. Subtract $7x^4 - 3x + 5$ from $2x^3 + 5x - 2x^2$

18. Subtract $-4y^9 + 7y - 6y^5$ from $12 + 7y$

19. $(-6a^3 + 7a^2 - 4a + 2) - (3a^3 - 6a^2 + 5a - 4)$

20. $(-2a^4 + 3a^2 - 7a) - (4a^3 + 2a^2 + 7a + 5)$

21. $(-3y - 17y^2 - 5) - (-6 + 2y^3 + 5y^2)$

22. $(6x - 4 + 2x^5 + 3x^2) - (-7 + 5x^4 + 2x^3)$

23. $$\begin{array}{r} -\ 7x^3 + 5x^2 - 7x + 5 \\ -(4x^3 + 7x^2 - 8x + 9) \\ \hline \end{array}$$

24. $$\begin{array}{r} -8x^3 + 7x^2 \qquad\qquad + 5 \\ -(-4x^3 \qquad\qquad + 5x - 9) \\ \hline \end{array}$$

25. Subtract: $-6x^2 + 2x - 4$ from the sum of $-7x^2 + 2x - 8$ and $5x^2 + 8x - 1$

26. Subtract $-8y^3 + 2y^2 + 6y - 8$ from the sum of $-3y^3 - 8y^2 + 2y - 8$ and $6y^3 + 5y - 1$

27. Subtract $-4x^4 + 2x^2 - 4$ from the sum of $5x^3 + 2x^2$ and $3x^4 + 9$

28. Subtract $6 + 3x^2 - 4x^4$ from the sum of $6x - 5x^4 + 2$ and $3x^2 - 9x^4 + 1$

Perform the indicated additions and subtractions and simplify:

29. $(6x^3 + 12x^2 - 4x + 1) + (2x^3 - 5x^2 + 2x - 1) - (6x^3 + 9x^2 - 3)$

30. $(-14 - 8x^3 + 2x) + (x^2 - 9x^3 + 1) - (x + 6x^3 + 9)$

31. $(-9y + 4) + (-7y - 9y^4 + 6y^2) - (5y^2 + 14 + 7y^3)$

32. $(6 - 3a^2 + 9a + 4a^3) - (6a^2 - 3 + 5a) + (-7 + 6a - 5a^3)$

33. Given polynomials $-3x^2 - 4x^3 + 8 - 5x$, $-x^3 + 4x - 6x^2$ and $17x - 15x^3 + 7 + x^2$, subtract the sum of the last two from the first.

34. Given polynomials $a^4 - a + 3a^2 - 14$, $9 + 6a^4 - a^3 - 7a$ and $2 - 4a^3 + 6a^4$, subtract the sum of the first two from the sum of the last two.

35. Given polynomials $y - 4y^3 + 2y^2 - 6$, $y^3 + 1$, $6y + 4y^3 - 8$ and $7y + 3y^2 - 6$, subtract the sum of the first two from the sum of the last two.

Answers to odd-numbered exercises on page 470.

6.3 MULTIPLICATION OF POLYNOMIALS

Multiplying a Monomial by a Monomial

Recall from Section 2.5 the rule for multiplying monomials.

> **To multiply monomials:**
> 1. Multiply the numerical coefficients.
> 2. Multiply the variables using the rule $x^a \cdot x^b = x^{a+b}$.

Example 1: Multiply the monomials.

(a) $(-6x^3)(5x^2)$
$= (-6)(5)\, x^3 \cdot x^2$
$= -30x^5$

(b) $(7x^4)(-6x^2)$
$= (7)(-6)\, x^4 \cdot x^2$
$= -42\, x^6$

(c) $(6mn^2)(3m^3n)$
$= (6)(3)\, m \cdot m^3 \cdot n^2 \cdot n$
$= 18\, m^4n^3$

Multiplying a Polynomial by a Monomial

Multiplying a polynomial by a monomial is an application of the distributive rule.

$$a(b + c + d + e + \ldots) = ab + ac + ad + ae + \ldots.$$

and

$$(b + c + d + e + \ldots)\, a = ba + ca + da + ea + \ldots.$$

> To multiply a polynomial by a monomial multiply each term of the polynomial by the monomial.

Example 2:

(a) $3x\,(6x^2 + 2x + 5)$

$= \boxed{3x} \cdot 6x^2 + \boxed{3x} \cdot 2x + \boxed{3x} \cdot 5$

$= 18x^3 + 6x^2 + 15x$

(b) $4x^4(3x^3 + 2x^2 - 4x + 9)$

$= \boxed{4x^4} \cdot 3x^3 + \boxed{4x^4} \cdot 2x^2 + \boxed{4x^4} \cdot (-4x) + \boxed{4x^4} \cdot 9$

$= 12x^7 + 8x^6 - 16x^5 + 36x^4$

(c) $-10x^2(-4x^2 + 2x - 9)$

$= \boxed{(-10x^2)}(-4x^2) + \boxed{(-10x^2)}(2x) + \boxed{(-10x^2)}(-9)$

$= 40x^4 - 20x^3 + 90x^2$

(d) $(-7n^3 + 2n^2 - 8)n^2$

$= -7n^3 \cdot \boxed{n^2} + 2n^2 \cdot \boxed{n^2} - 8 \cdot \boxed{n^2}$

$= -7n^5 + 2n^4 - 8n^2$

Multiplying a Polynomial by a Polynomial

Consider the product of two polynomials $(2x - 4)(3x^2 + 2x - 5)$. Applying the distributive rule we will multiply each term in one polynomial by each term in the other polynomial and combine similar terms. Arrange the problem vertically:

$3x^2 + 2x - 5$	Arrange the problem vertically.
$2x - 4$	
$6x^3 + 4x^2 - 10x$	← Product of $3x^2 + 2x - 5$ and $2x$
$-12x^2 - 8x + 20$	← Product of $3x^2 + 2x - 5$ and -4
$6x^3 - 8x^2 - 18x + 20$	← Combine like terms, which have been conveniently arranged in vertical columns.

To multiply a polynomial by a polynomial:

1. Multiply each term in one polynomial by each term in the other polynomial.
2. Arrange like terms under one another in the products.
3. Combine like terms in the products.

Example 3: Multiply $(6x^2 - 4x + 7)(3x + 2)$

Solution:

$6x^2 - 4x + 7$	Arrange the problem vertically.
$3x + 2$	
$18x^3 - 12x^2 + 21x$	$(3x)(6x^2 - 4x + 7)$
$12x^2 - 8x + 14$	$(2)(6x^2 - 4x + 7)$
$18x^3 + 0x^2 + 13x + 14$	Combine like terms.

$= 18x^3 + 13x + 14$

Example 4: $(3x^2 + 4x^2 - 5x + 6)(4x^2 - x + 1)$

Solution:

$$\begin{array}{rrrrrrl}
 & & 3x^3 & +\ 4x^2 & -\ 5x & +\ 6 & \\
 & & & 4x^2 & -\ x & +\ 1 & \\
\hline
12x^5 & +\ 16x^4 & -\ 20x^3 & +\ 24x^2 & & & (4x^2)\cdot\text{(top polynomial)} \\
 & -\ 3x^4 & -\ 4x^3 & +\ 5x^2 & -\ 6x & & (-x)\cdot\text{(top polynomial)} \\
 & & 3x^3 & +\ 4x^2 & -\ 5x & +\ 6 & (1)\cdot\text{(top polynomial)} \\
\hline
12x^5 & +\ 13x^4 & -\ 21x^3 & +\ 33x^2 & -\ 11x & +\ 6 & \text{Combine like terms.}
\end{array}$$

Example 5: Multiply.

$$\begin{array}{rrrrrl}
7x^3 & & +\ 3x & -\ 2 & & \\
 & -\ 6x^2 & & -\ 3 & & \\
\hline
-42x^5 & -\ 18x^3 & +\ 12x^2 & & & (-6x^2)\cdot\text{(top polynomial)} \\
 & -\ 21x^3 & & -\ 9x & +\ 6 & (-3)\cdot\text{(top polynomial)} \\
\hline
-42x^5 & -\ 39x^3 & +\ 12x^2 & -\ 9x & +\ 6 & \text{Combine like terms.}
\end{array}$$

We left spaces for missing terms so we could add like terms in columns.

EXERCISE 6.3

Multiply:

1. $(6x)(-5x^2)$
2. $(7x^3)(-5x^4)$
3. $(2y^2)(8y^2)$
4. $(3a^2b)(-4a^3b^4)$
5. $(-3x^2yz)(-2xyz^4)$
6. $(-3x^2y)(-2xy^2)(-7xy)$
7. $2x(3x + 5)$
8. $-4x(-x - 8)$
9. $3x^2(x^2 + 3x - 9)$
10. $-11x^2(-2x^2 + 2x - 1)$
11. $6x(4x^4 + 2x^3 + 3x^2 + 2x - 7)$
12. $7x^5(3x^3 + 4x^2 - 6x)$
13. $-10a^2(6a^3 - 7a^2 - 3a + 9)$
14. $-6a^2(-4a^4 + 2a^2 - 9)$
15. $(2x - 4)(x + 6)$
16. $(4x - 5)(9x + 2)$

17. $(x + 5)(x^2 + 2x + 7)$

18. $(y - 2)(y^2 + 2y + 4)$

19. $(6x - 1)(x^2 - 8x - 8)$

20. $(5x + 4)(6x^3 - 2x^2 + 3x - 4)$

21. $(7x^2 + 2x - 4)(3x + 7)$

22. $(4x^3 - 3x^2 + 3x + 8)(6x - 4)$

23. $(x^2 + 2x + 1)(x^2 - 5x + 6)$

24. $(2x^2 - 3x + 4)(x^2 - 4x - 4)$

25. $(x^3 + 2x^2 - 9x - 1)(x^2 + 2x + 3)$

26. $(2x^3 - 3x^2 - 3x + 5)(2x^2 - 3x - 1)$

27. $$\begin{array}{r} 4x^3 + 3x^2 - 6x - 6 \\ x^2 + 3 \\ \hline \end{array}$$

28. $$\begin{array}{r} 6y^3 + 7y^2 - 4 \\ - 3y^2 - 6 \\ \hline \end{array}$$

29. $$\begin{array}{r} 7a^3 - 7a^2 - a - 5 \\ 3a^2 + a + 5 \\ \hline \end{array}$$

30. $$\begin{array}{r} 7x^4 + 3x^3 + 2x^2 + 4x + 5 \\ x^2 + 2x - 3 \\ \hline \end{array}$$

Answers to odd-numbered exercises on page 470.

6.4 DIVISION OF POLYNOMIALS

Dividing a Polynomial by a Monomial

Our first problem will be to divide a polynomial by a monomial. Recall that when we multiplied a polynomial by a monomial we **multiplied** *each* term in the polynomial by the monomial. Since division is defined in terms of multiplication, it seems reasonable then, that to divide a polynomial by a monomial, we **divide** *each* term of the polynomial by the monomial.

To divide a polynomial by a monomial:

Divide *each* term of the polynomial by the monomial, using the rule $\frac{x^a}{x^b} = x^{a-b}$ to divide the variables.

Example 1: Divide the polynomial $(6x^3 + 12x^2 - 18x + 9)$ by the monomial 3.

Solution:

$$\frac{6x^3 + 12x^2 - 18x + 9}{3}$$

$$= \frac{6x^3}{\boxed{3}} + \frac{12x^2}{\boxed{3}} - \frac{18x}{\boxed{3}} + \frac{9}{\boxed{3}}$$ Divide each term by 3.

$$= 2x^3 + 4x^2 - 6x + 3$$

Example 2: Divide $(8x^4 - 24x^3 + 40x - 16) \div (-4x^2)$

Solution:

$$\frac{8x^4 - 24x^3 + 40x - 16}{-4x^2}$$

$$= \frac{8x^4}{\boxed{-4x^2}} - \frac{24x^3}{\boxed{-4x^2}} + \frac{40x}{\boxed{-4x^2}} - \frac{16}{\boxed{-4x^2}}$$ Divide each term by $-4x^2$.

$$= -2x^2 + 6x - 10x^{-1} + 4x^{-2}$$

$\left[\frac{40x}{-4x^2} = -10x^{1-2} = -10x^{-1}\right.$ and $\left.\frac{-16}{-4x^2} = \frac{-16x^0}{-4x^2} = 4x^{0-2} = 4x^{-2}\right]$

$$= -2x^2 + 6x - \frac{10}{x} + \frac{4}{x^2}$$

Write the answer with only positive exponents.

Example 3:

$$\frac{3x^3 + 7x^2 - 4x + 2}{2x}$$

$$= \frac{3x^3}{2x} + \frac{7x^2}{2x} - \frac{4x}{2x} + \frac{2}{2x}$$

$$= \frac{3x^2}{2} + \frac{7x}{2} - 2 + \frac{1}{x}$$

Division of a Polynomial by a Polynomial

To divide one polynomial by another is fairly complicated and is similar to long division in arithmetic. We will begin by reviewing a long division problem involving whole numbers.

Example 1: $34\overline{)726}$

Solution:

```
    2 ←———— Estimate by dividing 7 by 3 = 2 + .
34)726
   68 ←———— Multiply 34 × 2 = 68.
    4 ←———— Subtract 72 − 68 = 4.
```

```
    2
34)726
   68↓
   46 ←———— Bring down the 6.
```

Now repeat the four-step process again: estimate, multiply, subtract, bring down the next digit.

```
   21 ←———— Estimate by dividing 4 by 3 = 1 + .
34)726
   68
   46
   34 ←———— Multiply 34 × 1 = 34.
   12 ←———— Subtract 46 − 34 = 12.
     ↖———— The remainder = 12.
```

To check, we multiply the quotient (21) by the divisor (34) and add in the remainder (12) to obtain the dividend (726).

```
   21 ←———— quotient
 × 34 ←———— divisor
   84
  63
  714
+  12 ←———— remainder
  726 ←———— dividend
```

To divide one polynomial by a second polynomial, go through the following step-by-step procedure several times. You will see that we use the same basic four-step procedure that was used for division of whole numbers: estimate, multiply, subtract, bring down the next term. An example is worked out alongside each procedure to help you see exactly what is done in each step.

To divide a polynomial by a polynomial:

Example: $(-10 + 6x^2 + 11x) \div (2x + 5)$

1. Arrange the polynomials in descending order.

$$2x + 5\overline{)6x^2 + 11x - 10}$$

2. Divide the first term of the dividend by the first term of the divisor.

$$\begin{array}{r} \boxed{3x} \\ \boxed{2x} + 5\overline{)\boxed{6x^2} + 11x - 10} \end{array}$$

3. Multiply the entire divisor by the first term of the quotient.

$$\begin{array}{r} \boxed{3x} \\ \boxed{2x + 5}\overline{)6x^2 + 11x - 10} \\ \boxed{6x^2 + 15x} \end{array}$$

4. Subtract this product from the dividend. To subtract a polynomial we add its negative, or change the signs and add. Then bring down the next term from the dividend.

$$\begin{array}{r} 3x \\ 2x + 5\overline{)6x^2 + 11x - 10} \\ \underline{\boxed{\mp}6x^2 \boxed{\mp} 15x} \\ -4x - 10 \end{array}$$

5. Divide the first term of the new dividend by the first term of the divisor.

$$\begin{array}{r} 3x \boxed{-2} \\ \boxed{2x} + 5\overline{)6x^2 + 11x - 10} \\ \underline{\mp 6x^2 \mp 15x} \\ \boxed{-4x} - 10 \end{array}$$

6. Multiply the entire divisor by the second term of the quotient.

$$\begin{array}{r} 3x \boxed{-2} \\ \boxed{2x + 5}\overline{)6x^2 + 11x - 10} \\ \underline{\mp 6x^2 \mp 15x} \\ -4x - 10 \\ \boxed{-4x - 10} \end{array}$$

7. Subtract this product from the new dividend to obtain the remainder.

$$\begin{array}{r} 3x - 2 \\ 2x + 5\overline{)6x^2 + 11x - 10} \\ \underline{\mp 6x^2 \mp 15x} \\ -4x - 10 \\ \underline{\boxed{\pm} 4x \boxed{\pm} 10} \\ \boxed{0} \end{array}$$

Example 4: Divide $(-30 - x + 2x^2)$ by $(x - 4)$

Solution:

1. $x - 4\overline{)2x^2 - x - 30}$ Arrange in descending order.

2. $\begin{array}{r} \boxed{2x} \\ \boxed{x} - 4\overline{)\boxed{2x^2} - x - 30} \end{array}$ Divide $2x^2$ by x.

3. $$\begin{array}{r} \boxed{2x} \\ \boxed{x - 4}\,\overline{)\,2x^2 - x - 30} \\ 2x^2 - 8x \end{array}$$ Multiply $x - 4$ by $2x$.

4. $$\begin{array}{r} 2x \\ x - 4\,\overline{)\,2x^2 - x - 30} \\ \underline{\mp 2x^2 \pm 8x} \\ 7x - 30 \end{array}$$ Subtract $-x - (-8x) = -x + 8x = 7x$. Bring down the -30.

5. $$\begin{array}{r} 2x\boxed{+7} \\ \boxed{x} - 4\,\overline{)\,2x^2 - x - 30} \\ \underline{\mp 2x^2 \pm 8x} \\ \boxed{7x} - 30 \end{array}$$ Divide $7x$ by x.

6. $$\begin{array}{r} 2x\boxed{+7} \\ \boxed{x - 4}\,\overline{)\,2x^2 - x - 30} \\ \underline{\mp 2x^2 \pm 8x} \\ 7x - 30 \\ \boxed{7x - 28} \end{array}$$ Multiply $x - 4$ by 7.

7. $$\begin{array}{r} 2x + 7 \\ x - 4\,\overline{)\,2x^2 - x - 30} \\ \underline{\mp 2x^2 \pm 8x} \\ 7x - 30 \\ \underline{\mp 7x \pm 28} \\ \boxed{-2} \end{array}$$ Subtract $-30 - (-28) = -30 + 28 = -2$.

The remainder is -2.

Our answer is $2x + 7$ with a remainder of -2. To check, multiply the quotient by the divisor and add in the remainder and we should get the dividend.

$$\begin{array}{rrrl} 2x & +7 & & \leftarrow \text{quotient} \\ x & -4 & & \leftarrow \text{divisor} \\ \hline 2x^2 & +7x & & \\ & -8x & -28 & \\ \hline 2x^2 & -x & -28 & \\ & & -2 & \leftarrow \text{remainder} \\ \hline 2x^2 & -x & -30 & \leftarrow \text{dividend} \end{array}$$

Example 5: Divide $(5x + 5x^2 - 3 + x^3) \div (x + 3)$

Solution:

$$\begin{array}{r} x^2 + 2x - 1 \\ x + 3\,\overline{)\,x^3 + 5x^2 + 5x - 3} \\ \underline{\mp x^3 \mp 3x^2} \\ 2x^2 + 5x \\ \underline{\mp 2x^2 \mp 6x} \\ -x - 3 \\ \underline{\pm x \pm 3} \\ 0 \end{array}$$

Check:

$$\begin{array}{r} x^2 + 2x - 1 \\ x + 3 \\ \hline x^3 + 2x^2 - x \\ 3x^2 + 6x - 3 \\ \hline x^3 + 5x^2 + 5x - 3 \end{array}$$

Example 6: $(x^3 + 8) \div (x + 2)$

Solution:

$$x + 2\overline{)x^3 + 0x^2 + 0x + 8}$$

Always leave space for missing powers of the variable by using 0 as the coefficient.

$$\begin{array}{r} x^2 - 2x + 4 \\ x + 2\overline{)x^3 + 0x^2 + 0x + 8} \\ \mp x^3 \mp 2x^2 \\ \hline -2x^2 + 0x \\ \pm 2x^2 \pm 4x \\ \hline 4x + 8 \\ \mp 4x \mp 8 \\ \hline 0 \end{array}$$

Check:

$$\begin{array}{r} x^2 - 2x + 4 \\ x + 2 \\ \hline x^3 - 2x^2 + 4x \\ 2x^2 - 4x + 8 \\ \hline x^3 \qquad\qquad + 8 \end{array}$$

EXERCISE 6.4

Divide:

1. $(4x^3 + 6x^2 - 8x + 16) \div 2$
2. $(-6x^3 + 9x^2 - 12) \div (-3)$
3. $(14x^4 + 7x^3 - 21x + 35) \div 7x$
4. $(36y^4 - 6y^2 + 18y + 24) \div -6x^2$
5. $(8a^4 + 12a^3 - 16a^2 + 24) \div -4a^4$
6. $(7c^3 - 21c^2 + 14c - 42) \div 7c$

7. $(3x^4 + 2x^3 - 8x^2 + 22x - 11) \div 2x$

8. $(24x^5 + 16x^3 - 8x^2 + 6) \div -4x^2$

9. $(27x^3 + 2x^2 - 18x - 9) \div 9x^5$

10. $(33x^5 - 9x^3 + 6) \div 11x^4$

11. $(x^2 - 13x + 40) \div (x - 5)$

12. $(x^2 - 11x + 28) \div (x - 7)$

13. $(a^2 - 6a + 9) \div (a - 3)$

14. $(x^2 + 6x + 8) \div (x + 2)$

15. $(2y^2 + y - 6) \div (y + 2)$

16. $(2x^2 - 13x + 20) \div (2x - 5)$

17. $(6n^2 + 11n + 3) \div (3n + 1)$

18. $(8x^2 + 2x - 4) \div (2x + 1)$

19. $(6x^2 - 5x - 21) \div (3x + 5)$

20. $(11x + 4x^2 - 30) \div (4x - 5)$

21. $(17x + 1 + 12x^2) \div (5 + 3x)$

22. $(6 + 37t + 4t^2) \div (4t + 1)$

23. $(x^3 - 3x^2 + 100) \div (x + 4)$

24. $(x^3 - 8x + 3) \div (x + 3)$

25. $(6x^2 + 41x) \div (6x - 1)$

26. $(t^3 - 27) \div (t - 3)$

27. $(x^3 - x^2 - 5x - 3) \div (x - 3)$

28. $(-5x^2 - 5 - 13x + 2x^3) \div (2x + 1)$

29. $(y^4 + 2y^2 - 2y^3 - 8) \div (y - 2)$

30. $(z^3 - 8) \div (z - 2)$

31. $(x^4 + x^3 - 9x^2 - 14x - 4) \div (x^2 + 3x + 1)$

32. $(-7x^2 - 3 + 6x + x^4) \div (x^2 + 3 - 3x)$

Answers to odd-numbered exercises on page 471.

6.5 Chapter Summary

(6.1) **Terms:** The + and − signs in a mathematical expression divide it up into *terms*.

(6.1) **Numerical Coefficient:** The number part of a term.

(6.1) **Literal Part:** The letter, or variable, part of a term.

(6.1) **Like Terms:** Terms with exactly the same literal parts.

(6.1) **Unlike Terms:** Terms with different literal parts.

(6.1) **Polynomial:** An expression consisting of the sum or difference of one or more terms.

(6.1) **Monomial:** A polynomial containing *one* term.

(6.1) **Binomial:** A polynomial containing *two* terms.

(6.1) **Trinomial:** A polynomial containing *three* terms.

(6.1) **Degree of a Polynomial:** The highest power of the variable occurring in the polynomial.

(6.1) **Descending Order:** When the polynomial is written with the term containing the highest power of the variable first, then the next highest power of the variable, and so on.

(6.2) **To add polynomials horizontally:**
1. Write the polynomials in descending order.
2. Combine the like terms.

(6.2) **To add polynomials vertically:**
1. Write the polynomials in descending order with like terms under one another.
2. Combine the like terms.

(6.2) **To subtract polynomials:**
1. Write each polynomial in descending order.
2. Change the sign of *each* term in the polynomial being subtracted.
3. Combine the like terms.

(6.3) **To multiply monomials:**
1. Multiply the numerical coefficients.
2. Multiply the variables using the rule $x^a \cdot x^b = x^{a+b}$

Examples

(6.1) $\boxed{7x^3} + \boxed{3x^2}\boxed{-4x} + \boxed{5}$
terms

$\boxed{7}x^3 + \boxed{3}\,x^2\boxed{-4}x + \boxed{5}$
coefficients

$7\boxed{x^3} + 3\boxed{x^2} - 4\boxed{x} + 5$
literal parts

$\boxed{4xy^2} - 3x + \boxed{2xy^2} - 8$
like terms

$\boxed{3x^2} + \boxed{6x} + \boxed{2xy}$
unlike terms

$-6x^4 + 2x^3 - 2x + 1$
polynomial

$6x$
monomial

$3x + 2$
binomial

$6x^2 + 2x - 1$
trinomial

$7x^4 + 6x^{\boxed{5}} - 2x + 1$
degree = 5

$6x^{\boxed{5}} + 2x^{\boxed{4}} - 4x^{\boxed{2}} + \boxed{7}$
in descending order

(6.2) $(3x^3 + 2x^2 - 2x + 5) + (4x^3 - 6x^2 + 3x - 7)$

$= 7x^3 - 4x^2 + x - 2$

(6.2)
$$\begin{array}{r} 3x^3 + 2x^2 - 2x + 5 \\ + \; 4x^3 - 6x^2 + 3x - 7 \\ \hline 7x^3 - 4x^2 + x - 2 \end{array}$$

(6.2) $(5x^2 + 2x - 1) - (7x^2 + 3x - 5)$

$= 5x^2 + 2x - 1 - 7x^2 - 3x + 5$

$= -2x^2 - x + 4$

(6.3) $(-7x^4)(4x^6)$

$= (-7)(4)(x^4)(x^6)$

$= -28\,x^{10}$

(6.3) **To multiply a polynomial by a monomial:**
Multiply each term of the polynomial by the monomial.

(6.3) $5x^2(3x^2 + 2x - 1)$

$$= 5x^2 \cdot 3x^2 + 5x^2 \cdot 2x + 5x^2 \cdot (-1)$$
$$= 15x^4 + 10x^3 - 5x^2$$

(6.3) **To multiply a polynomial by a polynomial:**

1. Multiply each term in one polynomial by each term in the other polynomial.
2. Arrange like terms under one another in the products.
3. Combine the like terms.

(6.3)

$$\begin{array}{r} 7x^2 - 3x - 6 \\ 4x^2 - 7 \\ \hline 28x^4 - 12x^3 - 24x^2 \\ -49x^2 + 21x + 42 \\ \hline 28x^4 - 12x^3 - 73x^2 + 21x + 42 \end{array}$$

(6.4) **To divide a polynomial by a monomial:**
Divide *each* term of the polynomial by the monomial using the rule

$$\frac{x^a}{x^b} = x^{a-b}$$

to divide the variables.

(6.4) $(7x^3 + 4x^2 - 8x + 12) \div 2x$

$$= \frac{7x^3}{2x} + \frac{4x^2}{2x} - \frac{8x}{2x} + \frac{12}{2x}$$
$$= \frac{7}{2}x^2 + 2x - 4 + 6x^{-1}$$
$$= \frac{7}{2}x^2 + 2x - 4 + \frac{6}{x}$$

using only positive exponents.

(6.4) **To divide a polynomial by a polynomial:**
See the detailed explanation in Section 6.4.

(6.4)

$$\begin{array}{r} 3x + 2 \\ x - 2\overline{)3x^2 - 4x - 6} \\ \underline{\mp 3x^2 \pm 6x} \\ 2x - 6 \\ \underline{\mp 2x \pm 4} \\ -2 \end{array}$$

remainder ↗

Exercise 6.5 Chapter Review

(6.1) Fill in the following table:

	Polynomial	Descending Order	No. of Terms	Name	Degree
1.	$3x + 4x^2 - 8$				
2.	$5x^3 + 2x - 4$				
3.	$y^4 - 3y + 2y^2 - 1$				
4.	-8				
5.	$-14x$				

(6.2) Add the polynomials:

6. $7a^2 - 3a + 6$ and $-4a^2 - 4a - 5$

7. $-3y^3 + 4y^4 - 3y$ and $7y^3 + 3y^4 - 2y^2 - 7$

8. $-2x + 5x^3 + 2x^2 - 6$ and $x^3 + x^2 - 6 - 12x$

9. $-14x - 13x^3 + 7x^2$, $4x - 3x^2 + x^3 - 3$, and $6 + x^3 - 2x^2$

10.
$$\begin{array}{rrrr} -7x^3 & + 6x^2 & & - 8 \\ x^3 & & + 22x & - 5 \\ \hline \end{array}$$

11.
$$\begin{array}{rrrr} -3y^3 & + 4y^2 & & - 12 \\ & 2y^2 & + 8y & - 4 \\ 4y^3 & & + 2y & - 4 \\ \hline \end{array}$$

(6.2) Subtract the polynomials:

12. Subtract $4x^2 + 3x - 9$ from $2x^2 - 5x + 12$

13. Subtract $y^3 + 2y^2 + 6$ from $6y^3 + 7y - 4$

14. $(-5a^3 + 2a^2 - 6a + 3) - (2a^3 + 6a^2 + 3a + 5)$

15. $(12x^3 + 6x + 4) - (4x^2 + 3x)$

16.
$$\begin{array}{rrrr} -\ 6x^3 & +\ 2x^2 & -\ 4x & +\ 9 \\ \underline{-(3x^3} & & \underline{-\ 6x} & \underline{-\ 1)} \end{array}$$

17.
$$\begin{array}{rrrrr} -\ 4x^4 & & +\ 3x^2 & -\ 8x & +\ 1 \\ \underline{-(\quad} & \underline{-3x^3} & \underline{-\ 4x^2} & \underline{-\ 8x} & \underline{+\ 6)} \end{array}$$

(6.2) Add or subtract as indicated:

18. $(-6y - 8) + (4y^2 + 18y - 6) - (2 - 3y + 4y^2)$

19. $(a^4 + 3a^3 - 4 + 8a) - (6a + a^3 - a^4) + (2a^3 - 3a^4 + 9)$

20. Given the polynomials $3x + 14x^3 - 7x^2 + 3$, $4x^3 - 3x + 2$, and $-6x^3 + 3 - 4x^2 + x$, subtract the second polynomial from the sum of the first and the third.

21. Given the polynomials $-3y^3 + 2y^3 - 3 + 4y$, $6y^2 + 2y^4 + 6$, $3y^3 + 4y - 8y^2 + 7$, and $3y^4 + 2y^2 + 9y + 3y^3 - 5$, subtract the sum of the first two from the sum of the last two.

(6.3) Multiply:

22. $(-3x^4)(-2x^3)$

23. $(-4a^3y)(-2a^3y)(2ay^3)$

24. $-4x(6x - 1)$

25. $(6x + 5)(3x - 8)$

26. $(7x + 1)(2x + 5)$

27. $(x + 4)(2x^2 + 3x - 4)$

28. $(3x + 4)(4x^3 + 2x^2 - 3x + 5)$

29. $(2a + 3)(4a^2 - 6a + 9)$

30. $(2x^2 + 3x - 4)(x^2 + 3x + 2)$

(6.4) Divide:

31. $\dfrac{14x^3 - 21x^2 - 35x - 7}{7}$

32. $(-8y^4 - 12y^3 + 28y + 8) \div 4y$

33. $(8a^4 + 6a^3 - 8a^2 + 2a + 6) \div -2a$

34. $(6x^5 + 22x^4 - 8x^2 + 3x + 9) \div -3x^3$

35. $(x^2 - 6x - 27) \div (x - 9)$

36. $(6x^2 - 10x - 4) \div (2x - 4)$

37. $(15x^2 + x - 2) \div (5x + 2)$

38. $(x^3 - 7x + 6) \div (x - 3)$

39. $(4y^3 + y + 27) \div (3 + 2y)$

40. $(-4x + x^4 + 3) \div (x^2 - 2x + 1)$

Answers to odd-numbered exercises on page 471.

Chapter 6 Achievement Test

Name ______________________

Class ______________________

This test should be taken before you are tested in class on the material in Chapter 6. Solutions to each problem and the section where the type of problem is found are given on page 472.

Fill in the following table:

	Polynomial	Descending Order	No. of Terms	Name	Degree
1.	$-6x^2 + 4x - 3x^3$				
2.	$-26y$				
3.	$-7 - 3x$				

Add the polynomials:

4. $6x^3 + 4x^2 - 3x - 1$ and $2x^2 - 3 + 4x^3$ — 4. ______

5. $-6a + 4a^3 - 3$, $7a^4 + 2a^2 - 5$ and $2a^3 - 3 + 6a^2$ — 5. ______

Subtract the polynomials:

6. $(3x^4 - 6x^2 + 7x + 5) - (2x^4 - 3x^2 + x - 6)$ — 6. ______

7. $(8m + 2m^2 - 3m^3 - 1) - (7m^2 + 3m^4 - m)$ — 7. ______

8. Given polynomials $-2y^3 + 6y - 8 + y^2$, $7y^2 - 3y^3 + y - 4$, and $7 - y^3 + 7y$, subtract the last polynomial from the sum of the first two. — 8. ______

9. $(-3y^2)(4x^2y)(-5x^3y^2)$

10. $-3x^2(4x^2 + 2x - 5)$

11. $(7x - 2)(3x + 4)$

12. $(x - 4)(x^2 + 2x + 9)$

13. $(2x^2 + 3x - 1)(4x - 3)$

14. $(-27a^3 + 3a^2 - 6a + 18) \div 3a$

15. $(-6x^5 + 14x^4 - 12x^3 + 9) \div -4x^2$

16. $(x^2 - 4x - 12) \div (x - 6)$

17. $(x^3 - x^2 - 11x - 4) \div (x - 4)$

18. $(x^3 - 7x + 6) \div (x - 2)$

19. $(x^4 - 7x^2 - 18) \div (x - 3)$

9. ____________________

10. ____________________

11. ____________________

12. ____________________

13. ____________________

14. ____________________

15. ____________________

16. ____________________

17. ____________________

18. ____________________

19. ____________________

7 Special Products and Factoring

7.1 GREATEST COMMON FACTOR

The **greatest common factor** of two integers is the *largest* integer that is a factor of *each* integer.

Example 1: Find the greatest common factor of 12 and 18.

Solution:

$$12 = \boxed{2} \cdot 2 \cdot \boxed{3}$$

$$18 = \boxed{2} \cdot \boxed{3} \cdot 3$$

The largest integer common to both 12 and 18 is $2 \cdot 3 = 6$. We say that 6 is the greatest common factor of 12 and 18.

The **greatest common factor of a polynomial** is the monomial with (a) the largest coefficient and (b) the highest power of the variable, that divides *all* of the terms of the polynomial.

Example 2: Find the greatest common factor of the polynomial $6x^2 + 9x + 3x^2y$.

Solution: Find the factors of each term.

$$6x^2 = 2 \cdot \boxed{3} \cdot \boxed{x} \cdot x$$

$$9x = 3 \cdot \boxed{3} \cdot \boxed{x}$$

$$3x^2y = \boxed{3} \cdot \boxed{x} \cdot x \cdot y$$

The largest number that divides all terms is 3, and the highest power of the variable that divides all terms is x^1. Therefore, the greatest common factor of $6x^2 + 9x + 3x^2y$ is $3x$.

In Section 6.3 we multiplied monomials by polynomials by applying the distributive rule. For example:

$$3x(4x + 5) \underset{\text{factoring}}{\overset{\text{multiplying}}{\rightleftharpoons}} 12x^2 + 15x$$

Our next task is to reverse the process. Given $12x^2 + 15x$ we must write it in factored form. The process is called **factoring.**

Once we have found the greatest common factor, to find the other factor we divide each term of the given polynomial by the greatest common factor.

Example 3: Factor $6x^2 + 9x + 3x^2y$

Solution: In Example 2 we found that $3x$ is the greatest common factor. Now we must divide each term of the given polynomial by $3x$.

$$\frac{6x^2}{3x} + \frac{9x}{3x} + \frac{3x^2y}{3x}$$

$$= 2x + 3 + xy$$

Our final result is $6x^2 + 9x + 3x^2y = 3x(2x + 3 + xy)$ in factored form. All factoring problems should be checked by multiplying the factors together.

Example 4: Factor $27x^4 - 18x^2y^2 + 36x^3$

Solution:

$27x^4 - 18x^2y^2 + 36x^3$	
$= 3 \cdot 3 \cdot 3 \cdot x^4 - 2 \cdot 3 \cdot 3 \cdot x^2y^2 + 2 \cdot 2 \cdot 3 \cdot 3 \cdot x^3$	Write the coefficients as products of prime factors.
$3 \cdot 3 = 9$	The largest number that divides all the coefficients.
x^2	is the largest power of x that divides all terms.

Therefore the common factor is $3 \cdot 3 \cdot x^2$ or $9x^2$. The other factor is found by dividing each term of the given polynomial by $9x^2$.

$$\frac{27x^4}{9x^2} - \frac{18x^2y^2}{9x^2} + \frac{36x^3}{9x^2}$$

$$= 3x^2 - 2y^2 + 4x \qquad \text{the other factor.}$$

$$27x^4 - 18x^2y^2 + 36x^3 = 9x^2(3x^2 - 2y^2 + 4x) \text{ in factored form.}$$

Usually the coefficient of the common factor is found by inspection as the largest number that divides all of the coefficients of the given polynomial. However, if this leads to difficulty, you can use the method of prime factors that is illustrated in Example 4.

To find the greatest common factor:

1. Find the largest number that divides the coefficients of all terms in the polynomial. Do this by inspection or by prime factorization.
2. Find the highest power of the variable(s) that divides all the terms of the given polynomial.
3. The greatest common factor is the product of the numbers and variables found in steps 1 and 2.
4. The other factor is found by dividing each term of the polynomial by the greatest common factor found in step 3.
5. Check by multiplying the two factors together.

Example 5: Factor $-28x^5 - 14x^4 - 7x^3$

Solution: By inspection, -7 is the largest number that divides each of the coefficients and x^3 is the highest power of the variable that divides each term, so the greatest common factor is $-7x^3$. Now divide each term by $-7x^3$.

$$\frac{-28x^5}{-7x^3} - \frac{14x^4}{-7x^3} - \frac{7x^3}{-7x^3} = 4x^2 + 2x + 1$$

Therefore, the answer is $-28x^5 - 14x^4 - 7x^3 = -7x^3(4x^2 + 2x + 1)$ in factored form. Multiplication of the two factors yields the original polynomial, so the problem checks.

Example 6: Factor $14x^3y - 9a^2y + 6ax^4$

Solution: Factoring the coefficients indicates that there is no number except 1 that divides each of them.

$$14 = 2 \cdot 7, \qquad 9 = 3 \cdot 3, \qquad 6 = 2 \cdot 3$$

We can also see that there is no variable that is common to all of the terms. We are forced to conclude that there is no greatest common factor for the given polynomial, and we call it **prime**.

EXERCISE 7.1

Factor the following expressions. If not factorable, label the expression *prime*.

1. $3x + 9$
2. $10y + 30$
3. $12x - 14$
4. $36y - 27$
5. $6a^3 + 12a^2$
6. $4a^3 + 16a$
7. $18x^2 - 3x$
8. $32y^5 - 16y^3$
9. $27x^8 - 36x^4$
10. $24x^2 + 12y^2$
11. $3a^2b + 6x^2y$
12. $14a^2x - 7ay^2$
13. $16m^3 + 8m^2 - 24m$
14. $7x^2y - 3xy^2 + 9x^2y$
15. $y^6 - 3y^5 + y^2$
16. $x^5 - x^4 + x^2$
17. $2a^6 + 7a^4 - 3a$
18. $25x^6 + 15x^4 - 10x^2$
19. $9x^2 + 4y^3 - 7z$
20. $15a^2b - 60a^3b^2$
21. $-12x^3 - 16x^2 - 24x$
22. $-42x^2y^3 + 35x^4y^2 - 21x^5y^3$
23. $54m^2nq - 36m^4n^2q + 18m^2n^2q^2$
24. $30a^3x^3 + 45x^2y + 10b^2y^4$
25. $-27x^3y^8z^3 - 9xy^4z^{12} - 45xy^6z^9$
26. $-8x^4 + 9x^3y + 6y^5$
27. $48r^3t^2 - 32r^4t^4 + 24r^6$

28. $-120x^4y^3z^6 - 60x^5y^3z^{12} - 80x^5y^5z^6$.

Answers to odd-numbered exercises on page 472.

7.2 THE PRODUCT OF BINOMIALS

In order to factor trinomials in the next section, we first need to do more work on the product of two binomials. The product $(x + 3)(x + 4)$ was done in section 6.3 as follows:

$$\begin{array}{l} x\ +3 \\ \underline{x\ +4} \\ x^2 + 3x \\ \underline{\qquad 4x + 12} \\ x^2 + 7x + 12 \end{array}$$

We can obtain the same result horizontally by a method called the **FOIL** method.

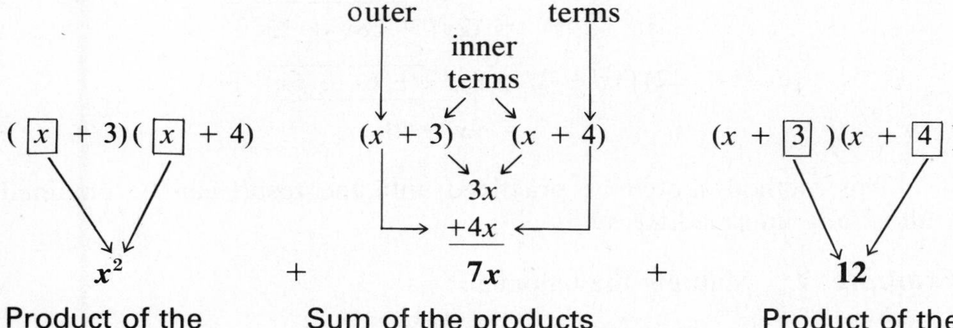

Product of the F irst terms of the binomials

Sum of the products of the O uter and I nner terms of the binomials

Product of the L ast terms

As you can see the result is the same. However, you will find that the **FOIL** method, once learned and practiced, is much faster.

> **To multiply two binomials by the *FOIL* method:**
>
> The first term = the product of the **F** irst terms of the binomials.
>
> The middle term = the sum of the **O** uter and **I** nner products.
>
> The last term = the product of the **L** ast terms of the binomials.

Example 1: Multiply the binomials $(x + 2)(x + 5)$

Solution:

	Product of **F**irst terms		**O**uter product		**I**nner product		Product of **L**ast terms
	↓		↓		↓		↓
$(x + 2)(x + 5) =$	x^2	$+$	$5x$	$+$	$2x$	$+$	10

$$(x + 2)(x + 5) = x^2 + \underbrace{5x + 2x}_{\text{sum}} + 10$$

$$= x^2 + 7x + 10$$

Example 2: Multiply the binomials:

(a) $(x + 6)(x - 2) = x^2 \boxed{-2x + 6x} - 12$
$= x^2 + 4x - 12$

(b) $(2x + 3)(x - 2) = 2x^2 \boxed{-4x + 3x} - 6$
$= 2x^2 - x - 6$

(c) $(3x - 4)(2x + 1) = 6x^2 + \boxed{3x - 8x} - 4$
$= 6x^2 - 5x - 4$

(d) $(6y - 5)(2y - 3) = 12y^2 \boxed{-18y - 10y} + 15$
$= 12y^2 - 28y + 15$

(e) $(3x + 2)(3x - 2) = 9x^2 \boxed{-6x + 6x} - 4$
$= 9x^2 - 4$

This method should be practiced until the result can be obtained without any intermediate steps.

Example 3: Multiply the binomials:

(a) $(4x - 3)(2x + 3) = 8x^2 + 6x - 9$
(b) $(3z + 5)(3z - 5) = 9z^2 - 25$
(c) $(a + 6)^2 = (a + 6)(a + 6) = a^2 + 12a + 36$
(d) $(2x - 7)(3x - 5) = 6x^2 - 31x + 35$
(e) $(7x + 9)(2x + 7) = 14x^2 + 67x + 63$

The procedure for finding the square of a binomial can be shortened even further. Consider $(2x + 3)^2$. We could apply the **FOIL** method, obtaining

$$(2x + 3)^2 = (2x + 3)(2x + 3) = 4x^2 + \boxed{6x} + \boxed{6x} + 9$$
$$= (2x)^2 + \boxed{2x \cdot 3} + \boxed{2x \cdot 3} + (3)^2$$
$$= 4x^2 + 12x + 9$$

Noting that the outer product and the inner product are the same, we state the following rule:

To square a binomial:

1. The square of the first term = the first term of the product.
2. Twice the product of the two terms = the middle term of the product.
3. The square of the last term = the last term of the product.

This will be best accomplished by memorizing the following statement.

"The square of a binomial equals the square of the first term, plus twice the product of the terms, plus the square of the last term."

Example 4: Square the binomial: $(3x - 5)^2$.

Solution:

$$(3x - 5)^2$$
$$= (3x)^2 \quad + (2)(3x)(-5) \quad + (-5)^2$$
$$= 9x^2 \quad - 30x \quad + 25$$

↑ square of the first term — ↑ twice the product of the terms — ↑ square of the last term

Example 5: Square the binomials:

(a) $(x + 6)^2 = x^2 + 12x + 36$
(b) $(5x - 1)^2 = 25x^2 - 10x + 1$
(c) $(2x + 7)^2 = 4x^2 + 28x + 49$
(d) $(2y - 3)^2 = 4y^2 - 12y + 9$
(e) $(4b + 5)^2 = 16b^2 + 40b + 25$

A common error, when squaring a binomial, is to leave out the middle term.

$(a + b)^2 \neq a^2 + b^2$

$(a + b)^2 = a^2 \boxed{+2ab} + b^2$

↖ This is frequently left out in error.

EXERCISE 7.2

Write the final answers to the following without writing down any intermediate steps:

1. $(x + 2)(x + 4)$
2. $(y + 3)(y + 2)$
3. $(a - 5)(a - 2)$
4. $(x - 4)(x - 5)$
5. $(x + 3)(x - 5)$
6. $(x + 7)(x - 6)$
7. $(n - 7)(n - 2)$
8. $(m - 3)(m - 7)$
9. $(x - 2)(x + 2)$
10. $(x - 8)(x + 8)$
11. $(3x + 4)(2x - 3)$
12. $(2x - 1)(4x + 1)$
13. $(3x + 2)(3x - 2)$
14. $(4y + 3)(y - 8)$
15. $(x + 3)^2$
16. $(x + 4)^2$
17. $(x - 7)^2$
18. $(x - 4)^2$
19. $(y - 9)^2$
20. $(y + 9)^2$
21. $(3y - 4)(4y + 3)$
22. $(2a - 5)(5a + 2)$
23. $(7x - 2)(7x + 2)$
24. $(4x - 7)(4x + 7)$
25. $(3x - 2)^2$
26. $(h + 4)(7h - 4)$
27. $(5n + 3)(5n - 3)$
28. $(3x + 2)^2$
29. $(7n - 1)^2$
30. $(4y - 3)^2$
31. $(7x + 2)^2$
32. $(2m - 9)^2$
33. $(x + y)(x + 3y)$
34. $(2a + b)(3a - b)$
35. $(3a + b)(3a - b)$
36. $(7x - 2y)(4x + 3y)$
37. $(10m - 3n)(5m + 7n)$
38. $(2x - 3y)^2$
39. $(4a - 5b)^2$
40. $(5x + 7y)^2$

Answers to odd-numbered exercises on page 473.

7.3 FACTORING TRINOMIALS WITH LEADING COEFFICIENT EQUAL TO 1

In the last section we multiplied two binomials together and in general we got a trinomial as a result. For example,

$$(x + 4)(x + 3) = x^2 + 7x + 12$$

$$\xrightarrow{\hspace{2cm}}$$

multiplying

Our task now will be to reverse the procedure and write the trinomial as the product of two binomials.

$$x^2 + 7x + 12 = (x + 4)(x + 3)$$

$$\xrightarrow{\hspace{2cm}}$$

factoring

The easiest type of trinomial to factor is one where the coefficient of the x^2 term is 1 (remember $x^2 = 1x^2$). Since $x \cdot x = x^2$, the first term of each binomial will always be x. We set up our work like this:

$$x^2 \quad + \quad 7x \quad + \quad 12 \quad = \quad (x \qquad)(x \qquad)$$

$$x^2 = x \cdot x$$

The product of the last two terms must equal 12, and the sum of these same last two terms must equal the coefficient of the middle term, 7. So what we're looking for are two integers whose product is 12 and whose sum is 7. A little thought tells us quickly that +4 and +3 are the integers.

Sum of +4 and +3 is 7

$$x^2 + 7x + 12 = (x + 4)(x + 3)$$

Product of +4 and +3 is 12

Example 1: Factor $x^2 + 6x + 8$

Solution: Set it up like this: $x^2 + 6x + 8 = (x \qquad)(x \qquad)$

Now we look for two integers whose product is 8 and whose sum is 6. The integers are +4 and +2. Our result, then, is $x^2 + 6x + 8 = (x + 4)(x + 2)$. As in any factoring problem, we should always check our answer by multiplying the factors together.

To factor a trinomial $x^2 + Bx + C = (x + p)(x + q)$, find integers p and q such that:

1. $p \cdot q = C$ (their product is C), and
2. $p + q = B$ (their sum is B).

Factoring trinomials is basically a trial and error procedure, but the following examples contain some hints on how to minimize the number of trials and the number of errors, so study them carefully. Notice that the order of the binomials in the answer doesn't matter since multiplication is commutative.

Example 2: Factor $x^2 + 3x - 10$

Solution: Find integers whose product is -10 and whose sum is $+3$. Since the product is negative they must have opposite signs. Since the sum is positive, the sign of the larger integer is positive. The only pair of integers that satisfies these conditions is $+5$ and -2.

$(+5)(-2) = -10$ and $(+5) + (-2) = +3$, so our result is

$$x^2 + 3x - 10 = (x + 5)(x - 2)$$

Check: $(x + 5)(x - 2) = x^2 + 3x - 10$ ✓

Example 3: Factor $x^2 - 8x + 12$

Solution: Find integers whose product is $+12$ and whose sum is -8. Since the product is positive, their signs must be alike, but since the sum is negative, they must both be negative. Possible pairs of factors are $(-12, -1)$, $(-3, -4)$ and $(-6, -2)$. Of these, only $(-6, -2)$ has a sum equal to -8. Our result is

$$x^2 - 8x + 12 = (x - 6)(x - 2)$$

Check: $(x - 6)(x - 2) = x^2 - 8x + 12$ ✓

Example 4: Factor $y^2 - 5y + 6$

Solution: Find integers whose product is $+6$ and whose sum is -5. Since the product is positive the signs are alike, but since the sum is negative, we know they are both negative. Possible factors are $(-6, -1)$ and $(-3, -2)$. Since the sum is -5, $(-3, -2)$ is the correct choice.

$$y^2 - 5y + 6 = (y - 3)(y - 2)$$

Check: $(y - 3)(y - 2) = y^2 - 5y + 6$ ✓

Example 5: Factor $x^2 + 9x + 20$

Solution: Find integers whose product is $+20$ and whose sum is $+9$. Since both the product and sum are positive they will both be positive. Choices of factors are $(+20, +1)$, $(+10, +2)$, and $(+4, +5)$. Since the sum is $+9$, $(+4, +5)$ is the correct choice, and the result is

$$x^2 + 9x + 20 = (x + 4)(x + 5)$$

Check: $(x + 4)(x + 5) = x^2 + 9x + 20$ ✓

Example 6: Factor $y^2 - 9x + 12$

Solution: Find the integers whose product is $+12$ and whose sum is -9. The positive product indicates that the signs are alike, but the negative sum tells us that they are both negative. Possible factors of 12 are $(-6, -2)$, $(-4, -3)$, and $(-12, -1)$. None of these factors, when added, have a sum of -9, so we conclude that the given trinomial is not factorable and we say that it is **prime.**

EXERCISE 7.3

Factor the following trinomials. If a trinomial is not factorable label it *prime*.

1. $x^2 + 5x + 4$
2. $x^2 + 6x + 8$
3. $x^2 + 4x + 4$
4. $x^2 - 2x - 3$
5. $x^2 - 4x + 3$
6. $y^2 + 7y + 6$
7. $y^2 + 12y + 35$
8. $y^2 - y - 6$
9. $y^2 + 11y - 10$
10. $y^2 - 3y - 10$
11. $a^2 + 13a - 14$
12. $a^2 - 12a + 35$
13. $a^2 - 9a + 14$
14. $a^2 + 15a + 12$
15. $a^2 - 10a + 16$
16. $b^2 - 8b + 16$
17. $b^2 - 5b - 6$
18. $b^2 + 5b + 6$
19. $b^2 - 5b + 6$
20. $b^2 + b - 6$
21. $h^2 - 12h + 27$
22. $h^2 - 9h + 10$
23. $h^2 + 7h + 10$
24. $h^2 - 7h + 10$

25. $h^2 - 9h + 18$

26. $u^2 + 11u + 24$

27. $u^2 + 4u + 24$

28. $u^2 - 12u + 20$

29. $u^2 - 11u + 18$

30. $u^2 - 15u + 14$

31. $t^2 - 12t - 8$

32. $t^2 + 7t - 30$

33. $t^2 + 13t - 30$

34. $t^2 + 11t + 30$

35. $t^2 - 11t + 30$

36. $z^2 - 7z - 18$

37. $z^2 - 14z + 49$

38. $z^2 + 10z + 25$

39. $z^2 + 6z - 35$

40. $z^2 - 10z + 25$

41. $m^2 - 10m + 24$

42. $m^2 + 2m + 24$

43. $m^2 + 11m + 24$

44. $m^2 + 16m - 24$

45. $m^2 - 5m - 24$

46. $x^2 - 30x - 64$

47. $y^2 + 15y + 56$

48. $u^2 + 10u - 75$

49. $a^2 - 11a - 42$

50. $z^2 + 20z + 51$

Answers to odd-numbered exercises on page 473.

7.4 FACTORING TRINOMIALS WITH LEADING COEFFICIENT GREATER THAN 1

Next we must concern ourselves with factoring trinomials when the coefficient of the squared term is greater than 1.

Example 1: Factor $3x^2 + 7x + 2$

Solution: Since all the signs in the trinomial are positive, all the signs in the factors will be positive. The products of the two first terms of our factors must equal $3x^2$, which gives us

$$(3x + \quad)(x + \quad)$$

It remains for us to find the two last terms of the binomials. Since their product must be equal to 2, and they are positive, our only choices are (2,1) or (1,2). Trying each of these to see which gives the correct middle term, $+7x$, we have

$$(3x + 2) \quad (x + 1)$$
$$\begin{array}{l} +2x \\ +3x \\ \hline +5x \leftarrow \text{Wrong middle term} \end{array}$$

$$(3x + 1) \quad (x + 2)$$
$$\begin{array}{l} +1x \\ +6x \\ \hline +7x \leftarrow \text{Correct middle term} \end{array}$$

Our result is $3x^2 + 7x + 2 = (3x + 1)(x + 2)$.

To factor trinomials of the form $Ax^2 + Bx + C$, $A > 1$:

1. The product of the first terms must equal Ax^2.
2. The product of the last terms must equal C.
3. The sum of the outer product and inner product must equal Bx.

Again, this is a trial-and-error procedure, but with sufficient practice you will learn to cut down on the number of trials by disregarding certain trials without actually trying them.

Example 2: Factor $2x^2 + 11x + 15$

Solution: Since all signs are positive, all signs in the factors will be positive.

Factors of 2	*Factors of 15*
2,1	3,5
	1,15

Wrong middle term ↙

Trial 1: $(2x + 3)(x + 5) = 2x^2 + \boxed{13}\,x + 15$ Incorrect

Trial 2: Reverse the order of factors 3,5 and try again.

$(2x + 5)(x + 3) = 2x^2 + 11x + 15$ Correct ✓

Example 3: Factor $3x^2 - 5x - 12$

Solution: Since the last term is negative, all pairs of factors of -12 will have different signs.

Factors of 3	*Factors of −12*
3,1	−2,6 or 2,−6
	−3,4 or 3,−4
	−1,12 or 1,−12

Trial 1: $(3x - 2)(x + 6) = 3x^2 + 16x - 12$ Incorrect

Trial 2: $(3x + 6)(x - 2) = 3x^2 - 12$ Incorrect

Trial 3: $(3x - 3)(x + 4) = 3x^2 + 9x - 12$ Incorrect

Trial 4: $(3x + 4)(x - 3) = 3x^2 - 5x - 12$ Correct ✓

Your should notice that in Trial 2, a common factor of 3 occurred in the first binomial that we tried $(3x + 6)$. If all common factors have been removed from the given trinomial then common factors will *never* occur in any of the binomial factors. This means that Trial 2 could have been eliminated immediately since the trial binomial $3x + 6$ contains a common factor. Similarly, we can see that Trial 3 will not work since the trial binomial $3x - 3$ contains a common factor 3. This hint can save you a lot of work, so it should be remembered.

Example 4: Factor $6x^2 - 7x + 2$

Solution: Since the last term of the trinomial is positive, the signs of the last terms of the binomial factors must be alike. However, since the middle term is negative, they will both be negative.

Factors of 6	*Factors of 2*
2,3	−2, −1
6,1	

Trial 1: $(2x - 2)(3x - 1)$ is immediately incorrect since $2x - 2$ contains a common factor of 2.

Trial 2: $(2x - 1)(3x - 2) = 6x^2 - 7x + 2$ Correct ✓

Example 5: Factor $4x^2 + 9x + 6$

Solution: Since all signs in the trinomial are positive all signs in both binomial factors will be positive.

Factors of 4	*Factors of 6*
2,2	2,3
4,1	6,1

Trial 1:	$(2x + 2)(2x + 3)$	contains common factor	Incorrect
Trial 2:	$(2x + 6)(2x + 1)$	contains common factor	Incorrect
Trial 3:	$(4x + 2)(x + 3)$	contains common factor	Incorrect
Trial 4:	$(4x + 3)(x + 2)$	$= 4x^2 + 11x + 6$	Incorrect
Trial 5:	$(4x + 6)(x + 1)$	contains common factor	Incorrect
Trial 6:	$(4x + 1)(x + 6)$	$= 4x^2 + 25x + 6$	Incorrect

None of the trials produce a correct result and since all of the possible combinations have been tried, we conclude that the trinomial is not factorable, or it is *prime*.

EXERCISE 7.4

Factor the following. If it is not possible to factor the trinomial, label it *prime*.

1. $2x^2 + 5x + 3$

2. $2x^2 - 7x + 3$

3. $3x^2 + 2x - 5$

4. $6x^2 + 5x + 1$

5. $2y^2 + y - 3$

6. $2y^2 + 7y + 6$

7. $3y^2 - 8y + 4$

8. $2y^2 + y - 6$

9. $4h^2 - 12h + 5$

10. $2h^2 + 11h + 5$

11. $3a^2 + 10a + 3$

12. $2a^2 - 7a + 5$

13. $5t^2 + 21t + 4$

14. $4m^2 + 13m + 3$

15. $5x^2 + 14x + 3$

16. $7y^2 - 4y + 5$

17. $3y^2 - 7y - 6$

18. $3x^2 - 22x + 7$

19. $5x^2 - 12x + 7$

20. $4x^2 - 8x - 5$

21. $2y^2 + 3y + 1$

22. $8y^2 + y - 7$

23. $2b^2 - 7b + 6$

24. $3t^2 - 22x + 7$

25. $4t^2 + 4t + 1$

26. $6n^2 - 7n - 10$

27. $11x^2 - 11x - 2$

28. $3a^2 + 14a + 8$

29. $9x^2 + 6x - 8$

30. $6t^2 + 5t - 6$

31. $4x^2 + 11x + 6$

32. $9y^2 + 14y - 8$

33. $12d^2 + 4d - 1$

34. $3a^2 - 17a + 20$

35. $4h^2 + 6h - 9$

36. $6y^2 - 13y - 28$

37. $2y^2 - y + 7$

38. $6t^2 + 19t + 10$

39. $7w^2 - 20w - 3$

40. $4n^2 - 7n - 15$

41. $6x^2 - 13x + 6$

42. $6h^2 - 11h + 6$

43. $25b^2 - 15b - 4$

44. $6m^2 - 5m - 14$

45. $8a^2 - 22a - 21$

46. $6x^2 - 31x + 40$

47. $9x^2 + 18x + 8$

48. $9x^2 + 9x - 16$

49. $8y^2 - 14y - 9$

50. $6a^2 + 14a - 7$

51. $4x^2 + 15x - 16$

52. $10h^2 - 21h - 10$

53. $15x^2 - 17x - 42$

54. $6y^2 - 23y + 21$

55. $8t^2 - 6t + 15$

56. $14x^2 - 11x + 12$

57. $9x^2 - 52x - 12$

58. $14y^2 - 33y - 5$

59. $8t^2 - 30t - 27$

60. $30x^2 - 37x + 10$

Answers to odd-numbered exercises on page 474.

7.5 FACTORING THE DIFFERENCE OF TWO SQUARES

When we multiply two binomials of the form $(x - y)(x + y)$ we get an interesting result.

Example 1:

(a) $(x + 2)(x - 2) = x^2 - 4$
(b) $(3x - 2)(3x + 2) = 9x^2 - 4$
(c) $(5y + 3)(5y - 3) = 25y^2 - 9$
(d) $(a + b)(a - b) = a^2 - b^2$

In every case the middle term is zero and the product consists of two *perfect squares* separated by a negative sign. We call this the **difference of two squares.**

To factor we reverse the procedure:

> **Factoring the difference of two squares:**
> A polynomial of the form $a^2 - b^2$ is factored as
> $a^2 - b^2 = (a + b)(a - b)$.

Example 1: Factor $4x^2 - 9$

Solution: Since $4x^2 = (2x)^2$ and $9 = 3^2$ and they are separated by a negative sign, we have the *difference of two squares*.

Square root of first term goes here

$$4x^2 - 9 = (2x + 3)(2x - 3)$$

Square root of second term goes here

The sign in one of the factors is + and the other −. Since multiplication is commutative they can be reversed.

$$(2x + 3)(2x - 3) = 4x^2 - 9 \checkmark$$

Example 2: Factor the following:

(a) $16x^2 - 25 = (4x + 5)(4x - 5)$
(b) $4y^2 - 1 = (2y + 1)(2y - 1)$
(c) $h^2 - 9g^2 = (h + 3g)(h - 3g)$
(d) $49x^2 - 16y^2 = (7x - 4y)(7x + 4y)$
(e) $9x^2 + 4$ is prime

STOP

The *sum* of two squares is *not* factorable. A common error is to factor $x^2 + y^2$ as $(x + y)(x + y)$. However, $(x + y)(x + y) = x^2 + 2xy + y^2$. $x^2 + y^2$ is not factorable.

Consider *even* powers of a variable such as y^4, x^6, and t^{14}. Since they can be written as

$$y^4 = (y^2)^2$$
$$x^6 = (x^3)^2$$
$$t^{14} = (t^7)^2$$

they are all perfect squares. In fact, the square root of any even-powered variable is equal to that variable to one-half the given exponent. This concept is used to factor the difference of two squares, as illustrated in the following example.

Example 3: Factor:

(a) $y^4 - 9 = (y^2)^2 - 9 = (y^2 + 3)(y^2 - 3)$
(b) $x^6 - 4 = (x^3)^2 - 4 = (x^3 + 2)(x^3 - 2)$
(c) $t^{14} - 25 = (t^7 + 5)(t^7 - 5)$
(d) $64x^6 - 49y^8 = (8x^3 + 7y^4)(8x^3 - 7y^4)$
(e) $16h^{10} + 9y^2$ is prime. Even though both terms are perfect squares, the *sum* of two squares is prime.

EXERCISE 7.5

Factor the following. If a problem is not factorable, label it *prime*.

1. $x^2 - 36$
2. $y^2 - 9$

3. $t^2 - 49$

4. $4p^2 - 1$

5. $16x^2 - 9$

6. $4x^2 - 25$

7. $9y^2 - 4$

8. $100t^2 - 1$

9. $4x^2 + 25$

10. $25x^2 - 36y^2$

11. $49m^2 - 4n^2$

12. $9y^2 - 25x^2$

13. $81a^2 - 49b^2$

14. $36t^2 - s^2$

15. $16x^2 + 9y^2$

16. $x^4 - 9$

17. $y^6 - 25$

18. $t^{10} - 36$

19. $4x^{12} - y^6$

20. $81x^6 - y^{12}$

21. $25x^2 - 36x^6y^4$

22. $x^8y^4 - 25z^6$

23. $144x^6 - 49m^6n^{18}$

24. $9x^6y^2 - 49z^{10}$

25. $100x^8 + 81y^8z^8$

Answers to odd-numbered exercises on page 474.

7.6 COMBINING DIFFERENT TYPES OF FACTORING

In many cases a polynomial may be factored more than once. Usually we are instructed to factor a polynomial completely, which means we must look at the factors to see if they may be factored further.

Always look for a greatest common factor first.

Example 1: Factor $3x^2 + 21x + 36$

Solution: We remove the greatest common factor, 3, first.

$$3x^2 + 21x + 36 = 3(x^2 + 7x + 12)$$

Next we factor the trinomial $x^2 + 7x + 12$.

$$x^2 + 7x + 12 = (x + 3)(x + 4)$$

Putting both of these steps together results in:

$$3x^2 + 21x + 36 = 3(x^2 + 7x + 12) = 3(x + 3)(x + 4)$$

As always, check by multiplying the factors together.

Example 2: Factor $24x^2 - 6y^2$

Solution:

$24x^2 - 6y^2 = 6(4x^2 - y^2)$ — The greatest common factor is 6.

$= 6(2x + y)(2x - y)$ — Difference of two squares: $4x^2 - y^2$

Example 3: Factor $3a^5 - 48a$

Solution:

$3a^5 - 48a = 3a(a^4 - 16)$ — The greatest common factor is $3a$.

$= 3a(a^2 + 4)(a^2 - 4)$ — Difference of two squares: $a^4 - 16$

$= 3a(a^2 + 4)(a + 2)(a - 2)$ — Difference of two squares: $a^2 - 4$

Example 4: Factor $18y^4 + 24y^3 - 24y^2$

Solution:

$18y^4 + 24y^3 - 24y^2 = 6y^2(3y^2 + 4y - 4)$ — The greatest common factor is $6y^2$.

$= 6y^2(3y - 2)(y + 2)$ — Factor the trinomial.

Example 5: Factor $-x^2 + 6x + 16$

Solution: When we are factoring any trinomial the squared term should always be positive. To accomplish this we will factor out -1 and then factor the resulting trinomial.

$-x^2 + 6x + 16 = -1(x^2 - 6x - 16)$	A common factor is -1.
$= -1(x - 8)(x + 2)$	Factor the trinomial.
$= -(x - 8)(x + 2)$	Usually only the negative sign is written instead of -1.

EXERCISE 7.6

Factor completely. If a problem is not factorable label it *prime*.

1. $3x^2 - 3y^2$
2. $4x^2 + 12x + 4$
3. $2x^2 + 6x + 4$
4. $5y^2 - 10y - 15$
5. $12a^2 - 3b^2$
6. $7x^2 - 14y^2$
7. $8y^2 - 12y - 8$
8. $6x^2 - 10x - 24$
9. $3x^2 - 6x - 24$
10. $4x^2 - 36$
11. $6x^2 - 19x + 15$
12. $4a^3 - 4ab^2$

13. $t^4 - 16$

14. $6t^2 + 18t + 6$

15. $a^3 - 4a$

16. $20m^2 + 6m - 2$

17. $4c^3 - 100c$

18. $-3x^2 - 7x - 2$

19. $6d^2 + 12c^2$

20. $18y^3 - 39y^2 - 15y$

21. $12h^4 - 14h^3 - 6h^2$

22. $xy^2 + xy - 12x$

23. $x^8 - 81$

24. $36c^2 - 9d^2$

25. $5x^{13} - 5$

26. $3x^2 + 12x + 4$

27. $32x^3 - 50xy^2$

28. $16a^2 - 36b^2$

29. $14x^2 + 21x - 35$

30. $48 - 3x^8$

31. $-2x^2 - x + 21$

32. $-15y^2 - 5y + 20$

Answers to odd-numbered exercises on page 474.

7.7 Chapter Summary

Examples

(7.1) To find the greatest common factor:
1. Find the largest number that divides the coefficient of all the terms in the polynomial using inspection or prime factorization.
2. Find the highest power of the variable(s) that divides all the terms of the given polynomial.
3. The greatest common factor is the product of the number and the variables found in steps 1 and 2.
4. The other factor is found by dividing each term of the polynomial by the greatest common factor found in step 3.
5. Check by multiplying the two factors together.

(7.1) Factor $12x^3 - 18x^4y + 24x^5y^2$. $6x^3$ is the greatest common factor.

$$\frac{12x^3}{6x^3} - \frac{18x^4y}{6x^3} + \frac{24x^5y^2}{6x^3}$$

$$= 2 - 3xy + 4x^2y^2$$

$$12x^3 - 18x^4y + 24x^5y^2 = 6x^3(2 - 3xy + 4x^2y^2)$$

(7.2) To multiply two binomials using the **FOIL** method:
1. The first term = the product of the **F**irst term of the binomial.
2. The middle term = the sum of the **O**uter and **I**nner products.
3. The last term = the product of the **L**ast term of the binomial.

(7.2) $(3x + 4)(2x - 5) = 6x^2 - 7x - 20$

(7.2) To square a binomial:
1. The square of the first term = the first term of the product.
2. Twice the product of the two terms = the middle term of the product.
3. The square of the last term = the last term of the product.

(7.2) $(2x - 3)^2 = 4x^2 - 12x + 9$

(7.3) To factor a trinomial, $x^2 + Bx + C = (x + p)(x + q)$, find integers p and q such that:
1. $p \cdot q = C$ (their product is C), and
2. $p + q = B$ (their sum is B).

(7.3) $x^2 + 3x - 28 = (x - 4)(x + 7)$

(7.4) When factoring a trinomial of the form $Ax^2 + Bx + C$, $A > 1$:
1. The product of the first terms must equal Ax^2.
2. The product of the last terms must equal C.
3. The sum of the outer products and inner products must equal Bx.

(7.4) $6x^2 - 13x + 5 = (2x - 1)(3x - 5)$

(7.5) Factoring the difference of two squares:
A polynomial of the form $a^2 - b^2$ is factored as

$$a^2 - b^2 = (a + b)(a - b)$$

(7.5) $36x^2 - 25 = (6x + 5)(6x - 5)$

(7.5) The square root of any even-powered variable is equal to that variable raised to one-half of the given exponent.

(7.5) $x^{10} - 49 = (x^5 + 7)(x^5 - 7)$

(7.6) Always look for a greatest common factor first.

(7.6) $20x^3 + 5x^2 - 15x$
$= 5x(4x^2 + x - 3)$
$= 5x(4x - 3)(x + 1)$

(7.1-7.6) A **prime** polynomial is one that is not factorable.

(7.1-7.6) $7x^4y - 8x^2z + 4yz^5$ is prime.
$x^2 - 6x + 2$ is prime.
$5x^2 - 3x + 4$ is prime.
$4x^2 + 9$ is prime.

Exercise 7.7 Chapter Review

(7.1) Factor completely. If a problem is not factorable, write *prime*.

1. $6x - 30$
2. $x^3 - 3x^2 + 7x$
3. $6x^2 + 9x$
4. $5y^2 + 14y^3$
5. $r^2 + R^2$
6. $16x^2y^6 - 18x^3y^4$
7. $14x^2y + 9yz^2 - 8xz^3$
8. $64a^4b + 24a^3b^2 - 12a^3b^3$
9. $-36y^4 - 16y^3 - 18y^2$
10. $28r^3t^3 - 35rt^5 + 49r^3t^5$

(7.2) Write the final answers without writing down any intermediate steps.

11. $(x + 4)(x - 3)$
12. $(2x + 1)(6x + 1)$
13. $(3x - 1)(2x + 3)$
14. $(7x - 2)(7x + 2)$
15. $(5x - 3)(2x - 3)$
16. $(6x + 5)(6x + 7)$
17. $(x + 4)^2$
18. $(2x + 3)^2$
19. $(2x - 5)^2$
20. $7x - 5y)^2$

(7.3) Factor. If a problem is not factorable, write *prime*.

21. $y^2 + 8x + 12$
22. $y^2 - 7y + 12$
23. $x^2 - 3x - 10$
24. $x^2 - 12x + 36$
25. $x^2 + 2x - 35$
26. $a^2 - 15a - 44$

27. $b^2 - 13b + 42$

28. $x^2 - 15x + 56$

29. $y^2 + 17x + 60$

30. $x^2 + 16x - 57$

(7.4) Factor. If a problem is not factorable, write *prime*.

31. $2x^2 - 5x - 12$

32. $12y^2 - 5y - 2$

33. $3b^2 + 4b - 7$

34. $16x^2 - 8x + 1$

35. $2y^2 - 13y + 15$

36. $6x^2 + 29x + 35$

37. $35c^2 - 52c + 12$

38. $14a^2 + 21a + 15$

39. $8x^2 + 42x + 49$

40. $21w^2 - 38w - 48$

(7.5) Factor. If a problem is not factorable, write *prime*.

41. $x^2 - 16$

42. $y^2 - 64$

43. $25c^2 - 1$

44. $4x^2 + 49$

45. $4a^2 - 25b^2$

46. $9y^2 - 64x^2$

47. $x^8 - 49$

48. $h^4 - 64g^6$

49. $25x^4 - 36y^2$

50. $64x^2y^4 - 9z^2$

(7.6) Factor completely. If a problem is not factorable, write *prime*.

51. $2a^2 - 8b^2$

52. $2x^2 + 2x - 24$

53. $6x^2 - 9x - 6$

54. $4y - y^3$

55. $4a^3 + 18a^2 + 20a$

56. $4x^2y + 20xy + 25y$

57. $2x^2y - 3xy - 5y$

58. $y^8 - 1$

59. $-2x^2 + 11x - 14$

60. $-28xy^2 - 46xy - 6x$

Answers to odd-numbered exercises on page 475.

Chapter 7 Achievement Test

Name ______________________

Class ______________________

This test should be taken before you are tested in class on the material in Chapter 7. Solutions to each problem and the section where the type of problem is found are given on page 475.

In exercises 1 through 4, write down the final answer without writing down any intermediate steps.

1. $(x - 4)(x - 5)$
2. $(3x + 2)(4x - 1)$
3. $(6x + 7)(3x - 5)$
4. $(3x - 5)^2$

In exercises 5 through 20, factor completely. If a problem is not factorable, write *prime*.

5. $6x^3 + 5x^6$
6. $12a^3b^2 + 9a^4b^2$
7. $9x^3y - 15x^2y^2 + 9x^4y$
8. $y^2 - y - 6$
9. $x^2 + 4x - 20$
10. $x^2 - 13x + 42$

1. ______________________
2. ______________________
3. ______________________
4. ______________________
5. ______________________
6. ______________________
7. ______________________
8. ______________________
9. ______________________
10. ______________________

11. $8a^2 + 2a - 1$

12. $4t^2 + 11t + 6$

13. $18x^2 - 27x - 5$

14. $8h^2 - 18h + 9$

15. $9y^2 - 25$

16. $4a^2 - 49b^2$

17. $9x^2 + 25y^2$

18. $2y^{12} - 32$

19. $16x^2y - 4xy - 6y$

20. $36t^2 - 57t + 15$

11. ______________________

12. ______________________

13. ______________________

14. ______________________

15. ______________________

16. ______________________

17. ______________________

18. ______________________

19. ______________________

20. ______________________

8 Algebraic Fractions

8.1 BASIC CONCEPTS

Your knowledge of arithmetic fractions is an excellent base on which to build the theory of algebraic fractions. You will find that the procedures for simplifying, multiplying, dividing, adding, and subtracting *algebraic* fractions are almost the same as they are for *arithmetic* fractions.

An algebraic fraction is an expression of the form

$$\frac{N}{D} \text{ where } N \text{ and } D \text{ are polynomials.}$$

Any value of the variable which makes the denominator of the fraction equal to zero cannot be allowed since division by zero is not defined.

For example, the algebraic fraction

$$\frac{2x+3}{x-4}$$

is not defined when $x = 4$ because it would make the value of the denominator equal to zero, and we call $x = 4$ an **excluded value.** You will frequently see

$$\frac{2x+3}{x-4} \qquad \boxed{x \neq 4}$$

and even if it is not explicitly stated, it is assumed that $x \neq 4$. It should be understood then, that any value(s) of the variable that make the denominator equal to zero must be excluded.

Example 1: Find the excluded values of each of the following algebraic fractions (if any).

(a) $\dfrac{4x-1}{x+7}$ — $x = -7$ is excluded because it would make the denominator zero.

(b) $\dfrac{x-5}{x^2+2x-3} = \dfrac{x-5}{(x+3)(x-1)}$ — x cannot equal either -3 or 1 since either value would make the denominator equal to zero.

(c) $\dfrac{3x+5}{7}$ — No value of x will make the denominator zero, so no value of x is excluded.

(d) $\dfrac{-9}{x}$ — $x \neq 0$

The Three Signs of a Fraction

Every fraction has three signs associated with it: the sign of the numerator, the sign of the denominator, and the sign of the fraction itself.

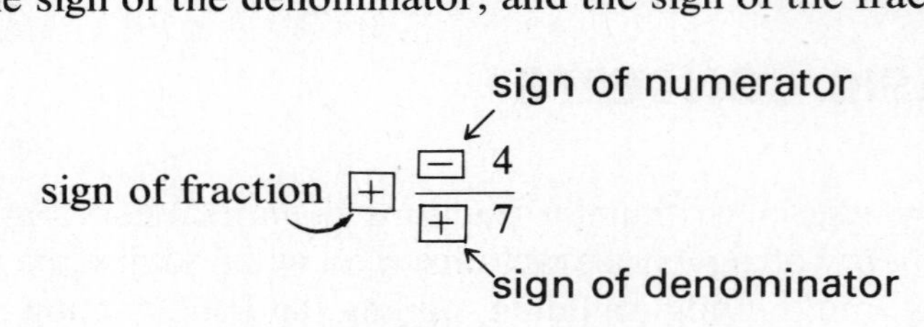

$$\text{sign of fraction } \boxed{+}\,\frac{\boxed{-}\,4}{\boxed{+}\,7}$$

From the rule of signs for division we know that

$$\frac{-a}{b} = \frac{a}{-b} = -\frac{a}{b}$$

This leads us to a rule of signs for fractions:

> If exactly two of the three signs of a fraction are changed, the value of the fraction remains the same.

Example 2:

$$+\,\frac{\boxed{-}\,6}{\boxed{-}\,2} = +(+3) = 3$$

Change the signs of the numerator and the denominator.

(a) $+\dfrac{+6}{+2} =$

$\boxed{-}\dfrac{\boxed{-}6}{+2} = -(-3) = 3$

↖ Change the signs of the fraction and the numerator.

$\boxed{-}\dfrac{+6}{\boxed{-}2} = -(-3) = 3$

↖ Change the signs of the fraction and the denominator.

(b) $-\dfrac{3}{-x} = \boxed{+}\dfrac{3}{\boxed{+}x} = \dfrac{3}{x}$

↑ Change the signs of the fraction and the denominator.

(c) $\dfrac{-3}{4-x} = \dfrac{\boxed{+}3}{\boxed{-}(4-x)} = \dfrac{3}{-4+x} = \dfrac{3}{x-4}$

↖ Change the signs of the numerator and the denominator.

This last example illustrates a property that we should look at further:

$$\boxed{(x - y) = -(y - x)}$$

Example 3:

(a) $\dfrac{3}{a-b} = \dfrac{?}{b-a}$

$\dfrac{3}{a-b} = \dfrac{\boxed{-}3}{\boxed{-}(a-b)} = \dfrac{-3}{b-a}$

↖ Change the signs of the numerator and the denominator.

(b) $-\dfrac{6}{2-x} = \boxed{+}\dfrac{6}{\boxed{-}(2-x)} = \dfrac{6}{x-2}$

↖ Change the signs of the fraction and the denominator.

EXERCISE 8.1

In exercises 1 through 12 find the value(s) of the variable (if any) for which the fraction is undefined.

1. $\dfrac{2x-1}{x-3}$

2. $\dfrac{-7}{2x}$

3. $\dfrac{x}{x+3}$

4. $\dfrac{x}{6}$

5. $\dfrac{4-x}{x-4}$

6. $\dfrac{3}{4}$

7. $\dfrac{x-3}{x^2+2x-3}$

8. $\dfrac{x-1}{x^2-x-2}$

9. $\dfrac{-3}{x}$

10. $\dfrac{2}{x^2-x}$

11. $\dfrac{-7}{x^2+3x-10}$

12. $\dfrac{3x}{2x-7}$

In exercises 13 through 21 find the missing term using the rule of signs for fractions.

13. $\dfrac{3}{5}=\dfrac{?}{-5}$

14. $\dfrac{-1}{4}=\dfrac{?}{-4}$

15. $-\dfrac{3}{4}=\dfrac{-3}{?}$

16. $\dfrac{x}{-y}=\dfrac{?}{y}$

17. $\dfrac{5}{1-x}=\dfrac{?}{-(1-x)}$

18. $\dfrac{5}{1-x}=\dfrac{?}{x-1}$

19. $\dfrac{2-y}{-3}=\dfrac{?}{3}$

20. $-\dfrac{3}{x-1}=\dfrac{?}{1-x}$

21. $\dfrac{1-x}{1-y}=\dfrac{?}{y-1}$

Answers to odd-numbered exercises on page 476.

8.2 REDUCING FRACTIONS TO LOWEST TERMS

In arithmetic we learned that we can divide both the numerator and the denominator of any fraction by the same number without changing the value of the fraction.

To reduce the arithmetic fraction $\frac{6}{9}$ to its lowest terms we factor both the numerator and the denominator completely and then divide both the numerator and the denominator by all factors common to both.

$$\frac{6}{9}=\frac{2\cdot\overset{1}{\cancel{3}}}{3\cdot\underset{1}{\cancel{3}}}=\frac{2}{3}$$

This procedure holds true in algebra, and is illustrated as follows:

$$\frac{ay}{by}=\frac{a\cdot\overset{1}{\cancel{y}}}{b\cdot\underset{1}{\cancel{y}}}=\frac{a}{b}$$

To reduce a fraction to lowest terms:

1. Factor both the numerator and denominator completely.
2. Divide both the numerator and denominator by any common factors.
3. Multiply the factors that remain.

Example 1: Reduce to lowest terms:

(a) $$\frac{\overset{1}{\cancel{6}} \cdot \cancel{a^2} \cdot \overset{b^2}{\cancel{b^3}}}{\underset{5}{\cancel{30}} \cdot \underset{a}{\cancel{a^3}} \cdot \cancel{b}} = \frac{b^2}{5a}$$

(b) $$\frac{\overset{2}{\cancel{10}} \cdot \overset{x}{\cancel{x^2}} \cdot \cancel{y} \cdot \cancel{z}}{\underset{7}{\cancel{35}} \cdot \cancel{x} \cdot \underset{y^3}{\cancel{y^4}} \cdot \cancel{z}} = \frac{2x}{7y^3}$$

(c) $$\frac{x-2}{x^2-4} = \frac{\overset{1}{\cancel{(x-2)}}}{\underset{1}{\cancel{(x-2)}}(x+2)} = \frac{1}{x+2}$$

The number 1 remains in the numerator after being divided by $(x - 2)$.

(d) $$\frac{2x+6}{x^2-9} = \frac{2\cancel{(x+3)}}{(x-3)\cancel{(x+3)}} = \frac{2}{x-3}$$

The 1's are usually not written when cancelling.

(e) $$\frac{a+b}{a}$$

cannot be reduced since there is no *factor* common to both numerator and denominator.

(f) $$\frac{a-b}{b-a} = \boxed{-}\ \frac{a-b}{\boxed{-}(b-a)} = -\frac{a-b}{a-b} = -\frac{\cancel{(a-b)}}{\cancel{(a-b)}} = -1$$

Change the signs of the fraction and denominator.

(g) $$\frac{x^2+2x-3}{x^2-2x-8} = \frac{(x-1)(x+3)}{(x-4)(x+2)}$$

There are no common factors; therefore the fraction is already in lowest terms.

(h) $$\frac{2x^2-5x-3}{2x^2-x-1} = \frac{(x-3)\cancel{(2x+1)}}{\cancel{(2x+1)}(x-1)} = \frac{x-3}{x-1}$$

(i) $$-\frac{x^2-2x-3}{6+x-x^2} = -\frac{x^2-2x-3}{-(x^2-x-6)}$$

Factor -1 out of the denominator and rearrange the terms.

$$= \frac{x^2-2x-3}{x^2-x-6}$$

Change the signs of the fraction and the denominator.

$$= \frac{\cancel{(x-3)}(x+1)}{(x+2)\cancel{(x-3)}}$$

Cancel $(x - 3)$ from the numerator and denominator.

$$= \frac{x+1}{x+2}$$

Only factors of both the numerator and denominator may be cancelled. For example:

(a) A common error is:

not a factor

$$\frac{\overset{1}{\not{2}} + 9}{\not{2}} = 10$$

is incorrect since 2 is *not* a factor of the numerator.

$$\frac{2+9}{2} = \frac{11}{2} = 5\frac{1}{2}$$

is correct.

(b) $$\frac{y + \overset{2}{\not{4}}}{y + \underset{1}{\not{2}}} = \frac{y+2}{y+1}$$

is incorrect since 2 is *not* a factor of either the numerator or the denominator.

$$\frac{\overset{1}{\not{y}} + 4}{\underset{1}{\not{y}} + 2} = \frac{5}{3}$$

is incorrect since y is *not* a factor of either the numerator or the denominator.

$$\frac{y+4}{y+2}$$

cannot be reduced further.

EXERCISE 8.2

Reduce the fractions to lowest terms:

1. $\frac{12}{18}$
2. $\frac{16}{36}$
3. $\frac{6x^3y^4}{12x^5y}$
4. $\frac{x^7y^3}{x^7y^4}$
5. $\frac{3xyz}{-9xy^2z}$
6. $\frac{24a^3bc}{6a^4c}$
7. $\frac{2xy}{4xy}$
8. $\frac{-72x^6y^3}{-6x^2y^2}$
9. $\frac{x^2y^4}{a^2b^3}$

10. $\frac{x^2 - 1}{2x + 2}$

11. $\frac{6}{3x + 9}$

12. $\frac{x + 3}{x - 3}$

13. $\frac{7}{7x + 14}$

14. $\frac{x + 6}{6}$

15. $\frac{3 + 6}{3}$

16. $\frac{x^2 - 2x}{x^3 - 3x^2 + 2x}$

17. $-\frac{3x - 2}{2 - 3x}$

18. $\frac{x^2 - 4x - 5}{x^2 + 5x + 4}$

19. $\frac{3x^2 + 5x + 2}{3x^2 - x - 2}$

20. $\frac{6 - 5h}{5h - 6}$

21. $\frac{x^2 + 3x}{x^3 + 9x^2 + 18x}$

22. $\frac{4x - 4y}{y^2 - x^2}$

23. $-\frac{2x^2 + 3x - 20}{16 - x^2}$

24. $\frac{x^2 - 5x + 6}{6 + x - x^2}$

25. $\frac{t^2 - 25}{t^2 - 2t - 15}$

26. $\frac{(x + y)^2}{x^2 + y^2}$

27. $\frac{x^3 + x^2 - 12x}{x^3 + 9x^2 + 20x}$

28. $\dfrac{3x^2 + 2xy - y^2}{6x^2 - 5xy + y^2}$ 29. $\dfrac{3a^2 - 3ab}{b^2 - a^2}$ 30. $\dfrac{x^2 - 4y^2}{2x^2 - 3xy - 2y^2}$

Answers to odd-numbered exercises on page 476.

8.3 MULTIPLICATION AND DIVISION OF ALGEBRAIC FRACTIONS

Multiplying Algebraic Fractions

In arithmetic we multiplied fractions as follows, making sure that all common factors were cancelled before multiplying.

$$\frac{3}{4} \cdot \frac{8}{7} = \frac{3 \cdot \overset{2}{\cancel{8}}}{\underset{1}{\cancel{4}} \cdot 7}$$ Cancel completely.

$$= \frac{6}{7}$$ Multiply the numerators together and multiply the denominators together.

In algebra we use exactly the same procedure.

To multiply fractions:

1. Factor all numerators and denominators completely.
2. Divide any numerator and any denominator by a factor that is common to both.
3. The answer is the product of the remaining numerators divided by the product of the remaining denominators.

Example 1: Multiply.

$$\frac{3x^2}{5y} \cdot \frac{2y}{9} = \frac{\cancel{3}x^2 \cdot 2\cancel{y}}{5\cancel{y} \cdot \underset{3}{\cancel{9}}}$$

$$= \frac{2x^2}{15}$$

Example 2: Multiply.

$$\frac{x^2+6x+5}{x^2-1}\cdot\frac{x^2-4x+3}{x^2-5x+6}$$

$$=\frac{(x+5)(x+1)}{(x-1)(x+1)}\cdot\frac{(x-1)(x-3)}{(x-3)(x-2)}$$

Factor all numerators and all denominators completely.

$$=\frac{(x+5)\cancel{(x+1)}\cdot\cancel{(x-1)}\cancel{(x-3)}}{\cancel{(x-1)}\cancel{(x+1)}\cdot\cancel{(x-3)}(x-2)}$$

Cancel.

$$=\frac{x+5}{x-2}$$

Example 3: Multiply.

$$\frac{x}{3x-6}\cdot\frac{5x-10}{x^3}$$

$$=\frac{x}{3(x-2)}\cdot\frac{5(x-2)}{x^3}$$

Factor.

$$=\frac{\cancel{x}\cdot 5\cancel{(x-2)}}{3\cancel{(x-2)}\cdot\underset{x^2}{\cancel{x^3}}}$$

Cancel.

$$=\frac{5}{3x^2}$$

Example 4: Multiply.

$$\frac{x^2-4}{x^2-x-6}\cdot\frac{x^2+9x+18}{x^2-x-2}\cdot\frac{x^2-4x-5}{x^2+x-30}$$

$$=\frac{(x-2)(x+2)}{(x-3)(x+2)}\cdot\frac{(x+6)(x+3)}{(x-2)(x+1)}\cdot\frac{(x-5)(x+1)}{(x+6)(x-5)}$$

Factor.

$$=\frac{\cancel{(x-2)}\cancel{(x+2)}\cdot\cancel{(x+6)}(x+3)\cdot\cancel{(x-5)}\cancel{(x+1)}}{(x-3)\cancel{(x+2)}\cdot\cancel{(x-2)}\cancel{(x+1)}\cdot\cancel{(x+6)}\cancel{(x-5)}}$$

Cancel.

$$=\frac{x+3}{x-3}$$

Example 5: Multiply.

$$\frac{x^2-4x-12}{x^2+4x+4}\cdot\frac{x^2-4}{6-x}$$

$$=\frac{(x-6)(x+2)}{(x+2)(x+2)}\cdot\frac{(x-2)(x+2)}{(-1)(x-6)}$$

Factor and write $(6-x)$ as $(-1)(x-6)$.

$$= \frac{\cancel{(x-6)}\cancel{(x+2)} \cdot (x-2)\cancel{(x+2)}}{\cancel{(x+2)}\cancel{(x+2)} \cdot (-1)\cancel{(x-6)}}$$

Cancel—don't forget -1 in the denominator.

$$= \frac{x-2}{-1}$$

$$= -x + 2 \text{ or } 2 - x$$

Either form is acceptable.

Division of Algebraic Fractions

In arithmetic we divide fractions by inverting the second fraction (the *divisor*) and multiplying. You may have learned an equivalent rule, "multiply by the reciprocal of the second fraction," which accomplishes the same result.

$$\frac{3}{4} \div \frac{5}{8} = \frac{3}{\underset{1}{\cancel{4}}} \times \frac{\overset{2}{\cancel{8}}}{5} = \frac{6}{5}$$

The same procedure applies to division of algebraic fractions.

To divide fractions:

Invert the second fraction (divisor) and multiply.

$$\frac{a}{b} \div \frac{c}{d} = \frac{a}{b} \cdot \frac{d}{c}$$

Example 6: Divide.

$$\frac{3x+6}{x^2} \div \frac{x^2-4}{x^3}$$

Invert the divisor.

$$= \frac{3x+6}{x^2} \cdot \frac{x^3}{x^2-4}$$

Multiply.

$$= \frac{3(x+2)}{x^2} \cdot \frac{x^3}{(x-2)(x+2)}$$

Factor.

$$= \frac{3\cancel{(x+2)} \cdot \overset{x}{\cancel{x^3}}}{\cancel{x^2} \cdot (x-2)\cancel{(x+2)}}$$

Cancel.

$$= \frac{3x}{x-2}$$

Example 7: Divide.

$$\frac{2x^2-5x-3}{x^2-2x-3} \div 2x^2-9x-5$$

$2x^2 - 5x - 3$ can be written as $\frac{2x^2-5x-3}{1}$

$$= \frac{2x^2 - 5x - 3}{x^2 - 2x - 3} \cdot \frac{1}{2x^2 - 9x - 5}$$ Invert and multiply.

$$= \frac{(2x+1)(x-3)}{(x-3)(x+1)} \cdot \frac{1}{(2x+1)(x-5)}$$ Factor.

$$= \frac{\cancel{(2x+1)}\cancel{(x-3)} \cdot 1}{\cancel{(x-3)}(x+1) \cdot \cancel{(2x+1)}(x-5)}$$ Cancel.

$$= \frac{1}{(x+1)(x-5)}$$

Example 8: Divide.

$$\frac{y^2 - x^2}{6xy - 6y^2} \div \frac{x - y}{3x - 3y}$$

$$= \frac{y^2 - x^2}{6xy - 6y^2} \cdot \frac{3x - 3y}{x - y}$$ Invert and multiply.

$$= \frac{(y-x)(y+x)}{6y(x-y)} \cdot \frac{3(x-y)}{(x-y)}$$ Factor.

$$= \frac{(-1)\cancel{(x-y)}(y+x) \cdot \overset{1}{\cancel{3}}\cancel{(x-y)}}{\underset{2}{\cancel{6}}y\cancel{(x-y)} \cdot \cancel{(x-y)}}$$ Write $y - x$ as $(-1)(x - y)$ in the numerator and cancel.

$$= \frac{-(y+x)}{2y}$$ Don't forget the (-1).

$$= \frac{-(x+y)}{2y}$$ Variables are usually written in alphabetical order.

EXERCISE 8.3

Multiply or divide as indicated.

1. $\frac{9}{16} \cdot \frac{2}{3}$

2. $\frac{32x^3}{24y} \cdot \frac{10y^4}{12x^5}$

3. $\frac{12x^2y^3}{21x} \div \frac{4xy^2}{7x^3}$

4. $\frac{6a}{8b} \div \frac{3ab}{4a^3b}$

5. $\frac{14x^2y}{7xy^3} \div \frac{2xy^2}{y^4}$

6. $\frac{36x^3y^4z}{18xyz^4} \cdot \frac{4x}{15z^2}$

7. $\dfrac{b^2}{a - b} \cdot \dfrac{a^2 - b^2}{3ab + 3b^2}$

8. $\dfrac{y - 4}{y + 4} \div \dfrac{3y^2 - 12y}{y^2 + 2y - 8}$

9. $\dfrac{4x + 32}{2x + 10} \div \dfrac{x^2 + 3x - 40}{x^2 - 25}$

10. $\dfrac{4n - 12}{4n} \cdot \dfrac{2n^2}{16}$

11. $\dfrac{a^3}{a^2b - ab} \div \dfrac{a^2}{ab}$

12. $\dfrac{(x + 1)^2}{x^2 - 64} \cdot \dfrac{x^2 - 6x - 16}{x^2 + 3x + 2}$

13. $\dfrac{x^2 - 9}{x^2 - 5x^2 + 6x} \cdot \dfrac{16x^2}{8x + 24}$

14. $\dfrac{4x^2 - 36}{2x^2 - 2x - 12} \cdot \dfrac{x^2 + 8x + 15}{x^2 + 6x + 65}$

15. $\dfrac{x^2 - 2x - 63}{x^2 - 4x - 45} \div \dfrac{2x^2 - 5x - 12}{2x^2 + 13x + 15}$

16. $\dfrac{5}{a + 6} \div \dfrac{5a + 30}{a^2 + 12x + 36}$

17. $\dfrac{9x^2 - 1}{6x^2 + 17x + 5} \div (1 - 3x)$

18. $\dfrac{x - y}{2x - y} \cdot \dfrac{2x^2 + xy - y^2}{2y^2 - 3xy + x^2}$

19. $\dfrac{x^2 - 7x + 12}{x^2 + 6x - 7} \cdot \dfrac{x^2 - 2x - 8}{x^2 - x - 6} \cdot \dfrac{x^2 + 4x - 21}{x^2 - 8x + 16}$

20. $\dfrac{x^2 + 10x + 21}{x^2 - 2x - 15} \div (x^2 + 2x - 35)$

21. $\dfrac{(x - 5)^2}{(x - 5)^3} \div \dfrac{x^2 - x - 6}{x^2 - 3x - 10}$

22. $(3 - x) \div \dfrac{x^2 - x - 6}{x^2 - 5x - 14}$

23. $\dfrac{x^2 + 3x - 10}{x^2 - x - 2} \cdot \dfrac{x^2 - 10x + 9}{x^2 - 4x - 45} \cdot \dfrac{x^2 + 2x + 1}{1 - x^2}$

24. $\dfrac{x^2 - x - 2}{x^2 - 2x - 3} \cdot \dfrac{x^2 + 10x + 24}{x^2 + 4x - 12} \div \dfrac{2x^2 + 5x - 12}{x^2 - 8x + 15}$

25. $\dfrac{6x^2+x-1}{x^2+5x+6}\cdot\dfrac{x^3-2x^2-8x}{4x^2-1}\cdot\dfrac{2x^2-7x+3}{3x^3-13x^2+4x}$

Answers to odd-numbered exercises on page 476.

8.4 ADDING AND SUBTRACTING LIKE FRACTIONS

The same basic procedure is followed for adding or subtracting algebraic fractions as is used for adding or subtracting fractions in arithmetic.

Like fractions are fractions which have the same denominators. We can add fractions only when they have the same denominators.

To add or subtract like fractions:

1. Add or subtract the numerators.
2. Write this sum or difference over the same denominator.
3. Reduce the resulting fraction to lowest terms (when necessary).

Example 1: Add or subtract the following fractions:

(a) $\dfrac{2}{7}+\dfrac{3}{7}=\dfrac{2+3}{7}=\dfrac{5}{7}$

(b) $\dfrac{4}{9}-\dfrac{1}{9}=\dfrac{4-1}{9}=\dfrac{3}{9}=\dfrac{1}{3}$

(c) $\dfrac{3}{x^2}+\dfrac{7}{x^2}=\dfrac{10}{x^2}$

(d) $\dfrac{4}{x-1}+\dfrac{3}{x-1}=\dfrac{7}{x-1}$

(e) $\dfrac{6a}{2a-b}-\dfrac{3b}{2a-b}=\dfrac{6a-3b}{2a-b}$

$=\dfrac{3\cancel{(2a-b)}}{\cancel{(2a-b)}}$ — Factor out 3 in the numerator and cancel.

$=3$

(f) $\frac{4}{x-2} - \frac{2x}{x-2} = \frac{4-2x}{x-2}$

$= \frac{2(2-x)}{x-2}$

$= \frac{-2\cancel{(x-2)}}{\cancel{(x-2)}}$ Factor out −1.

$= -2$

(g) $\frac{x+8}{(x+2)(x-2)} \boxed{-} \frac{x+5}{(x+2)(x-2)}$

The entire numerator is being subtracted here.

$= \frac{x+8 \boxed{-} (x+5)}{(x+2)(x-2)}$

This affects the sign of each term in the numerator being subtracted. Students frequently make errors here.

$= \frac{x+8 \boxed{-} x \boxed{-} 5}{(x+2)(x-2)}$

$= \frac{3}{(x+2)(x-2)}$

(h) $\frac{x^2-2x+1}{(x-1)(x+5)} - \frac{x^2-3x+2}{(x-1)(x+5)}$

$= \frac{x^2-2x+1-(x^2-3x+2)}{(x-1)(x+5)}$ Subtract the numerators.

$= \frac{x^2-2x+1-x^2+3x-2}{(x-1)(x+5)}$ Each sign in the numerator being subtracted has been changed.

$= \frac{x-1}{(x-1)(x+5)}$ Combine the similar terms.

$= \frac{\cancel{(x-1)}}{\cancel{(x-1)}(x+5)}$ Cancel.

$= \frac{1}{x+5}$

The Least Common Denominator

Since we are able to add or subtract fractions only when their denominators are the same, what should we do if we are asked to add or subtract fractions with different denominators (*unlike* fractions)? In this case we find a *least common denominator* (LCD) and rewrite each fraction as an equivalent fraction with the LCD as its denominator.

To find the LCD of a set of fractions:

1. Factor each denominator.
2. Write down each factor the greatest number of times it appears in any one factorization.
3. The LCD is the product of the factors which you wrote down in step 2.

Example 2: Find the LCD for $\frac{1}{24}$ and $\frac{5}{36}$.

(1) Factor each denominator.

$$24 = \boxed{2 \cdot 2 \cdot 2} \cdot 3$$

$$36 = 2 \cdot 2 \cdot \boxed{3 \cdot 3}$$

(2) 2 occurs three times in the first factorization and 3 occurs twice in the second factorization. This is the *greatest* number of times that they occur in any denominator.

(3) LCD = $\boxed{2 \cdot 2 \cdot 2} \cdot \boxed{3 \cdot 3} = 72.$

Example 3: Find the LCD for $\frac{2}{9x^2y}$, $\frac{7}{18xy}$ and $\frac{1}{12xy^2}$.

Factor each denominator
$$\begin{cases} 9x^2y = \boxed{3 \cdot 3} \cdot \boxed{x^2} \cdot y \\ 18xy = 2 \cdot 3 \cdot 3 \cdot x \cdot y \\ 12xy^2 = \boxed{2 \cdot 2} \cdot 3 \cdot x \cdot \boxed{y^2} \end{cases}$$

Note that the greatest number of times that a variable occurs will always be the highest power of the variable.

$$\text{LCD} = 2 \cdot 2 \cdot 3 \cdot 3 \cdot x^2 \cdot y^2 = 36x^2y^2$$

Example 4: Find the LCD for $\frac{x}{x^2 - 1}$ and $\frac{5}{x^2 + 3x + 2}$.

$$x^2 - 1 = \boxed{(x - 1)(x + 1)}$$

$$x^2 + 3x + 2 = \boxed{(x + 2)}(x + 1)$$

$$\text{LCD} = (x - 1)(x + 1)(x + 2)$$

Example 5: Find the LCD for $\frac{a}{a+2}$, $\frac{b}{a^2-a-6}$ and $\frac{c}{a^2-6a+9}$.

$\boxed{a+2}$ is already factored.

$a^2 - a - 6 = (a + 2)(a - 3)$

$a^2 - 6a + 9 = \boxed{(a - 3)}\boxed{(a - 3)}$ The factor $(a - 3)$ occurs twice here.

LCD $= (a + 2)(a - 3)(a - 3)$

Changing to Equivalent Fractions

In order to add or subtract with different denominators we must first express each fraction as an equivalent fraction using the LCD as the new denominator.

> **To change a fraction to an equivalent fraction with the LCD as the new denominator:**
>
> 1. Divide the LCD by the denominator of the original fraction.
> 2. Multiply both the numerator and the denominator of the original fraction by the quotient obtained in step 1.

Example 6: $\frac{3}{4} = \frac{?}{12}$

(1) Divide 12 by 4, which gives us $\frac{12}{4} = 3$.

(2) Now multiply numerator and denominator of the original fraction $\frac{3}{4}$ by 3.

$$\frac{3}{4} \cdot \boxed{\frac{3}{3}} = \frac{9}{12} \text{ so } \frac{3}{4} = \frac{9}{12}$$

Note that we multiplied the original fraction by $\frac{3}{3}$ or 1, so we would expect our result to be *equivalent* to the original fraction.

Example 7: $\frac{5}{8x^2y} = \frac{?}{48x^4y^4}$

(1) $\frac{48x^4y^4}{8x^2y} = 6x^2y^3$ Divide the LCD by the denominator.

(2) $\frac{5}{8x^2y} \cdot \boxed{\frac{6x^2y^3}{6x^2y^3}} = \frac{30x^2y^3}{48x^4y^4}$

Example 8: $\dfrac{x-5}{x^2+2x-15} = \dfrac{?}{(x+5)(x-3)(x+1)}$

(1) Factor the original denominator in order to divide.

$$x^2 + 2x - 15 = (x+5)(x-3)$$

$$\frac{\text{LCD}}{\text{original denominator}} = \frac{\cancel{(x+5)}\cancel{(x-3)}(x+1)}{\cancel{(x+5)}\cancel{(x-3)}} = x+1$$

(2) $\dfrac{x-5}{x^2+2x-15} = \dfrac{(x-5)\boxed{(x+1)}}{(x+5)(x-3)\boxed{(x+1)}} = \dfrac{x^2-4x-5}{(x+5)(x-3)(x+1)}$

Example 9: $\dfrac{a+6}{(a+2)(a+2)} = \dfrac{?}{(a+2)(a+2)(a-1)}$

(1) $\dfrac{\cancel{(a+2)}\cancel{(a+2)}(a-1)}{\cancel{(a+2)}\cancel{(a+2)}} = a-1$

(2) $\dfrac{a+6}{(a+2)(a+2)} = \dfrac{(a+6)\boxed{(a-1)}}{(a+2)(a+2)\boxed{(a-1)}} = \dfrac{a^2+5a-6}{(a+2)(a+2)(a-1)}$

EXERCISE 8.4

Add or subtract the fractions as indicated, and reduce them to their lowest terms.

1. $\dfrac{5}{12} + \dfrac{1}{12}$
2. $\dfrac{a}{6} - \dfrac{4a}{6}$
3. $\dfrac{2}{3x} + \dfrac{4}{3x}$
4. $\dfrac{4}{a+b} - \dfrac{3}{a+b}$
5. $\dfrac{6}{5x} + \dfrac{7}{5x} - \dfrac{3}{5x}$
6. $\dfrac{3y}{y-4} - \dfrac{12}{y-4}$
7. $\dfrac{12a}{3a-b} - \dfrac{4b}{3a-b}$
8. $\dfrac{x+1}{6x^2+3} - \dfrac{x+4}{6x^2+3}$
9. $\dfrac{a}{x} + \dfrac{b}{x} + \dfrac{c}{x}$
10. $\dfrac{4}{z-1} - \dfrac{4z}{z-1}$
11. $\dfrac{3m}{m-2n} - \dfrac{2n}{m-2n} - \dfrac{2m}{m-2n}$
12. $\dfrac{x+2}{2x^2+4x} + \dfrac{3x-5}{2x^2+4x} - \dfrac{6x-1}{2x^2+4x}$

Find the LCD for each group of fractions.

13. $\frac{5}{8}, \frac{1}{12}, \frac{1}{10}$

14. $\frac{1}{x^2y}, \frac{3}{x^3y}$

15. $\frac{1}{6xy^3}, \frac{5}{8x^2}$

16. $\frac{3}{a}, \frac{4}{a+1}$

17. $\frac{6}{x}, \frac{-2}{x-4}$

18. $\frac{x}{3a^2bc}, \frac{y}{12a^3b}, \frac{z}{15ab^4c^2}$

19. $\frac{5}{x+1}, \frac{3x}{x^2+2x+1}$

20. $\frac{-4}{x^2-2x}, \frac{-7}{x^2-2x+1}$

21. $\frac{3x}{7x+1}, \frac{4}{7x}, \frac{-5}{7x-1}$

22. $\frac{2x-1}{x^2-x-12}, \frac{x-5}{x^2+6x+9}$

23. $\frac{a-1}{a^2-4}, \frac{-3}{a^2+4a+4}, \frac{a^2+1}{a^2+2a}$

24. $\frac{6n}{6n+18}, \frac{1}{n}, \frac{3n+1}{n^2+6n+9}$

25. Under what circumstances is the LCD equal to the product of the denominators of the original fractions?

For each of the following, find a new fraction which is equivalent to the fraction on the left by finding the missing numerator.

26. $\frac{1}{3} = \frac{?}{12}$

27. $\frac{2}{5} = \frac{?}{15}$

28. $\frac{3}{4x^3y} = \frac{?}{16x^5y^2}$

29. $\frac{5}{6xy^2} = \frac{?}{30x^3y^5}$

30. $\frac{2}{5ab} = \frac{?}{15a^2bc}$

31. $\frac{7}{4ab^2} = \frac{?}{28ab^3c^2}$

32. $\frac{3}{x+2} = \frac{?}{(x+2)(x-5)}$

33. $\frac{4}{x+3} = \frac{?}{(x+3)(x-4)}$

34. $\frac{x-3}{x-4} = \frac{?}{(x-4)(x+1)}$

35. $\frac{x-2}{x-7} = \frac{?}{(x-7)(x+2)}$

36. $\frac{2}{x+3} = \frac{?}{(x+1)(x+2)(x+3)}$

37. $\frac{4}{x-2} = \frac{?}{(x-4)(x-2)(x+1)}$

38. $\frac{x+4}{(x-2)(x+3)} = \frac{?}{(x-2)(x+3)(x-1)}$

39. $\frac{x-4}{(x+1)(x+5)} = \frac{?}{(x+1)(x+5)(x-4)}$

40. $\frac{x-1}{(x+1)(x+1)} = \frac{?}{(x+1)(x+2)(x+1)}$

Answers to odd-numbered exercises on page 477.

8.5 ADDING AND SUBTRACTING UNLIKE FRACTIONS

We have now learned everything that is needed to add or subtract fractions with different denominators.

To add or subtract unlike fractions:

1. Find the LCD for all the given fractions.
2. Change all the given fractions to equivalent fractions with the LCD as the denominator.
3. Add or subtract the like fractions as before.
4. Reduce the result to its lowest terms.

Example 1: $\dfrac{2}{x-1}+\dfrac{5}{x+2}$

(1) $\text{LCD} = (x-1)(x+2)$

(2) $\dfrac{2}{x-1}+\dfrac{5}{x+2}$

$$=\frac{2\,\boxed{(x+2)}}{(x-1)\,\boxed{(x+2)}}+\frac{5\,\boxed{(x-1)}}{(x+2)\,\boxed{(x-1)}}$$ Form equivalent fractions.

(3) $\dfrac{2(x+2)+5(x-1)}{(x-1)(x+2)}$ Add the numerators.

$$=\frac{2x+4+5x-5}{(x-1)(x+2)}$$ Simplify and combine.

$$=\frac{7x-1}{(x-1)(x+2)}$$

Example 2: $\dfrac{5}{3x^2y}-\dfrac{1}{4xy^2}$

(1) $\text{LCD} = 12x^2y^2$

(2) $\dfrac{5}{3x^2y}-\dfrac{1}{4xy^2}$

$$\frac{12x^2y^2}{3x^2y}=4y \downarrow \qquad \frac{12x^2y^2}{4xy^2}=3x \downarrow$$

$$=\frac{5\cdot\boxed{4y}}{3x^2y\cdot\boxed{4y}}-\frac{1\cdot\boxed{3x}}{4xy^2\cdot\boxed{3x}}$$ Form equivalent fractions.

$$=\frac{20y}{12x^2y^2}-\frac{3x}{12x^2y^2}$$ Subtract the numerators.

(3) $=\dfrac{20y-3x}{12x^2y^2}$

Example 3: $\frac{3}{x} + \frac{x}{x-2}$

(1) LCD $= x(x - 2)$

(2) $\frac{3}{x} + \frac{x}{x-2} = \frac{3\,\boxed{(x-2)}}{x\,\boxed{(x-2)}} + \frac{x \cdot \boxed{x}}{(x-2)\,\boxed{x}}$

(3) $= \frac{3(x-2) + x \cdot x}{x(x-2)} = \frac{3x - 6 + x^2}{x(x-2)} = \frac{x^2 + 3x - 6}{x(x-2)}$

Example 4: $\frac{20}{a^2 - 4} + \frac{5}{a+2}$

(1) $a^2 - 4 = (a + 2)(a - 2)$ so LCD is $(a + 2)(a - 2)$

(2) $\frac{20}{(a+2)(a-2)} + \frac{5}{a+2} = \frac{20}{(a+2)(a-2)} + \frac{5\,\boxed{(a-2)}}{(a+2)\,\boxed{(a-2)}}$

(3) $= \frac{20 + 5a - 10}{(a+2)(a-2)} = \frac{5a + 10}{(a+2)(a-2)}$

(4) $= \frac{5\cancel{(a+2)}}{\cancel{(a+2)}(a-2)} = \frac{5}{a-2}$ Factor and reduce to lowest terms.

Example 5: $\frac{x-2}{x^2 + 6x + 9} - \frac{x+4}{x^2 - 9}$

(1) Since $x^2 + 6x + 9 = (x + 3)(x + 3)$
and $x^2 - 9 = (x - 3)(x + 3)$
the LCD is $(x + 3)(x + 3)(x - 3)$

(2) $\frac{x-2}{(x+3)(x+3)} - \frac{x+4}{(x+3)(x-3)}$

$= \frac{(x-2)\,\boxed{(x-3)}}{(x+3)(x+3)\,\boxed{(x-3)}} - \frac{(x+4)\,\boxed{(x+3)}}{(x+3)(x-3)\,\boxed{(x+3)}}$

$= \frac{x^2 - 5x + 6}{(x+3)(x+3)(x-3)} \boxed{-} \frac{x^2 \boxed{+} 7x \boxed{+} 12}{(x+3)(x+3)(x-3)}$

The entire numerator is being subtracted, which affects the sign of each term in it.

(3) $= \frac{x^2 - 5x + 6 \boxed{-} x^2 - 7x \boxed{-} 12}{(x+3)(x+3)(x-3)}$

(4) $= \frac{-12x - 6}{(x+3)(x+3)(x-3)} = \frac{-6(2x - 1)}{(x+3)(x+3)(x-3)}$

EXERCISE 8.5

Perform the indicated operations:

1. $\frac{1}{3} + \frac{2}{9} + \frac{5}{12}$

2. $\frac{3}{x} + \frac{4}{x^2}$

3. $\frac{2}{x^2y} - \frac{1}{xy^2}$

4. $\frac{7}{6x^2y^2z} - \frac{1}{1xy^3z}$

5. $\frac{5}{6y^2} + \frac{9}{4y}$

6. $\frac{2}{3x} + \frac{4}{2x} - \frac{5}{9x^2}$

7. $\frac{a}{bc} + \frac{b}{ac} + \frac{2c}{ab}$

8. $4x - \frac{2}{x}$

9. $\frac{3}{x} + \frac{x}{x+3}$

10. $\frac{y}{y-6} - \frac{6}{y}$

11. $\frac{x+y}{y} + \frac{y}{x-y}$

12. $\frac{2}{x} + \frac{1}{x+1}$

13. $\dfrac{2a}{5a + 10} - \dfrac{3a}{15a + 30}$

14. $\dfrac{1}{x + 2} + \dfrac{5}{x^2 - x - 6}$

15. $\dfrac{3}{x + 2} - \dfrac{5}{x - 5}$

16. $\dfrac{7}{x - 5} - \dfrac{3}{5 - x}$

17. $\dfrac{2y}{5y + 10} + \dfrac{4}{y^2 - 3y - 10}$

18. $\dfrac{5}{x^2 + 6x + 9} - \dfrac{4}{x^2 - 9}$

19. $\dfrac{6}{x - 1} - \dfrac{4}{x - 2} + \dfrac{4}{x^2 - 3x + 2}$

20. $\dfrac{b}{a^2 - ab} - \dfrac{a}{ab - b^2}$

21. $\dfrac{x + 2}{3 - x} + \dfrac{x - 1}{x^2 - 9}$

22. $\dfrac{3x - 1}{2x^2 + x - 3} - \dfrac{x + 1}{x - 1}$

23. $\dfrac{3}{x + 1} + \dfrac{2}{x^2 - 1} - \dfrac{1}{x^2 + x - 2}$

24. $\dfrac{x - 3}{x + 5} - \dfrac{x - 5}{x + 3}$

25. $\dfrac{2x + 3}{2x - 6} + \dfrac{x - 3}{x^2 - 9}$

26. $\dfrac{4}{x^2 + 2x - 3} + \dfrac{2x}{x^2 + 5x + 6}$

More than one operation may occur in a problem. Use the correct order of operations and the meaning of parentheses in the following.

27. $\frac{2}{a} + \frac{2}{b} \cdot \frac{3b}{a}$

28. $\left(\frac{2}{a} + \frac{2}{b}\right) \cdot \frac{3b}{a}$

29. $\frac{3}{x^2 - 2x - 8} \div \frac{1}{x - 4} - \frac{1}{x + 2}$

30. $\frac{3}{x^2 - 2x - 8} \div \left(\frac{1}{x - 4} - \frac{1}{x + 2}\right)$

31. $\frac{2x}{x^2 - 4} + \frac{2x - 6}{x^2 - 4x + 3} \cdot \frac{x^2 - 5x + 4}{x^2 - 2x - 8} - \frac{x - 3}{x^2 + 3x - 10} \div \frac{x - 3}{3x + 15}$

Answers to odd-numbered exercises on page 477.

8.6 COMPLEX FRACTIONS

A fraction whose numerator or denominator (or both) is itself a fraction is called a *complex fraction*. The following are all examples of complex fractions:

$$\frac{\frac{2}{3} - \frac{1}{2}}{\frac{1}{8}}, \quad \frac{x - y}{1 - \frac{x}{y}}, \quad \frac{\frac{a}{b} + \frac{b}{a}}{\frac{b}{a} - \frac{a}{b}}$$

A fraction in which neither the numerator nor the denominator contains a fraction is called a **simple fraction.** Examples of simple fractions are the following:

$$\frac{3}{4}, \quad \frac{x - 5}{x}, \quad \frac{-17}{x^2 + 2x - 1}, \quad \frac{x - 2}{x^2 + 6x + 5}$$

To simplify a complex fraction we change it to a simple fraction and then reduce it if possible. There are two methods for accomplishing this task, both of which we will investigate. We will find that in some cases Method I will be easier to use while in other problems Method II is superior.

To simplify complex fractions:

Method I

1. Find the LCD of all the fractions in the problem.
2. Multiply both the numerator and the denominator by this LCD.
3. Simplify the result.

Method II

1. Add or subtract, wherever necessary, to obtain a single fraction in both the numerator and the denominator.
2. Divide the simplified fraction in the numerator by the simplified fraction in the denominator.

Example 1: Simplify the complex fraction.

$$\frac{\frac{2}{3}-\frac{1}{4}}{1+\frac{1}{6}}$$

Method I

(1) The LCD of all the fractions is 12.

(2) Multiply the entire numerator and the entire denominator by the LCD, 12.

$$\frac{\boxed{12}\left(\frac{2}{3}-\frac{1}{4}\right)}{\boxed{12}\left(1+\frac{1}{6}\right)} = \frac{(\overset{4}{\boxed{\not{12}}})\frac{2}{\not{3}}-(\overset{3}{\boxed{\not{12}}})\frac{1}{\not{4}}}{(\boxed{12})\,1+(\overset{2}{\boxed{\not{12}}})\frac{1}{\not{6}}} = \frac{8-3}{12+2} = \frac{5}{14}$$

Method II

(1) Obtain a single fraction in both the numerator and the denominator.

Numerator: $\frac{2}{3}-\frac{1}{4}=\frac{8}{12}-\frac{3}{12}=\frac{5}{12}$

Denominator: $1+\frac{1}{6}=1\frac{1}{6}=\frac{7}{6}$

(2) Divide the simplified numerator by the simplified denominator.

$$\frac{\frac{2}{3}-\frac{1}{4}}{1+\frac{1}{6}} = \frac{\frac{5}{12}}{\frac{7}{6}} = \frac{5}{12} \div \frac{7}{6} = \frac{5}{\underset{2}{\not{12}}} \cdot \frac{\overset{1}{\not{6}}}{7} = \frac{5}{14}$$

Example 2: Simplify.

$$\frac{\dfrac{3}{x}-\dfrac{4}{x^2}}{\dfrac{3}{x}+2}$$

Method I

The LCD is x^2.

$$\frac{\boxed{x^2}\left(\dfrac{3}{x}-\dfrac{4}{x^2}\right)}{\boxed{x^2}\left(\dfrac{3}{x}+2\right)}=\frac{\overset{x}{\boxed{\cancel{x^2}}}\,\dfrac{3}{\cancel{x}}-\boxed{\cancel{x^2}}\,\dfrac{4}{\cancel{x^2}}}{\overset{x}{\boxed{\cancel{x^2}}}\,\dfrac{3}{\cancel{x}}+\boxed{x^2}\,2}=\frac{3x-4}{3x+2x^2}\quad\text{or}\quad\frac{3x-4}{x\,(2x+3)}$$

Method II

$$\frac{\dfrac{3}{x}-\dfrac{4}{x^2}}{\dfrac{3}{x}+2}=\frac{\dfrac{3\,\boxed{x}}{x\,\boxed{x}}-\dfrac{4}{x^2}}{\dfrac{3}{x}+\dfrac{2\,\boxed{x}}{1\,\boxed{x}}}=\frac{\dfrac{3x-4}{x^2}}{\dfrac{3+2x}{x}}$$

$$=\frac{3x-4}{\underset{x}{\cancel{x^2}}}\cdot\frac{\cancel{x}}{3+2x}=\frac{3x-4}{x\,(2x+3)}$$

Example 3: Simplify the complex fraction.

$$\frac{\dfrac{1}{x^2}-\dfrac{1}{y^2}}{\dfrac{1}{x^2y^2}}$$

Method I

The LCD is x^2y^2.

$$\frac{\boxed{x^2y^2}\left(\dfrac{1}{x^2}-\dfrac{1}{y^2}\right)}{\boxed{x^2y^2}\;\dfrac{1}{x^2y^2}}=\frac{\boxed{\cancel{x^2}y^2}\,\dfrac{1}{\cancel{x^2}}-\dfrac{1}{\cancel{y^2}}\,\boxed{x^2\cancel{y^2}}}{\boxed{\cancel{x^2y^2}}\;\dfrac{1}{\cancel{x^2y^2}}}=\frac{y^2-x^2}{1}=y^2-x^2$$

Method II

$$\frac{\dfrac{1}{x^2}-\dfrac{1}{y^2}}{\dfrac{1}{x^2y^2}}=\frac{\dfrac{y^2}{x^2y^2}-\dfrac{x^2}{x^2y^2}}{\dfrac{1}{x^2y^2}}=\frac{\dfrac{y^2-x^2}{x^2y^2}}{\dfrac{1}{x^2y^2}}$$

$$=\frac{y^2-x^2}{\cancel{x^2y^2}}\cdot\frac{\cancel{x^2y^2}}{1}=y^2-x^2$$

EXERCISE 8.6

Simplify the complex fractions:

1. $\dfrac{\dfrac{a}{b}}{\dfrac{a^3}{b^2}}$

2. $\dfrac{\dfrac{1}{x}}{\dfrac{1}{y}}$

3. $\dfrac{\dfrac{a}{b}}{a}$

4. $\dfrac{\dfrac{3}{4}+\dfrac{5}{8}}{\dfrac{1}{2}+\dfrac{3}{8}}$

5. $\dfrac{1-\dfrac{3}{5}}{4-\dfrac{2}{15}}$

6. $\dfrac{\dfrac{1}{8}+2}{5-\dfrac{3}{4}}$

7. $\dfrac{\dfrac{2x^4y}{x^3y}}{\dfrac{4x}{xy^2}}$

8. $\dfrac{\dfrac{12a^2}{3b}}{\dfrac{4a^3}{5b^2}}$

9. $\dfrac{\dfrac{1}{a}+\dfrac{1}{b}}{a}$

10. $\dfrac{\dfrac{1}{x}+\dfrac{1}{y}}{\dfrac{1}{x}-\dfrac{1}{y}}$

11. $\dfrac{\dfrac{x}{y} - 2}{2 - \dfrac{y}{x}}$

12. $\dfrac{\dfrac{x}{y} - \dfrac{y}{x}}{x + y}$

13. $\dfrac{\dfrac{1}{x+3} - \dfrac{1}{x}}{\dfrac{2}{x+3}}$

14. $\dfrac{\dfrac{x}{x+y}}{\dfrac{y}{x-y}}$

15. $\dfrac{\dfrac{1}{x+y} - \dfrac{1}{x-y}}{\dfrac{1}{x^2 - y^2}}$

16. $\dfrac{\dfrac{1}{3a} - \dfrac{1}{4b}}{\dfrac{1}{2a} + \dfrac{1}{3b}}$

17. $\dfrac{\dfrac{1}{a} + \dfrac{2}{a^2}}{\dfrac{3}{a^2} + 4}$

18. $\dfrac{\dfrac{x}{y} + 3}{5 - \dfrac{x^2}{y^2}}$

19. $\dfrac{\dfrac{x}{y} + \dfrac{y}{x}}{\dfrac{1}{xy}}$

20. $\dfrac{\dfrac{1}{4x} + \dfrac{2}{3y}}{\dfrac{3}{4y} - \dfrac{1}{3x}}$

21. $\dfrac{\dfrac{2}{x} - \dfrac{1}{x^2 - x}}{\dfrac{3}{x-1}}$

22. $\dfrac{\dfrac{x}{x+1} - \dfrac{2}{x}}{\dfrac{1}{x} + \dfrac{2}{x+1}}$

23. $\dfrac{\dfrac{1}{x^2 - x - 2} + \dfrac{1}{x + 1}}{\dfrac{1}{x - 2} + \dfrac{1}{x^2 - x - 2}}$

24. $\dfrac{a}{1 + \dfrac{a - 1}{a + 1}}$

25. $\dfrac{1 - a + \dfrac{2}{a^2}}{\dfrac{4}{a} - a}$

26. $\dfrac{x + 1}{1 + \dfrac{1}{x}}$

27. $\dfrac{\dfrac{3}{x^2 - 3x + 2} - \dfrac{1}{x - 1}}{\dfrac{1}{x - 1} + 1}$

28. $\dfrac{\dfrac{x + 2}{x - 2} - \dfrac{x}{x + 2}}{3 + \dfrac{2}{x}}$

29. $\dfrac{\dfrac{x}{x + 1} - \dfrac{x^2}{x^2 - 1}}{1 + \dfrac{1}{x + 1}}$

30. $\dfrac{3 - \dfrac{2}{x + 1}}{\dfrac{3}{x + 1} - 4}$

Answers to odd-numbered exercises on page 478.

8.7 FRACTIONAL EQUATIONS

In section 3.4 we found that we could clear the fractions from a simple equation by multiplying both sides of the equation by the LCD of all the fractions in the equation. Now we are going to solve more complicated equations using essentially the same procedure.

To solve an equation containing fractions:

1. Clear of fractions by multiplying each term on both sides of the equation by the LCD of all the fractions.
2. Solve the resulting equation.
3. Check the solution in the original equation.

Example 1: Solve for x: $\frac{x}{2}+\frac{x}{4}=6$

(1) LCD = 4
(2) Multiply both sides of the equation by the LCD.

$$\frac{x}{2}+\frac{x}{4}=6$$

$$\overset{2}{\boxed{\not 4}}\,\frac{x}{\underset{1}{\not 2}}+\overset{1}{\boxed{\not 4}}\,\frac{x}{\underset{1}{\not 4}}=\boxed{4}\cdot 6$$

$$2x+x=24 \qquad \text{The fractions are gone!}$$

$$3x=24$$

$$x=8$$

(3) Check:

$$\frac{8}{2}+\frac{8}{4}\overset{?}{=}6$$

$$4+2=6 \checkmark$$

Example 2: $\frac{x+2}{2}+\frac{x+3}{4}=4$

(1) LCD = 4 by inspection.
(2) Multiply both sides of the equation by the LCD, 4.

$$\frac{x+2}{2}+\frac{x+3}{4}=4$$

$$\overset{2}{\boxed{\not 4}}\left(\frac{x+2}{\not 2}\right)+\overset{1}{\boxed{\not 4}}\left(\frac{x+3}{\not 4}\right)=\boxed{4}\cdot 4$$

$$2(x+2)+1(x+3)=16$$

$$2x+4+x+3=16$$

$$3x+7=16$$

$$3x=9$$

$$x=3$$

(3) Check:

$$\frac{3+2}{2}+\frac{3+3}{4}\overset{?}{=}4$$

$$\frac{5}{2}+\frac{6}{4}\overset{?}{=}4$$

$$\frac{10}{4}+\frac{6}{4}=\frac{16}{4}=4 \checkmark$$

An equation that contains a variable in the denominator must be checked for excluded values. There can be apparent solutions that yield zero in a denominator even though no error was made in the procedure. These are called **extraneous solutions** and they do not satisfy the equation. This is illustrated in the next example.

Example 3: $\frac{1}{x+1} = \frac{-2}{x^2-1}$

(1) LCD $= (x-1)(x+1)$

(2) $\frac{1}{x+1} = \frac{-2}{(x-1)(x+1)}$

$$\frac{\boxed{(x-1)\cancel{(x+1)}}\ 1}{\cancel{(x+1)}} = \frac{\boxed{\cancel{(x-1)}\cancel{(x+1)}}\ (-2)}{\cancel{(x-1)}\cancel{(x+1)}}$$

Multiply both sides by the LCD.

$$x-1 = -2$$
$$x = -1$$

(3) Check: We can see by inspection that $x = -1$ is an excluded value. If we check $x = -1$ in the original equation we see that it cannot be a solution.

$$\frac{1}{-1+1} \stackrel{?}{=} \frac{-2}{(-1)^2-1} \text{ or } \boxed{\frac{1}{0}} \stackrel{?}{=} \boxed{\frac{-2}{0}}$$

Zero as a denominator is meaningless.

Since $x = -1$ is an excluded value, it is an extraneous solution. Therefore, the original equation has **no solution.**

Example 4: Solve for y: $\frac{3}{y+5} + \frac{2}{y-2} = \frac{4}{y^2+3y-10}$

(1) Factoring $y^2 + 3y - 10$ gives us $(y+5)(y-2)$ so the LCD is $(y+5)(y-2)$.

(2) $\frac{3}{y+5} + \frac{2}{y-2} = \frac{4}{(y+5)(y-2)}$

$$\frac{\boxed{\cancel{(y+5)}(y-2)}\ 3}{\cancel{(y+5)}} + \frac{\boxed{(y+5)\cancel{(y-2)}}\ 2}{\cancel{(y-2)}} = \frac{\boxed{\cancel{(y+5)}\cancel{(y-2)}}\ 4}{\cancel{(y+5)}\cancel{(y-2)}}$$

Multiply by the LCD.

$$3y - 6 + 2y + 10 = 4$$
$$5y + 4 = 4$$
$$5y = 0$$
$$y = 0$$

(3) Check:

$$\frac{3}{0+5}+\frac{2}{0-2}\stackrel{?}{=}\frac{4}{0^2+3\cdot 0-10}$$

$$\frac{3}{5}+\frac{2}{-2}\stackrel{?}{=}\frac{4}{-10}$$

$$\frac{6}{10}-\frac{10}{10}=\frac{-4}{10}\ \checkmark$$

Example 5: Solve for x: $\dfrac{3}{x-2}-\dfrac{4}{x+1}=\dfrac{5}{x^2-x-2}$

(1) Factoring x^2-x-2 yields $(x-2)(x+1)$, so the LCD is $(x-2)(x+1)$

(2) $\dfrac{3}{x-2}-\dfrac{4}{x+1}=\dfrac{5}{(x-2)(x+1)}$

$$\frac{\boxed{\cancel{(x-2)}(x+1)}\ 3}{\cancel{(x-2)}}\ \boxed{-}\ \frac{\boxed{(x-2)\cancel{(x+1)}}\ 4}{\cancel{(x+1)}}=\frac{\boxed{\cancel{(x-2)}\cancel{(x+1)}}\ 5}{\cancel{(x-2)}\cancel{(x+1)}}$$ Multiply by the LCD.

Be careful with this negative sign.

$$3x+3\ \boxed{-}\ (4x-8)=5$$

The signs of both terms have been changed.

$$3x+3\ \boxed{-}\ 4x\ \boxed{+}\ 8=5$$

$$-x+11=5$$

$$-x=-6$$

$$x=6$$

(3) Check:

$$\frac{3}{6-2}-\frac{4}{6+1}\stackrel{?}{=}\frac{5}{6^2-6-2}$$ Substitute 6 for x.

$$\frac{3}{4}-\frac{4}{7}\stackrel{?}{=}\frac{5}{28}$$

$$\frac{21}{28}-\frac{16}{28}=\frac{5}{28}$$

$$\frac{5}{28}=\frac{5}{28}\ \checkmark$$

Some word problems result in equations which contain algebraic fractions.

Example 6: The sum of the reciprocals of two consecutive even integers is equal to 7 divided by 4 times the first integer. Find the integers.

Solution: Let x = the first even integer
then $x + 2$ = the second even integer.

The *reciprocal* of the *first* integer $= \frac{1}{x}$.

The *reciprocal* of the *second* integer $= \frac{1}{x+2}$.

The *sum* of the reciprocals is $\frac{1}{x} + \frac{1}{x+2}$.

The complete equation is $\frac{1}{x} + \frac{1}{x+2} = \frac{7}{4x}$.

Now solve for x using the methods of this section.

(1) By inspection, the LCD is $4x(x + 2)$.

(2) $\boxed{\cancel{4x}(x+2)} \cdot \frac{1}{\cancel{x}} + \boxed{4x\,\cancel{(x+2)}} \cdot \frac{1}{\cancel{x+2}} = \boxed{\cancel{4x}(x+2)} \cdot \frac{7}{\cancel{4x}}$ Multiply by the LCD.

$$4x + 8 + 4x = 7x + 14$$
$$8x + 8 = 7x + 14$$
$$x = 6$$

Therefore the integers are 6 and 8.

(3) Check:

$$\frac{1}{6} + \frac{1}{8} \stackrel{?}{=} \frac{7}{4 \cdot 6}$$
$$\frac{4}{24} + \frac{3}{24} = \frac{7}{24}$$
$$\frac{7}{24} = \frac{7}{24} \checkmark$$

EXERCISE 8.7

Solve the following equations and check:

1. $\frac{x}{2} - \frac{x}{3} = 5$
2. $\frac{x}{3} = \frac{5}{30}$
3. $\frac{y}{5} + \frac{y}{2} = 14$

4. $\frac{5}{6}x + \frac{1}{3}x = \frac{1}{2}$

5. $\frac{2}{3}a - \frac{2}{5} = \frac{2}{5}a$

6. $\frac{1}{x} + \frac{1}{2x} = 3$

7. $\frac{1}{b} = \frac{5}{2b}$

8. $\frac{x}{5} + \frac{x}{3} = 8$

9. $\frac{x+1}{5} - \frac{x-1}{2} = 1$

10. $\frac{a}{5} + \frac{a+2}{4} = \frac{1}{4}$

11. $\frac{x+2}{4} + \frac{x+6}{2} = 3$

12. $\frac{6a-3}{9} - \frac{2a+6}{4} = \frac{1}{3}$

13. $\frac{2}{x} = \frac{3}{x} - 1$

14. $\frac{2y-1}{3} - \frac{3y}{4} = \frac{5}{6}$

15. $\frac{n-3}{3n+2} = \frac{1}{4}$

16. $\frac{x-6}{x-9} = \frac{3}{x-9}$

17. $\frac{2}{x+2} + \frac{3}{x-2} = \frac{12}{x^2-4}$

18. $\frac{2y}{4y-4} = \frac{7}{8}$

19. $\frac{x}{x-3} = \frac{3}{x-3} + 4$

20. $\frac{x}{x+1} = 2 - \frac{1}{x+1}$

21. $\frac{1}{x+5} + \frac{1}{x-5} = \frac{1}{x^2-25}$

22. $\frac{4}{y-3} + \frac{2y}{y^2-9} = \frac{1}{y+3}$

23. $\dfrac{x}{x+4} + \dfrac{4}{4-x} = \dfrac{x^2+16}{x^2-16}$

24. $\dfrac{a-1}{a} = 1 + \dfrac{1}{a-1}$

25. $\dfrac{x-1}{x+3} + \dfrac{8}{x^2+x-6} = \dfrac{x+1}{x-2}$

26. $\dfrac{2x-3}{x-1} + \dfrac{1}{x-1} = 3$

27. $\dfrac{x+2}{x^2+3x} = \dfrac{3}{x+3} + \dfrac{1}{x}$

28. $\dfrac{2}{3x+6} + \dfrac{5}{4x+8} = \dfrac{1}{3}$

29. $\dfrac{x-1}{x-2} - \dfrac{2x+1}{x-3} = \dfrac{3-x^2}{x^2-5x+6}$

30. $\dfrac{4x-3}{2x-1} = \dfrac{6x}{3x+1}$

31. Two divided by a certain number is the same as 2 more than twice the number divided by four times the number. What is the number?

32. One number is 8 more than a second number. If the larger number is divided by the smaller number, the result is $\frac{7}{3}$. Find the number.

33. Find the integer which when added to both the numerator and denominator of $\frac{7}{9}$ will yield a fraction equivalent to $\frac{6}{7}$.

34. The numerator of a fraction is 4 more than the denominator. If this fraction is added to $\frac{2}{3}$, the result is 3. Find the fraction.

35. One number is equal to four times another number. The sum of the reciprocals of the two numbers is $\frac{10}{3}$. Find the two numbers.

36. One more than three times a certain number, divided by 9 more than the number is equal to 1. Find the number.

37. One number is 2 less than a second number. If the second number is divided by twice the first number, the result is $\frac{5}{6}$. Find the numbers.

Answers to odd-numbered exercises on page 478.

8.8 Chapter Summary

Examples

(8.1) An excluded value is a value of the variable that makes the denominator of a fraction equal to zero.

(8.1) x may not equal 4 in $\dfrac{3}{x-4}$ since the denominator would be zero.

(8.1) If exactly two of the three signs of a fraction are changed, the value of the fraction remains the same.

(8.1) $-\dfrac{2}{3-x} = \boxed{+}\dfrac{2}{\boxed{-}(3-x)}$

$= \dfrac{2}{x-3}$

(8.2) To reduce a fraction to lowest terms:

1. Factor both the numerator and the denominator completely.
2. Divide both the numerator and the denominator by any common factor.
3. Multiply the factors that remain.

(8.2) $\dfrac{3x+6}{x^2-4} = \dfrac{3\cancel{(x+2)}}{(x-2)\cancel{(x+2)}}$

$= \dfrac{3}{x-2}$

(8.3) To multiply fractions:

1. Factor all numerators and denominators completely.
2. Divide any numerator and any denominator by a factor that is common to both.
3. The answer is the product of the remaining numerators divided by the product of the remaining denominators.

(8.3)

$\dfrac{x^2-3x-4}{4x+8} \cdot \dfrac{2x-2}{x^2-1}$

$= \dfrac{(x-4)\cancel{(x+1)}}{\cancelto{2}{4}(x+2)} \cdot \dfrac{\cancelto{1}{2}\cancel{(x-1)}}{\cancel{(x-1)}\cancel{(x+1)}}$

$= \dfrac{x-4}{2(x+2)}$

(8.3) To divide fractions:
Invert the second fraction (divisor) and multiply.

(8.3)

$\dfrac{x^2+x-6}{x^2+2x-3} \div \dfrac{x^2+2x-8}{x^2-2x+1}$

$= \dfrac{x^2+x-6}{x^2+2x-3} \cdot \dfrac{x^2-2x+1}{x^2+2x-8}$

$= \dfrac{\cancel{(x+3)}\cancel{(x-2)}}{\cancel{(x-1)}\cancel{(x+3)}} \cdot \dfrac{\cancel{(x-1)}(x-1)}{(x+4)\cancel{(x-2)}}$

$= \dfrac{x-1}{x+4}$

(8.4) To add or subtract *like* fractions:

1. Add or subtract the numerators.
2. Write their sum or difference over the same denominator.
3. Reduce the resulting fraction to lowest terms.

(8.4) $\dfrac{3}{x-2} + \dfrac{x+1}{x-2} = \dfrac{x+4}{x-2}$

(8.4) To find the LCD (least common denominator) of a set of fractions:

1. Factor each denominator.
2. Write down each factor the greatest number of times that it appears in any one factorization.
3. The LCD is the product of the factors that you wrote down in step 2.

(8.4) Find the LCD for the fractions.

$$\frac{2}{x^2-9}, \frac{6}{x^2-x-12},$$

$$x^2 - 9 = (x - 3)(x + 3)$$

$$x^2 - x - 12 = (x + 3)(x - 4)$$

$$\text{LCD} = (x - 3)(x + 3)(x - 4)$$

(8.4) To change a fraction to an equivalent fraction with the LCD as the new denominator:

1. Divide the LCD by the denominator of the original fraction.
2. Multiply both the numerator and the denominator of the original fraction by the quotient obtained in step 1.

(8.4)

$$\frac{x-2}{(x+3)(x+4)} = \frac{?}{(x+3)(x+4)(x+2)}$$

$$\frac{x-2}{(x+3)(x+4)} = \frac{(x-2)\boxed{(x+2)}}{(x+3)(x+4)\boxed{(x+2)}} = \frac{x^2-4}{(x+3)(x+4)(x+2)}$$

(8.5) To add or subtract *unlike* fractions:

1. Find the LCD for all the given fractions.
2. Change all the given fractions to equivalent fractions with the LCD as the denominator.
3. Add or subtract the like fractions as before.
4. Reduce the result to lowest terms.

(8.5) $\frac{3}{x-4} + \frac{6}{x-2}$

1. LCD is $(x - 4)(x - 2)$
2. $\frac{3}{x-4} + \frac{6}{x-2}$
$$= \frac{3\boxed{(x-2)}}{(x-4)\boxed{(x-2)}} + \frac{6\boxed{(x-4)}}{(x-2)\boxed{(x-4)}}$$
3. $$= \frac{3x-6}{(x-4)(x-2)} + \frac{6x-24}{(x-4)(x-2)}$$
$$= \frac{9x-30}{(x-4)(x-2)}$$

(8.6) To simplify complex fractions:
Method I

1. Find the LCD of all the fractions in the problem.
2. Multiply both the numerator and the denominator by this LCD.

(8.6) Simplify $\dfrac{\frac{2}{x}+3}{4-\frac{5}{x}}$ by Method I.
LCD is x.

3. Simplify the result.

$$\frac{\boxed{\cancel{x}}\ \frac{2}{\cancel{x}} + \boxed{x}\ 3}{\boxed{x}\ 4 - \boxed{\cancel{x}}\ \frac{5}{\cancel{x}}} = \frac{2 + 3x}{4x - 5}$$

(8.6)

Method II

1. Add or subtract, wherever necessary, to obtain a single fraction in both the numerator and denominator.
2. Divide the simplified fraction in the numerator by the simplified fraction in the denominator.

(8.6) Simplify $\dfrac{\frac{3}{2x}}{1 + \frac{1}{x}}$ by Method II.

$$\frac{\frac{3}{2x}}{\frac{x}{x} + \frac{1}{x}} = \frac{\frac{3}{2x}}{\frac{x+1}{x}}$$

$$= \frac{3}{2\cancel{x}} \cdot \frac{\cancel{x}}{x+1}$$

$$= \frac{3}{2(x+1)}$$

(8.7) To solve an equation containing fractions:

1. Clear of fractions by multiplying each term on both sides of the equation by the LCD of all the fractions.
2. Solve the resulting equation.
3. Check the solution in the original equation.

(8.7) Solve for x.

$$\frac{5}{x-1} = \frac{4}{x} + \frac{2}{3x}$$

LCD is $3x(x - 1)$.

$$\boxed{3x\cancel{(x-1)}}\ \frac{5}{\cancel{(x-1)}} = \boxed{3\cancel{x}(x-1)}\ \frac{4}{\cancel{x}} + \boxed{\cancel{3x}(x-1)}\ \frac{2}{\cancel{3x}}$$

$$15x = 12x - 12 + 2x - 2$$

$$15x = 14x - 14$$

$$x = -14$$

Check:

$$\frac{5}{-14-1} \stackrel{?}{=} \frac{4}{-14} + \frac{2}{3(-14)}$$

$$\frac{5}{-15} \stackrel{?}{=} -\frac{2}{7} - \frac{1}{21}$$

$$= -\frac{1}{3} \stackrel{?}{=} \frac{-6}{21} - \frac{1}{21}$$

$$-\frac{1}{3} = -\frac{7}{21} \checkmark$$

Exercise 8.8 Chapter Review

(8.1) Find the value(s) of the variable (if any) for which the fraction is undefined:

1. $\dfrac{3}{x-1}$

2. $\dfrac{x-4}{3}$

3. $\dfrac{-16}{x^2-4}$

(8.1) Find the missing term using the rule of signs for fractions:

4. $\dfrac{5}{8}=\dfrac{?}{-8}$

5. $\dfrac{y+4}{-2}=\dfrac{?}{2}$

6. $-\dfrac{x-5}{3}=\dfrac{5-x}{?}$

(8.2) Reduce to lowest terms:

7. $\dfrac{12}{32}$

8. $\dfrac{-24x^3yz}{-6xy^4z}$

9. $\dfrac{3x+1}{3x-1}$

10. $\dfrac{3-4x}{4x-3}$

11. $\dfrac{2x^2+5x-3}{2x^2-7x+3}$

(8.3) Multiply or divide as indicated:

12. $\dfrac{5}{15}\cdot\dfrac{9}{16}$

13. $\dfrac{12x^2y}{3x^4}\div\dfrac{6xy^4}{9xy}$

14. $\dfrac{3x-12}{x}\cdot\dfrac{8x^2}{x-4}$

15. $\dfrac{2x^2+9x+4}{3x^4}\cdot\dfrac{x^3-5x^2}{2x+1}$

16. $\dfrac{x^2-5x+6}{x^2-7x+10}\div\dfrac{x^2-3x-4}{x^2-4x-5}$

17. $\dfrac{2+x-x^2}{3x^2-2x-8}\div\dfrac{x^2-4x-5}{3x^2-11x-20}$

(8.4) Add or subtract as indicated and reduce to lowest terms:

18. $\frac{3}{16} + \frac{5}{16}$

19. $\frac{14}{3y} - \frac{5}{3y}$

20. $\frac{3a}{x+5} + \frac{7b}{x+5}$

(8.4) Find the LCD for each group of fractions:

21. $\frac{3}{8}, \frac{1}{4}, \frac{3}{5}$

22. $\frac{7}{x^3y}, \frac{2}{xy^3}$

23. $\frac{6}{x-5}, \frac{-2}{x^2-8x+15}$

24. $\frac{9}{x^2-2x+1}, \frac{3}{5x-5}, \frac{-7}{2-2x}$

(8.5) Add or subtract as indicated and reduce to lowest terms:

25. $\frac{5}{3x^2} + \frac{9}{2x^3}$

26. $\frac{x}{x-4} - \frac{2}{x+1}$

27. $\frac{3x-1}{x^2-x-2} - \frac{2}{x-2}$

28. $\frac{x+1}{x-2} + \frac{3}{x+3} - \frac{4x-3}{x^2+x-6}$

(8.5) Combine using the correct order of operations:

29. $\frac{2x+6}{x^2-2x-3} + \frac{x-2}{x^2+x} \cdot \frac{x^2+3x}{x^2+x-6}$

(8.6) Simplify the complex fractions:

30. $\dfrac{\frac{2}{3}+\frac{3}{4}}{\frac{5}{6}-\frac{1}{2}}$

31. $\dfrac{\frac{a}{b}-3}{3-\frac{b}{a}}$

32. $\dfrac{\frac{3}{x-y}+\frac{2}{x+y}}{\frac{1}{x^2-y^2}}$

(8.7) Solve and check:

33. $\frac{x}{9} + \frac{x}{12} = \frac{3}{4}$

34. $\frac{1}{4x} - \frac{1}{3x} = \frac{1}{x^2}$

35. $\frac{2}{y+4} + \frac{5}{y-2} = \frac{3y}{y^2+2y-8}$

36. $\frac{2x}{4-x} + \frac{8}{x-4} = 1$

37. $\frac{2x+1}{x+3} - \frac{x-1}{x-4} = \frac{x^2+8}{x^2-x-12}$

Answers to odd-numbered exercises on page 479.

Chapter 8 Achievement Test

Name ______________________

Class ______________________

This test should be taken before you are tested in class on the material in Chapter 8. Solutions to each problem and the section where the type of problem is found are given on page 479.

For what value of x is the fraction not defined?

1. $\dfrac{2x}{x+1}$

2. $\dfrac{x-2}{x^2+6x+5}$

Find the missing term using the rule of signs for fractions:

3. $-\dfrac{2}{3}=\dfrac{2}{?}$

4. $\dfrac{2}{2-x}=\dfrac{?}{x-2}$

Reduce to lowest terms:

5. $\dfrac{x^2-1}{2x-2}$

6. $\dfrac{x^2-3x-4}{x^2+3x-10}$

7. $\dfrac{x^2-2x+1}{x^2+3x-4}$

Perform the indicated operations. Reduce to lowest terms.

8. $\dfrac{x^2}{x-y}\cdot\dfrac{x^2-y^2}{5x^3y+x^2}$

1. ______________________

2. ______________________

3. ______________________

4. ______________________

5. ______________________

6. ______________________

7. ______________________

8. ______________________

9. $\dfrac{x^2 - 9}{x^2 + 2x - 3} \div \dfrac{x^2 + x - 12}{x^2 + 3x - 4}$

9. ____________________

10. $\dfrac{7x}{3x - y} - \dfrac{x + 2y}{3x - y}$

10. ____________________

11. $3x + \dfrac{4}{x}$

11. ____________________

12. $\dfrac{3}{a - 2} + \dfrac{a}{2 - a}$

12. ____________________

13. $\dfrac{6}{x^2 + 2x + 1} - \dfrac{4}{x^2 - 1}$

13. ____________________

14. $\dfrac{x - 2}{x + 1} - \dfrac{x^2 + 6x + 8}{x^2 - 3x - 4} \div \dfrac{x^2 + 8x + 12}{x^2 + 2x - 24}$

14. ____________________

Simplify the complex fractions:

15. $\dfrac{\dfrac{1}{x - 2} - \dfrac{3}{x}}{\dfrac{4}{x - 2}}$

15. ____________________

16. $$\frac{\dfrac{a^2}{a^2-1}+\dfrac{a}{a-1}}{\dfrac{3}{a+1}-\dfrac{1}{a-1}}$$

16. ______________

Solve for x and check:

17. $\frac{2}{3}x-\frac{1}{4}x=\frac{5}{12}$

17. ______________

18. $\frac{3}{x-4}+\frac{3}{x+2}=\frac{6}{x^2-2x-8}$

18. ______________

19. $\frac{a}{a-2}+\frac{3}{2-a}=\frac{a^2-1}{a^2-4}$

19. ______________

20. One number is 6 greater than a second number. If the smaller number is divided by the larger number, the result is $\frac{6}{5}$. Find the numbers.

20. 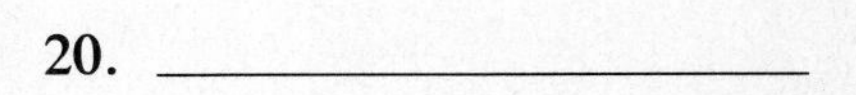

9 Graphing

9.1 THE RECTANGULAR COORDINATE SYSTEM

In large cities, streets and avenues are often numbered to make it easy to find a particular place. In a similar way, finding points on a piece of paper is made convenient by a numbering system called a **rectangular coordinate system.**

In previous chapters we made extensive use of the number line. Each point on the number line has some specific number associated with it. Now let's draw another line which intersects the number line at an angle of 90° (a right angle) and passes through the point zero. This point of intersection is called the **origin.** These two lines are called the **axes,** and they determine a flat surface called a **plane.** The horizontal line is called the **x-axis,** and the vertical line is called the **y-axis.** The y-axis is numbered using the same scale as is used on the number line. Starting with zero at the point of intersection (origin), we use positive numbers as we travel in an upward direction and negative numbers in the downward direction.

There is a scheme for associating each point in the plane with a pair of numbers. For example, consider the **ordered pair** of numbers (4, −2). The first number (4), called the **x-coordinate** or **abscissa,** tells us how far to go to the left (−) or to the right (+), and the second number (−2), called

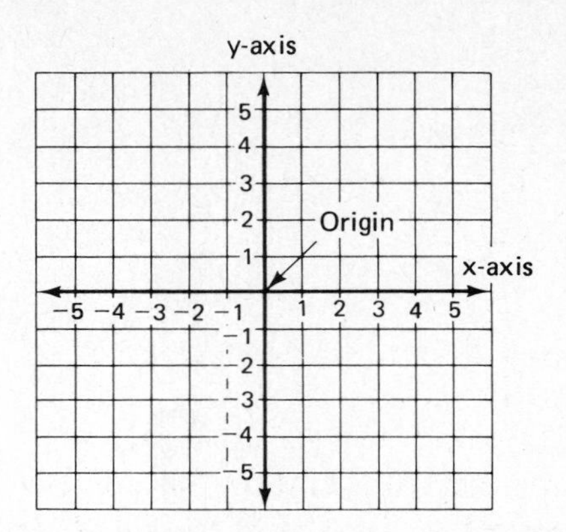

Figure 1.

the **y-coordinate** or **ordinate,** tells us how far to go upward (+) or downward (−).

So, starting from the origin we travel 4 units to the right, since 4 is positive. From there, the second number, −2, tells us to go 2 units in a downward direction, since −2 is negative. Study the location of this point in Figure 2.

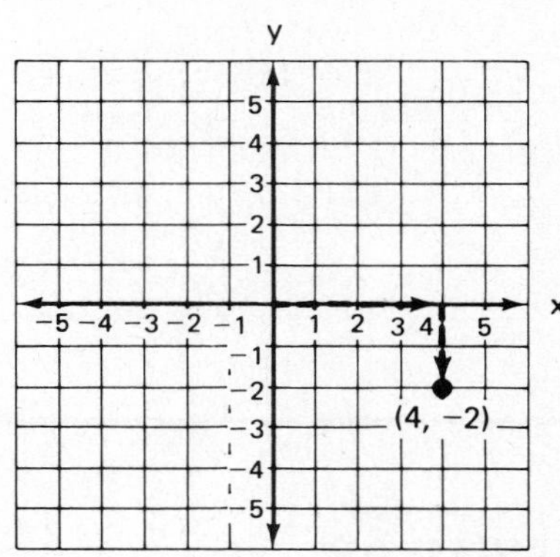

Figure 2.

Example 1: Plot the point (−3, 4).

Solution: Start at the origin. The x-coordinate is a negative number (−3), so we move 3 units to the left. From there, the y-coordinate (4) is positive, so we travel 4 units upward to find the location of the point (−3, 4).

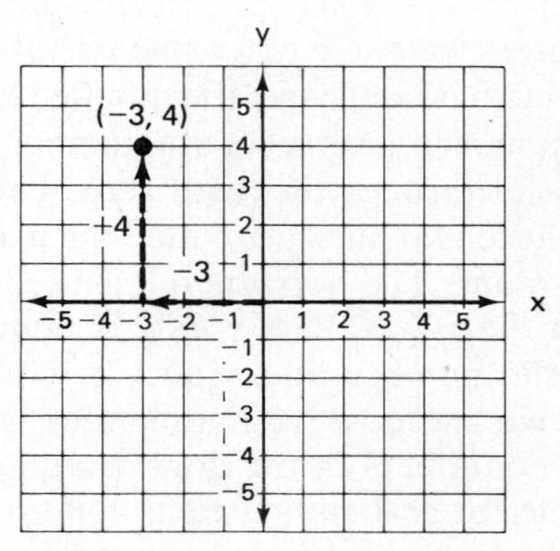

Figure 3.

The following examples are illustrated in Figure 4.

Example 2:

(a) Locate the point A (2,3) on a rectangular coordinate system.
(b) Plot the point B (0,−3) in the plane.
(c) Graph the point C (−3,−2) in the plane.

Solution:

(a) In the ordered pair (2,3) the x-coordinate (2) indicates that we move 2 units to the right of zero. From there the y-coordinate (3) directs us to go 3 units upward.
(b) In the ordered pair (0,−3) the abscissa (0) tells us that we do not move right or left at all. The ordinate (−3) tells us to move downward 3 units.
(c) To graph the ordered pair (−3,−2), we start at the origin, move 3 units to the left (−3), and then downward 2 units (−2).

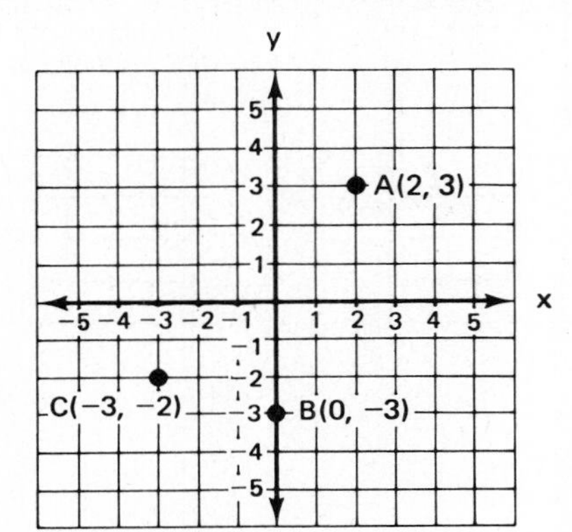

Figure 4.

EXERCISE 9.1

Draw a set of axes, label them properly, and locate the following points.

1. a. (1,4)	2. a. (−3,4)
b. (3,5)	b. (−2,−5)
c. (4,−2)	c. (1,−4)
d. (1,−1)	d. (5,5)
e. (−3,−3)	e. (3,−3)

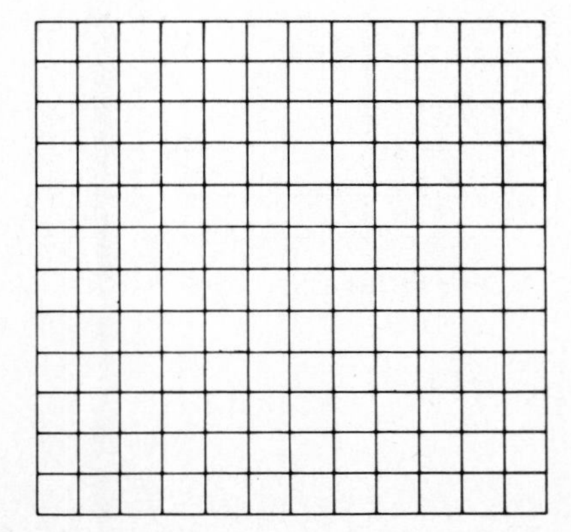

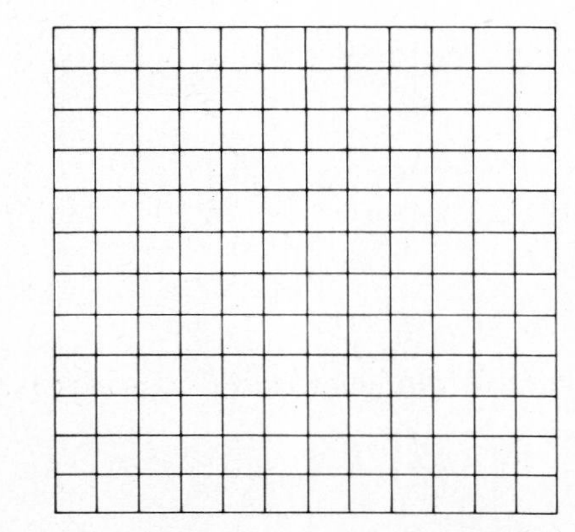

3. a. $(-4,0)$
   b. $(0,0)$
   c. $(0,5)$
   d. $(-5,-5)$
   e. $(2\frac{1}{2},3)$ (estimate)

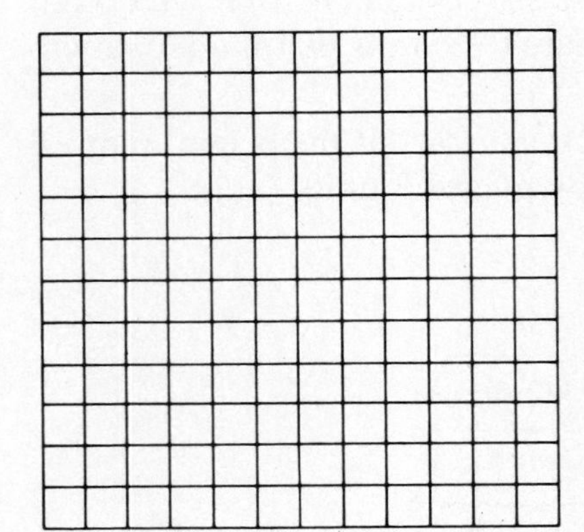

4. a. $(\frac{1}{2}, \frac{1}{2})$ (estimate)
   b. $(4.5,3)$ (estimate)
   c. $(-1\frac{1}{2}, -2\frac{1}{2})$ (estimate)
   d. $(-1.5,2.5)$ (estimate)
   e. $(3.4,-1.8)$ (estimate)

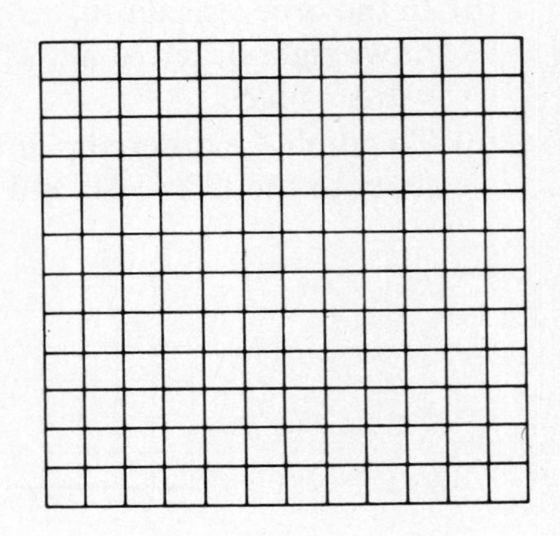

In exercises 21 through 28 give the ordered pair which represents each of the points in Figure 5.

5. A (,)
   B (,)
   C (,)
   D (,)
   E (,)
   F (,)
   G (,)
   H (,)

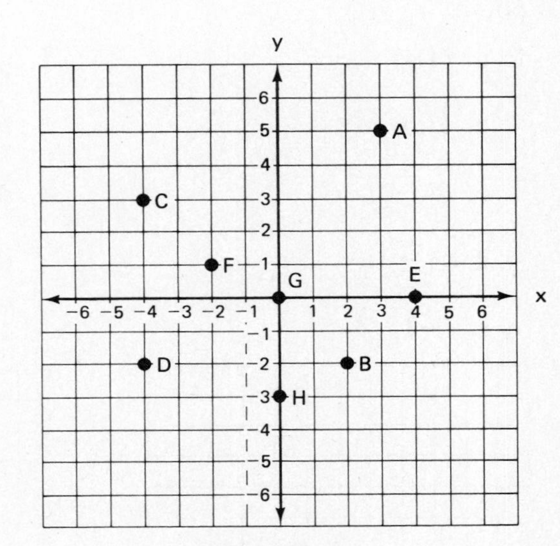

Figure 5.

Answers to odd-numbered exercises on page 480.

9.2 GRAPHING STRAIGHT LINES

Now that we know how to graph points in the plane, the next step is to graph straight lines in the plane.

The graph of a linear equation with two variables, x and y, is the set of points representing the set of ordered pairs which are the solutions to the equation.

Consider the equation $y = 2x - 1$. The points we are to graph are represented by the values for x and y which make the equation a true statement, that is, the solutions. We find these values by arbitrarily choosing a value for x and calculating a corresponding value for y. For example, if $x = 1$ we have

$$\begin{aligned} y &= 2x - 1 \\ &= 2 \cdot \boxed{1} - 1 \quad \text{Substitute 1 for } x. \\ &= 2 - 1 \\ &= 1 \end{aligned}$$

So the ordered pair $(x,y) = (1,1)$ in a solution to the given equation. If $x = 0$, we have

$$\begin{aligned} y &= 2x - 1 \\ &= 2 \cdot \boxed{0} - 1 \quad \text{Substitute 0 for } x. \\ &= 0 - 1 \\ &= -1 \end{aligned}$$

So $(0,-1)$ is also a solution. Usually, we tabulate these values in a table such as the following:

Choose x	*Calculate y*	x	y	*Ordered Pairs*
$x = 0$	$y = 2 \cdot \boxed{0} - 1 = 0 - 1 = -1$	0	−1	(0,−1)
$x = 1$	$y = 2 \cdot \boxed{1} - 1 = 2 - 1 = 1$	1	1	(1,1)
$x = 2$	$y = 2 \cdot \boxed{2} - 1 = 4 - 1 = 3$	2	3	(2,3)
$x = 3$	$y = 2 \cdot \boxed{3} - 1 = 6 - 1 = 5$	3	5	(3,5)
$x = -1$	$y = 2 \cdot \boxed{-1} - 1 = -2 - 1 = -3$	−1	−3	(−1,−3)

Remember we picked any value for x that we wanted and calculated the value of y that corresponded to it. If we graph the five ordered pairs that we found, we notice an interesting fact.

When the points are connected, they all lie on a straight line. This is true of any **linear equation,** which is an equation of degree 1. Although only two points are absolutely necessary to determine a straight line, we always graph three points as a check.

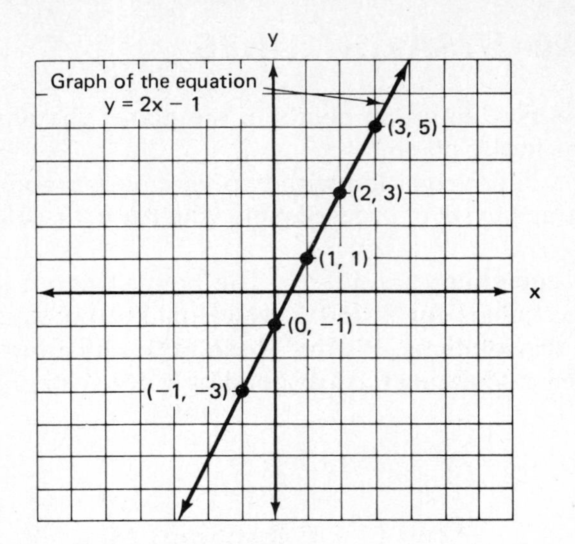

Figure 6.

Example 1: Graph $x + y = 4$

Solution: It is easier in many cases to solve for y before making our substitution for x. Since we are free to choose any values we like for x, it is best to choose values that give easy calculations and will not run off the graph paper.

Choose x	*Calculate y*	x	y	*Ordered pairs*
0	$y = 4 - x = 4 - \boxed{0} = 4$	0	4	(0,4)
2	$y = 4 - x = 4 - \boxed{2} = 2$	2	2	(2,2)
4	$y = 4 - x = 4 - \boxed{4} = 0$	4	0	(4,0)

Now plot the points and connect them.

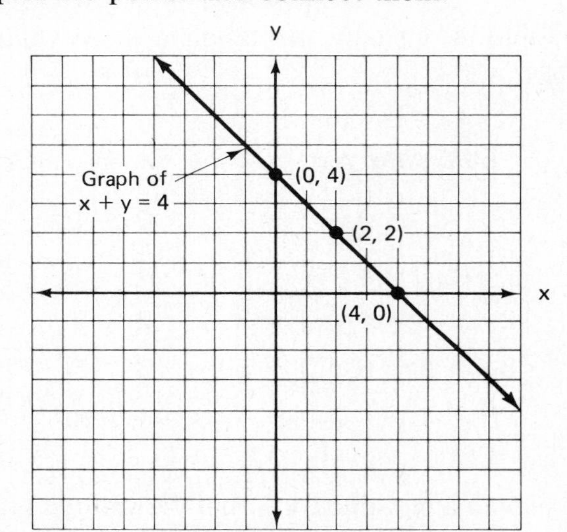

Figure 7.

Example 2: Graph $y = 4$

Solution: As you can see immediately, the x-term is missing. Think of this as having the coefficient of the x-term equal to zero.

$$y = 4 \text{ is the same as } y = 0 \cdot x + 4$$

Now make the table as before.

Choose x	*Calculate y*	x	y	*Ordered pairs*
0	$y = 0 \cdot x + 4 = 0 \cdot \boxed{0} + 4 = 4$	0	4	(0,4)
2	$y = 0 \cdot x + 4 = 0 \cdot \boxed{2} + 4 = 4$	2	4	(2,4)
−2	$y = 0 \cdot x + 4 = 0 \cdot \boxed{-2} + 4 = 4$	−2	4	(−2,4)

As you can see, no matter what value we choose for x, the value of y is always equal to 4. Graphing the ordered pairs yields a horizontal line on which the value of y is 4, no matter what the value of x.

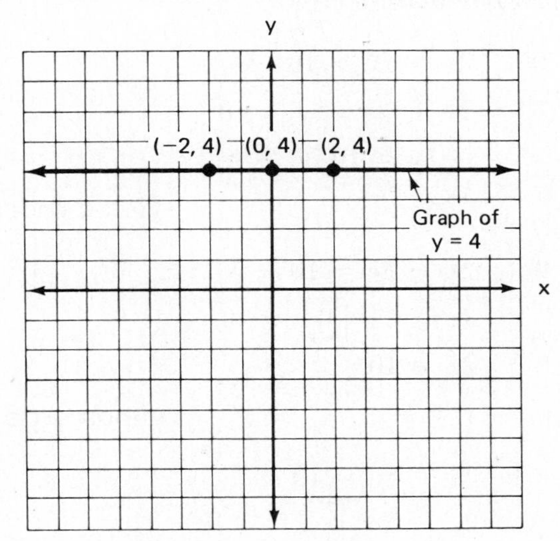

Figure 8.

The graph of an equation of the form $y = b$ (b = some constant) will be a horizontal line through all points where $y = b$.

Example 3: Graph $x = -2$

Solution: In this case the y-term is considered to have a zero coefficient, and we can say that

$$x = -2 \text{ is the same as } x + 0 \cdot y = -2$$

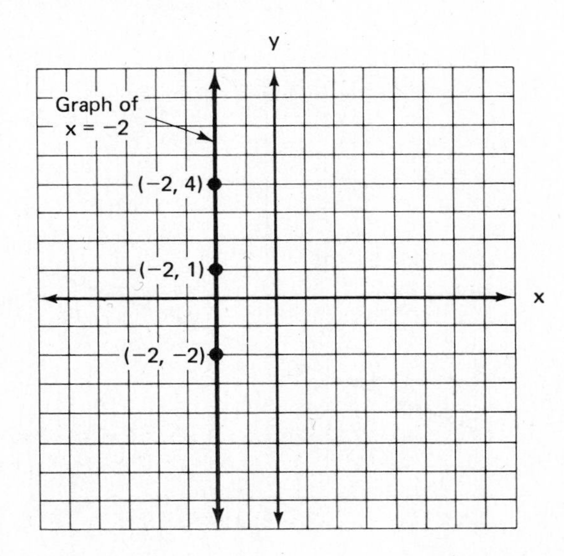

Figure 9.

x always equals -2 regardless of the value of y. Some typical ordered pairs would be $(-2,-2)$, $(-2,4)$, and $(-2,1)$. Graphing them gives a vertical line, and $x = -2$ at every point on this line.

The graph of an equation of the form $x = c$ (c = some constant) will be a vertical line through all the points where $x = c$.

Example 4: Graph the equation $2x - 5y = 10$

Solution: Sometimes instead of choosing values for x and calculating y, it is easier to choose values for y and calculate corresponding values for x. Frequently, choosing $x = 0$ and $y = 0$ results in the simplest calculation.

If $x = 0$

$$2x - 5y = 10$$
$$2 \cdot \boxed{0} - 5y = 10$$
$$-5y = 10$$
$$y = -2$$

yields ordered pair $(0,-2)$.

If $y = 0$

$$2x - 5y = 10$$
$$2x - 5 \cdot \boxed{0} = 10$$
$$2x = 10$$
$$x = 5$$

yields ordered pair $(5,0)$.

If $y = -1$

$$2x - 5y = 10$$
$$2x - 5 \cdot \boxed{-1} = 10$$
$$2x + 5 = 10$$
$$2x = 5$$
$$x = \frac{5}{2} \text{ or } 2\frac{1}{2}$$

yields ordered pair $(2\frac{1}{2},-1)$.

Graphing these ordered pairs gives us the graph of the equation.

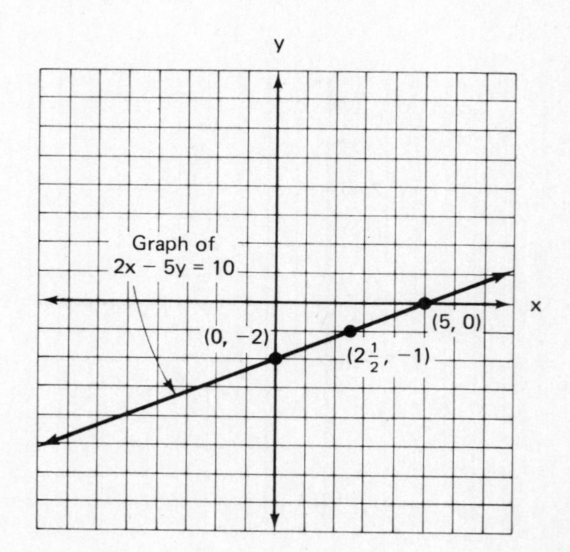

Figure 10.

To graph a linear equation:

1. Choose an arbitrary value for one of the variables and calculate the corresponding value for the remaining variable. Repeat the procedure three times yielding three ordered pairs.
2. Plot the three ordered pairs obtained in step 1.
3. Draw a straight line through the points.

Vertical and horizontal lines:

1. The graph of the equation $x = c$ (c = constant) is a vertical line through all points where $x = c$.
2. The graph of the equation $y = b$ (b = constant) is a horizontal line through all points where $y = b$.

EXERCISE 9.2

Graph each of the following equations:

1. $y = x - 3$

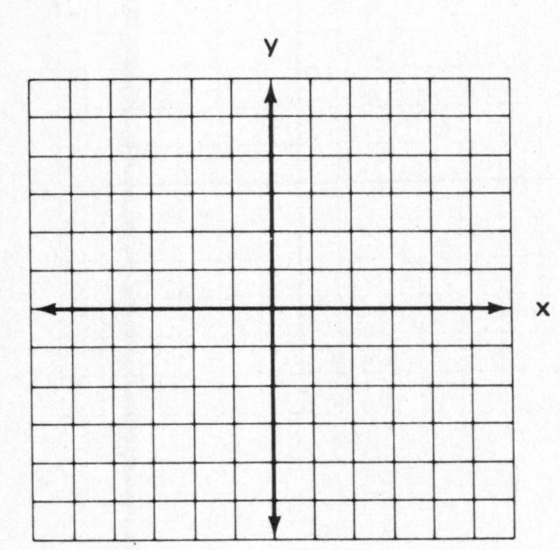

2. $x - y = 2$

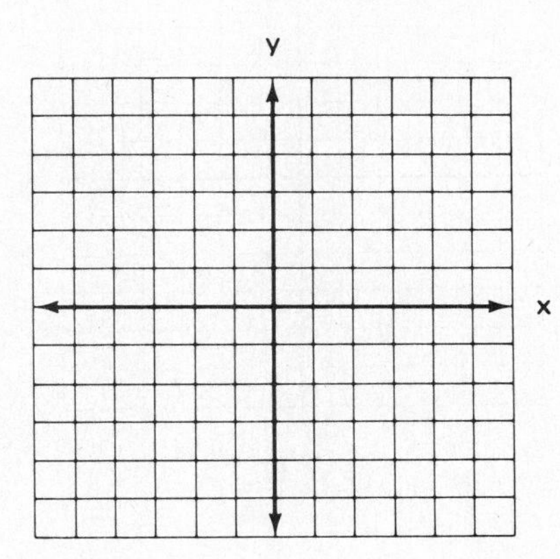

3. $2x + y = 3$

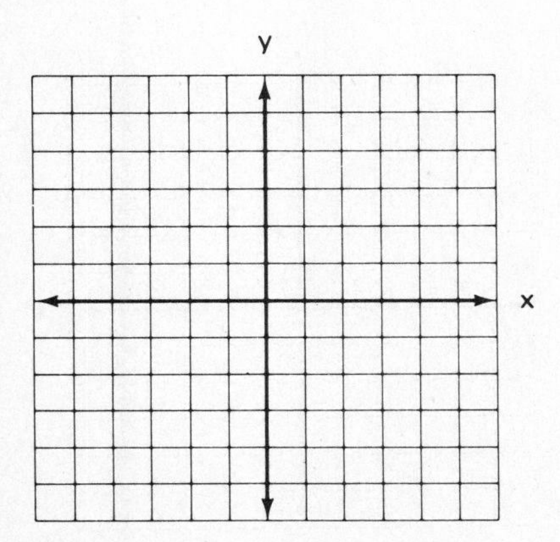

4. $y - x = 0$

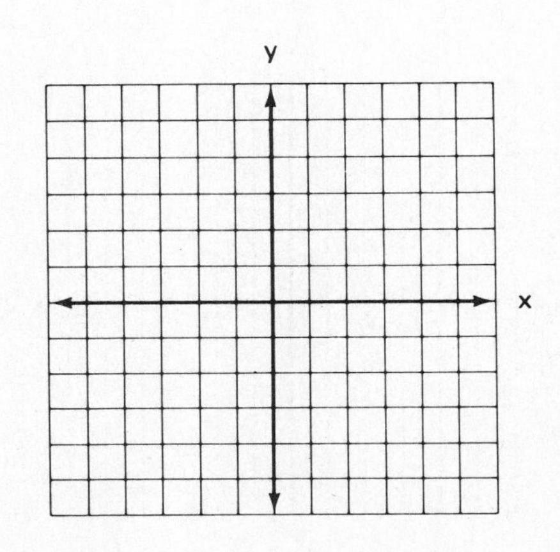

5. $y = 3x - 1$

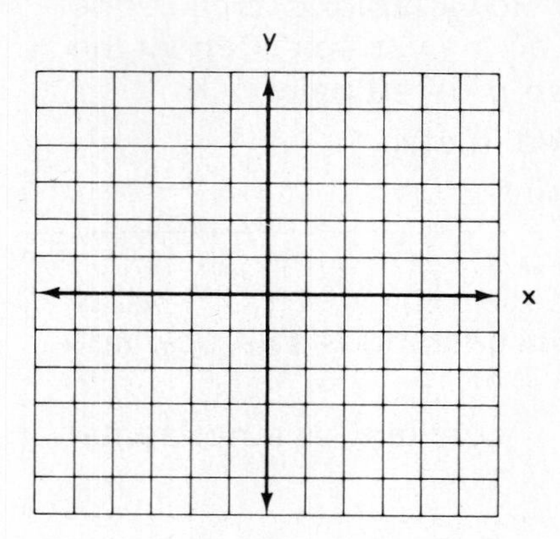

6. $2y - 3x = 6$

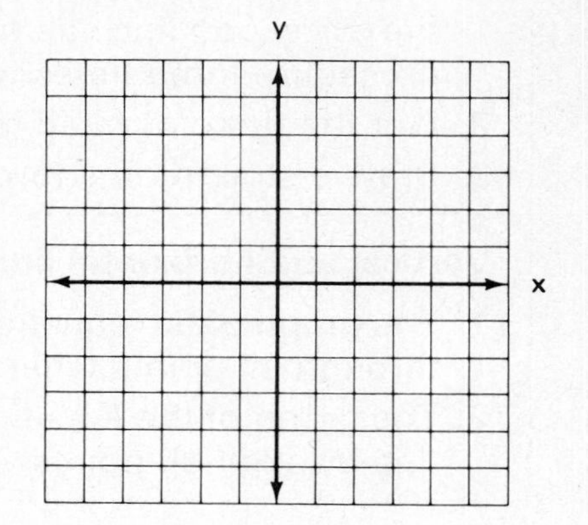

7. $y + 2x = 6$

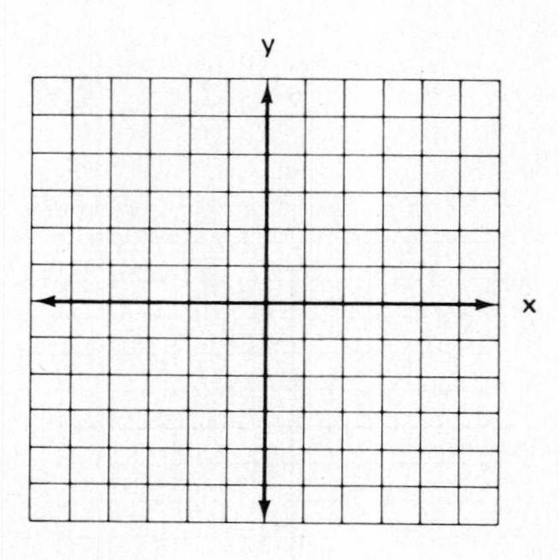

8. $y = -3$

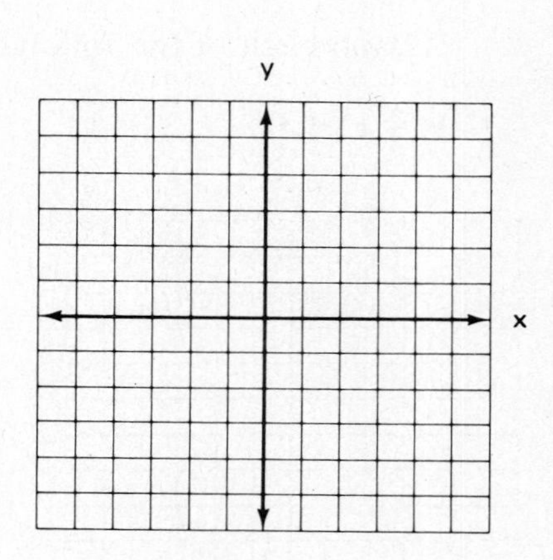

9. $5y = 2x - 5$

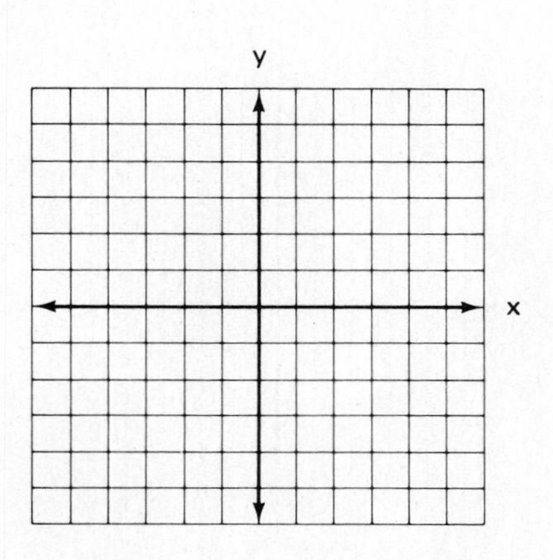

10. $x = 4$

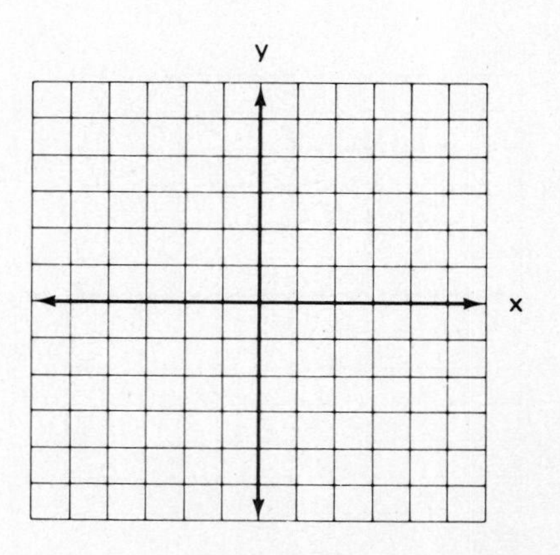

11. $3 - y = 2x$

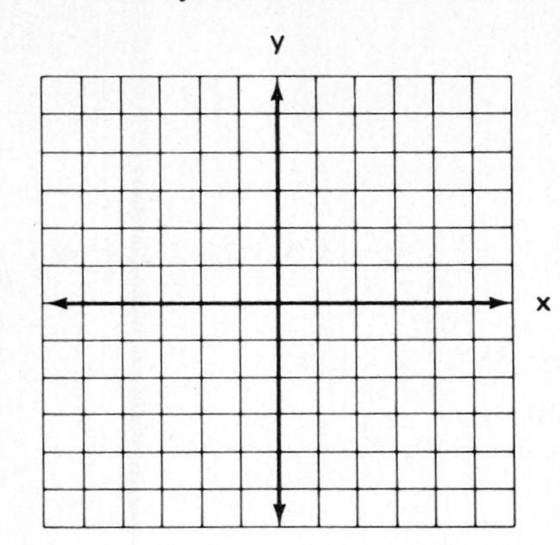

12. $y = 2$

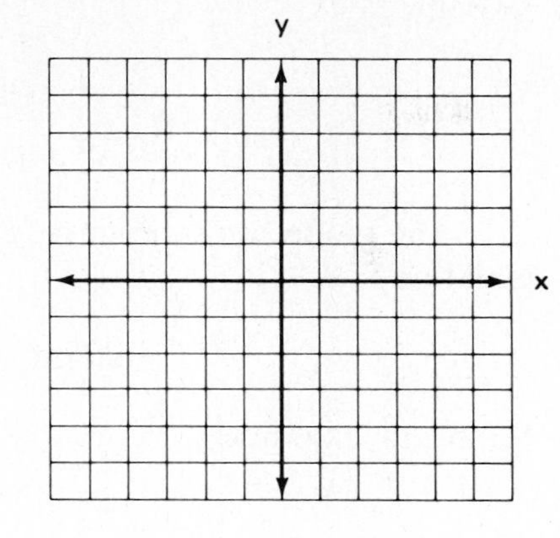

13. $x = -1$

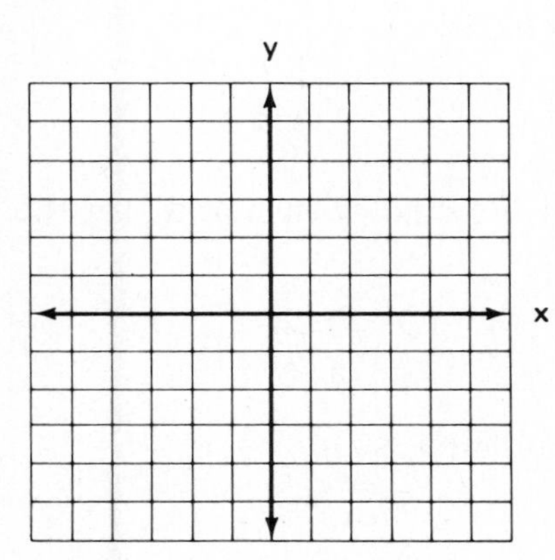

14. $y = \frac{1}{2}x + 1$

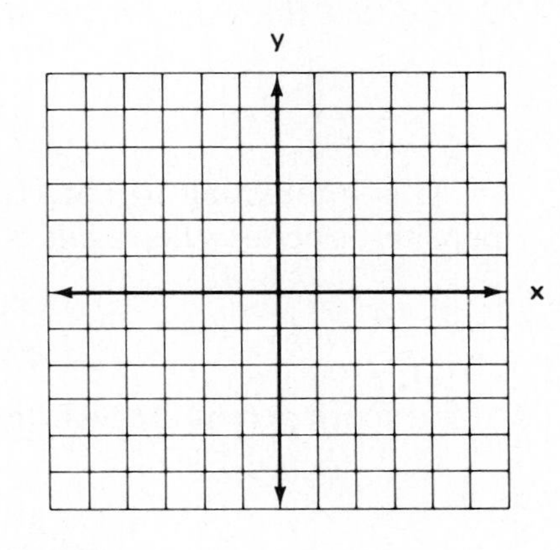

Answers to odd-numbered exercises on page 480.

9.3 SLOPE OF STRAIGHT LINES

Consider the straight line drawn through the points (1,1) and (5,4).

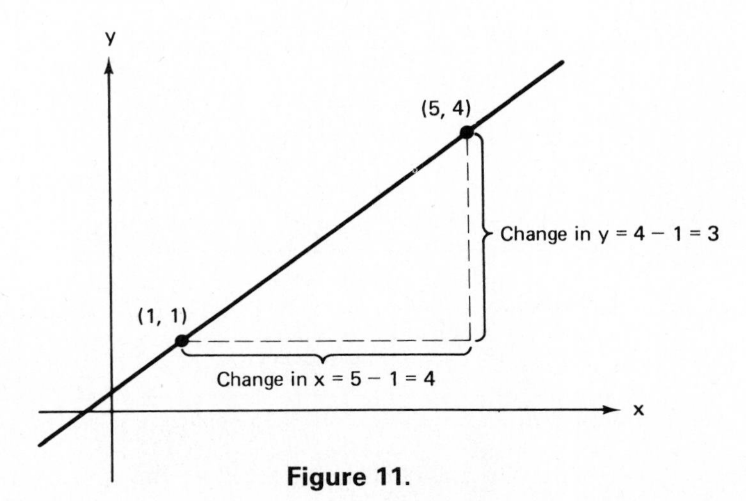

Figure 11.

We can calculate the change in both y and x as we go from point (1,1) to point (5,4).

$$\text{change in } y = 4-1 = 3$$
$$\text{change in } x = 5-1 = 4$$

The slope, or steepness, of a line (which we denote by the letter m) is defined as:

$$\text{slope} = m = \frac{\text{change in } y}{\text{change in } x}$$

So in our example,

$$\text{slope} = m = \frac{\text{change in } y}{\text{change in } x} = \frac{4-1}{5-1} = \frac{3}{4}$$

We say that the slope of the line through points (1,1) and (5,4) is

$$m = \frac{3}{4}$$

It is interesting to note that if we find the change in y and x in the opposite direction the result is the same:

$$\text{slope} = \frac{\text{change in } y}{\text{change in } x} = \frac{1-4}{1-5} = \frac{-3}{-4} = \frac{3}{4}$$

In more general terms, the definition of slope is:

> If (x_1, y_1) and (x_2, y_2) are any two points on a line, then the slope of the line is given by:
>
> $$\text{slope} = m = \frac{\text{change in } y}{\text{change in } x} = \frac{y_2 - y_1}{x_2 - x_1}$$

Example 1: Find the slope of a line drawn through points (4,2) and (−3,−3).

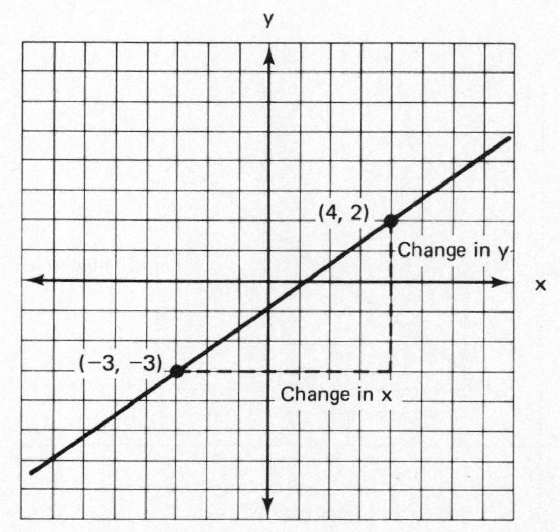

Figure 12.

Solution:

$$\text{slope} = m = \frac{\text{change in } y}{\text{change in } x} = \frac{y_2 - y_1}{x_2 - x_1} = \frac{2-(-3)}{4-(-3)} = \frac{2+3}{4+3} = \frac{5}{7}$$

Remember, we could have found the changes in x and y in the reverse direction:

$$m = \frac{y_1 - y_2}{x_1 - x_2} = \frac{-3\ -2}{-3\ -4} = \frac{-5}{-7} = \frac{5}{7}$$

which is the same result as before.

Example 2: Find the slope of the line containing points (−3,2) and (3,−1).

Solution:

$$m = \frac{y_2 - y_1}{x_2 - x_1}$$

$$= \frac{-1\ -2}{3-(-3)} = \frac{-3}{6} = -\frac{1}{2}$$

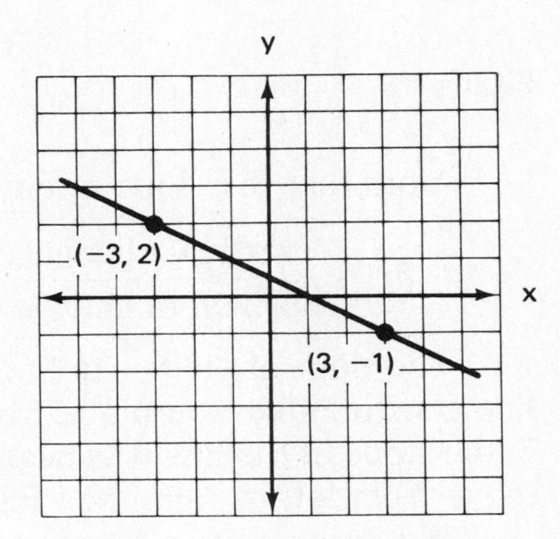

Figure 13.

Example 3: Find the slope of the line containing points (−2,2) and (3,2).

Solution:

$$m = \frac{y_2 - y_1}{x_2 - x_1} = \frac{2-2}{3-(-2)} = \frac{0}{5} = 0$$

The slope is 0.

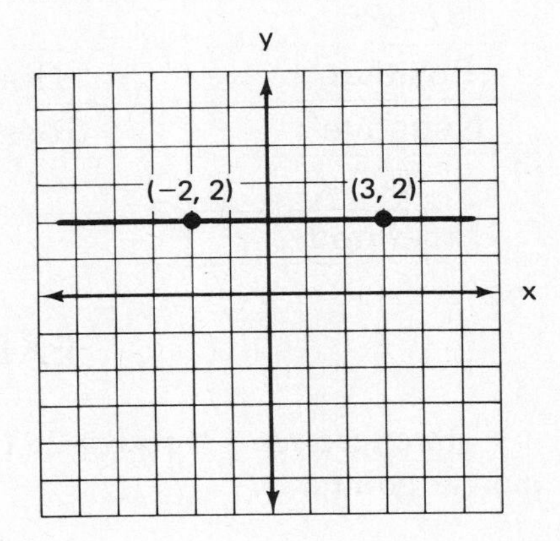

Figure 14.

Note that the slope of *any* horizontal line is 0:

$$m = \frac{\text{vertical change}}{\text{horizontal change}} = \frac{0}{\text{horizontal change}} = 0$$

Example 4: Find the slope of the line through the points (−2,3) and (−2,−1).

Solution:

$$m = \frac{y_2 - y_1}{x_2 - x_1} = \frac{-1 - 3}{-2 - (-2)} = \frac{-4}{0}$$

The slope is *undefined* since we have 0 in the denominator.

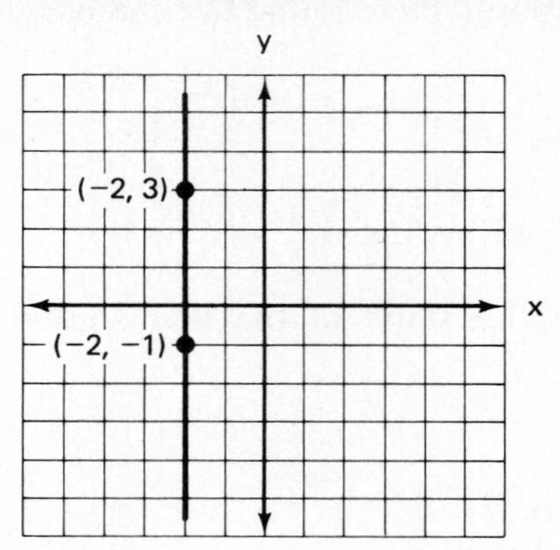

Figure 15.

Note that the slope of any vertical line is undefined since

$$m = \frac{\text{vertical change}}{\text{horizontal change}} = \frac{\text{vertical change}}{0} \quad \text{which is undefined.}$$

The slope of the line in example 1 was a positive number ($+\frac{5}{7}$), and the line goes uphill as we move from left to right (see Figure 12). In example 2, the slope of the line was negative ($-\frac{1}{2}$), and the line goes downhill as we move from left to right (see Figure 13).

We know quite a bit about the nature of a line just by looking at the number that represents its slope. This is summarized as follows:

SLOPE	*LINE*
Positive	Goes uphill as it goes from left to right
Negative	Goes downhill as it goes from left to right
0	Horizontal line
Undefined	Vertical line

EXERCISE 9.3

In exercises 1 through 20 find the slope of the line through the given pair of points.

1. (2,3), (6,5)
2. (0,0), (2,3)
3. (3,4), (1,5)
4. (1,2), (−1,3)
5. (4,6), (1,3)
6. (−2,1), (−4,−1)

7. (−3,−2), (−5,−6) 8. (2,−1), (2,5) 9. (−3,4), (−3,7)

10. (−5,2), (−5,−6) 11. (3,−2), (7,−2) 12. (4,4), (−3,4)

13. (7,−1), (−5,−1) 14. (4,4), (−2,−2) 15. (−3,−5), (−6,−10)

16. (−2,4), (0,6) 17. (−7,0), (4,0) 18. (0,−6), (0,5)

19. $(4,\frac{1}{2})$, $(3,2\frac{1}{2})$ 20. $(2\frac{1}{2},3)$, $(6,-2\frac{1}{2})$

In exercises 21 and 22, find the slope of each equation by first finding two points on the line.

21. $x + y = 6$ 22. $3x - 2y = 6$

Answers to odd-numbered exercises on page 481.

9.4 EQUATION OF STRAIGHT LINES

The Point-Slope Form of the Equation of a Straight Line

Consider a line having slope m and containing a known point with coordinates (x_1,y_1). If we let (x,y) be *any* other point on the line we can find the slope as follows:

$$\text{slope} = \frac{y - y_1}{x - x_1} = m$$

$$\frac{\boxed{\cancel{(x - x_1)}}\,(y - y_1)}{\cancel{x - x_1}} = m\,\boxed{(x - x_1)}$$

$$y - y_1 = m\,(x - x_1)$$

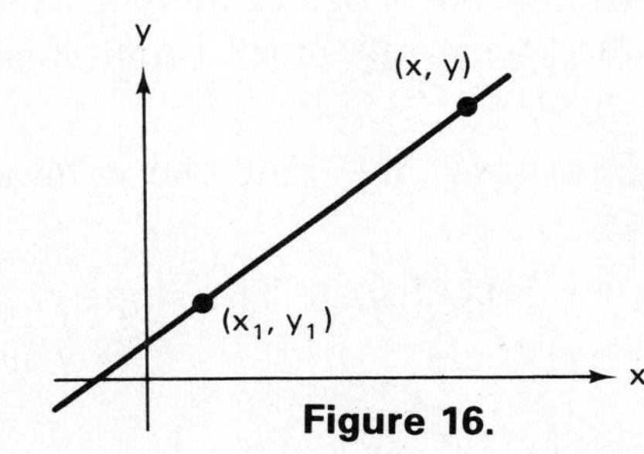

Figure 16.

Point-slope form of the equation of a line:

$$y - y_1 = m(x - x_1)$$

where m = slope and (x_1, y_1) is a known point on the line.

Equations are frequently written in a form called the **standard form** of an equation of a straight line:

Standard form of an equation of a line:

$$Ax + By + C = 0$$

where A, B, and C are integers and A and B are not both equal to zero. A should be positive if possible.

Example 1: Find the equation of the line passing through points (2,−1) having a slope of 3. Write the equation in standard form.

Solution:

$y - y_1 = m(x - x_1)$	Point-slope form.
$y - (\boxed{-1}) = 3\ (x - \boxed{2})$	Substitute $x_1 = 2$, $y_1 = -1$, $m = 3$.
$y + 1 = 3x - 6$	
$3x - y - 7 = 0$	Standard form of the equation.

Example 2: Find the equation of the line in standard form which has a slope of −5 and passes through the point (0,−4).

Solution:

$y - y_1 = m(x - x_1)$	Point-slope form
$y - (\boxed{-4}) = \boxed{-5}(x - \boxed{0})$	$x_1 = 0$, $y_1 = -4$, $m = -5$
$y + 4 = -5x$	
$5x + y + 4 = 0$	Standard form

To find the equation of a line when we are given *two* points on the line, we find the slope of the line as before and then use this slope and either of the two points in the point-slope formula.

Example 3: Find the equation of the line through points (1,−2) and (3,4).

Solution: The slope is

$$m = \frac{y_2 - y_1}{x_2 - x_1} = \frac{4 - (-2)}{3 - 1} = \frac{6}{2} = 3$$

Now use the point-slope formula and *either* of the given points to find the equation. Use the point (1,−2).

$y - y_1 = m(x - x_1)$ Point-slope form

$y - (\boxed{-2}) = 3(x - \boxed{1})$ $x = 1, y = -2, m = 3$

$y + 2 = 3x - 3$

$3x - y - 5 = 0$ Standard form

We could have also used the other point (3,4).

$y - y_1 = m(x - x_1)$ Point-slope form

$y - \boxed{4} = \boxed{3}(x - \boxed{3})$ $x = 3, y = 4, m = 3$

$y - 4 = 3x - 9$

$3x - y - 5 = 0$ This is the same result.

Example 4: Find the equation of a horizontal line that passes through the point (2,4).

Solution: A horizontal line has a slope equal to 0.

$y - y_1 = m(x - x_1)$ Point-slope form

$y - \boxed{4} = \boxed{0}(x - \boxed{2})$ $x_1 = 2, y_1 = 4, m = 0$

$y - 4 = 0$

$y = 4$

Example 5: Find the equation of a vertical line that passes through the point (1,−5).

Solution: Since all vertical lines have an undefined slope we cannot use the point-slope form. However, we can write the equation directly if we realize that all points on any vertical line have the same x-coordinate. In this case, since the line passes through the point (1,−5), x is always equal to 1 and the equation is $x = 1$.

Slope-Intercept Form of the Equation of a Straight Line.

The point where a line crosses the y-axis is called the **y-intercept.** Since every point on the y-axis has an x-coordinate equal to zero, the y-intercept will be of the form $(0,b)$. Now let's find the equation of a line with y-intercept $(0,b)$ and slope m.

$y - y_1 = m(x - x_1)$

$y - \boxed{b} = m(x - \boxed{0})$ $x_1 = 0, y_1 = b$

$y - b = mx$

$y = \boxed{m}x + \boxed{b}$

$\boxed{m}$ = slope

$\boxed{b}$ = y-intercept

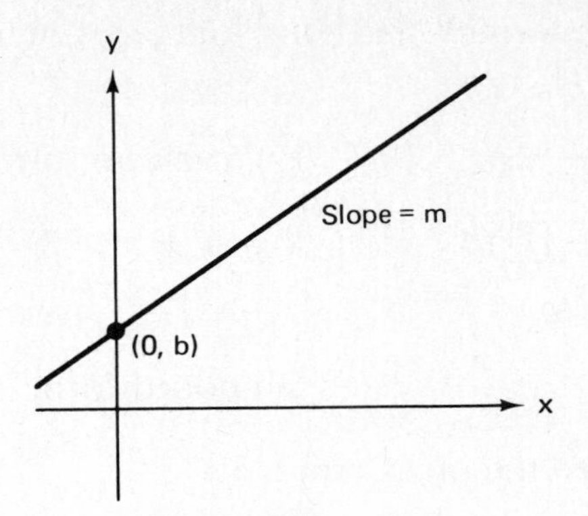

Figure 17.

> **The slope-intercept form of the equation of a line is:**
>
> $$y = mx + b$$
>
> where m = slope and b = y-intercept.

Example 6: Write the equation, in slope-intercept form, of a line with a slope of $\frac{1}{2}$ that crosses the y-axis at the point (0,3).

Solution: Since $m = \frac{1}{2}$ and $b = 3$ we can write the equation immediately.

$y = mx + b$ Slope-intercept form

$y = \boxed{\frac{1}{2}} x + \boxed{3}$ Substitute $m = \frac{1}{2}$, $b = 3$.

Example 7: Find the slope and y-intercept of a line with the following equation:

$$3x - 2y + 8 = 0$$

Solution: We must put the given equation in the slope-intercept form $y = mx + b$.

$3x - 2y + 8 = 0$

$-2y = -3x - 8$ Add $-3x$ and -8 to each side.

$y = \frac{-3x}{-2} - \frac{8}{-2}$ Divide each side by -2.

$y = \boxed{\frac{3}{2}} x + \boxed{4}$ Slope-intercept form $y = mx + b$.

m ↑ ↑ b

$m = \frac{3}{2}$ and the y-intercept is (0,4)

EXERCISE 9.4

Write the equation of the line with the given slope passing through the given point. Write the equation in standard form.

1. $(1,3)$, $m = 2$
2. $(-1,4)$, $m = 4$
3. $(-3,-2)$, $m = -3$
4. $(4,-5)$, $m = -5$
5. $(-2,-7)$, $m = 1$
6. $(-6,3)$, $m = \frac{2}{3}$
7. $(-1,-4)$, $m = \frac{3}{4}$
8. $(2,0)$, $m = 0$
9. $(0,4)$, $m = 0$
10. $(2,-5)$, $m = -\frac{3}{2}$

Write the equation, in standard form, of the line containing the two points.

11. $(0,0)$, $(3,4)$
12. $(1,3)$, $(5,6)$
13. $(2,3)$, $(1,4)$
14. $(5,-2)$, $(2,-2)$
15. $(0,3)$, $(6,0)$
16. $(2,0)$, $(2,-6)$
17. $(4,-2)$, $(-1,-8)$
18. $(3,0)$, $(4,0)$
19. $(-5,2)$, $(-5,5)$
20. $(-5,3)$, $(-8,-6)$

Find the slope and y-intercept of the line having the given equation.

21. $4x - 3y + 6 = 0$

22. $3x - 5y - 12 = 0$

23. $x - 3y + 6 = 0$

24. $3x - 5 = -2y$

25. $5x - 3y = 9$

26. $17 = 3x - 4y$

27. $\frac{3}{5}x - \frac{1}{2}y + \frac{7}{2} = 0$

28. $\frac{1}{2}x + \frac{2}{3}y - 2 = 0$

29. $y = -4$

30. $x = 3$

31. Write the equation of a horizontal line containing the point $(-2,3)$.

32. Write the equation of a horizontal line containing the point $(2,1)$.

33. Write the equation of a vertical line containing the point $(-2,-3)$.

34. Write the equation of a vertical line containing the point $(0,4)$.

Answers to odd-numbered exercises on page 481.

9.5 GRAPHING CURVES

To graph a straight line required plotting only two points (we plotted a third as a check). Graphing curves uses the same procedure except that more than two points need to be plotted.

To graph a curve:

1. Make a table of ordered pairs by choosing values for one variable and calculating a corresponding value for the remaining variable.
2. Plot a sufficient number of points to determine the graph.
3. Connect the points with a smooth curve.

Example 1: Graph $y = x^2 - 2$

Solution:

Choose x	Calculate y	x	y	Ordered pairs
0	$y = 0^2 - 2 = 0 - 2 = -2$	0	−2	(0,−2)
1	$y = 1^2 - 2 = 1 - 2 = -1$	1	−1	(1,−1)
−1	$y = (-1)^2 - 2 = 1 - 2 = -1$	−1	−1	(−1,−1)
2	$y = 2^2 - 2 = 4 - 2 = 2$	2	2	(2,2)
−2	$y = (-2)^2 - 2 = 4 - 2 = 2$	−2	2	(−2,2)
3	$y = 3^2 - 2 = 9 - 2 = 7$	3	7	(3,7)
−3	$y = (-3)^2 - 2 = 9 - 2 = 7$	−3	7	(−3,7)

Plot the ordered pairs (points) and connect the points using a smooth curve (see Figure 18).

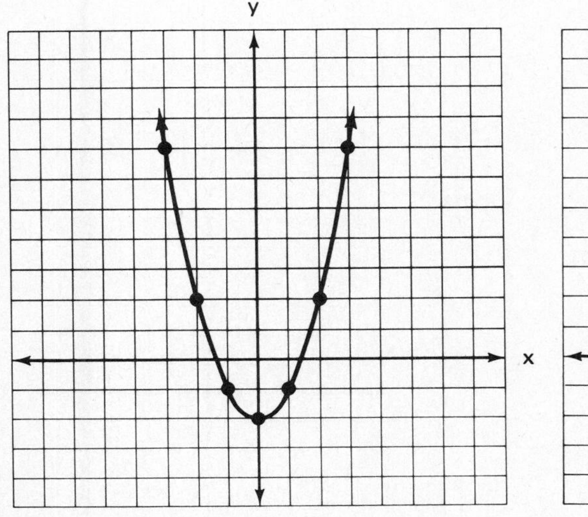

Figure 18.

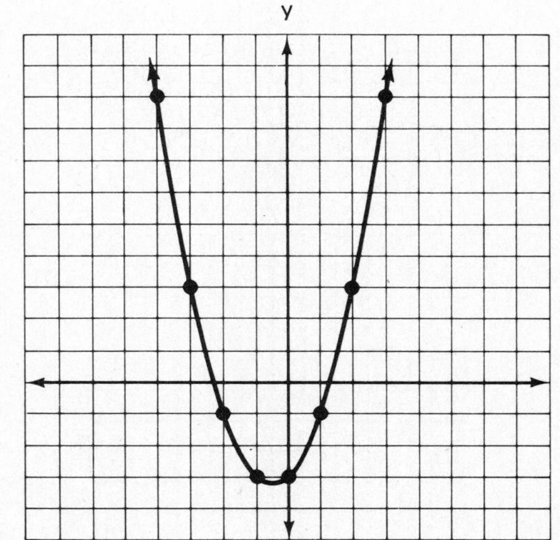

Figure 19.

Example 2: Graph $y = x^2 + x - 3$

Solution:

Choose x	*Calculate y*	x	y	*Ordered pairs*
0	$y = 0^2 + 0 - 3 = -3$	0	−3	(0,−3)
1	$y = 1^2 + 1 - 3 = 1 + 1 - 3 = -1$	1	−1	(1,−1)
−1	$y = (-1)^2 - 1 - 3 = 1 - 1 - 3 = -3$	−1	−3	(−1,−3)
2	$y = 2^2 + 2 - 3 = 4 + 2 - 3 = 3$	2	3	(2,3)
−2	$y = (-2)^2 - 2 - 3 = 4 - 2 - 3 = -1$	−2	−1	(−2,−1)
3	$y = 3^2 + 3 - 3 = 9 + 3 - 3 = 9$	3	9	(3,9)
−3	$y = (-3)^2 - 3 - 3 = 9 - 3 - 3 = 3$	−3	3	(−3,3)
−4	$y = (-4)^2 - 4 - 3 = 16 - 4 - 3 = 9$	−4	9	(−4,9)

Now plot the points and connect them with a smooth curve (see Figure 19).

EXERCISE 9.5

Graph the following curves:

1. $y = x^2$

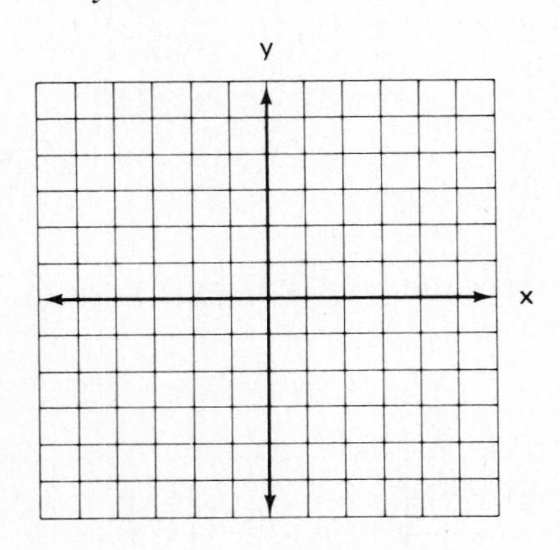

2. $y = x^2 - 4$

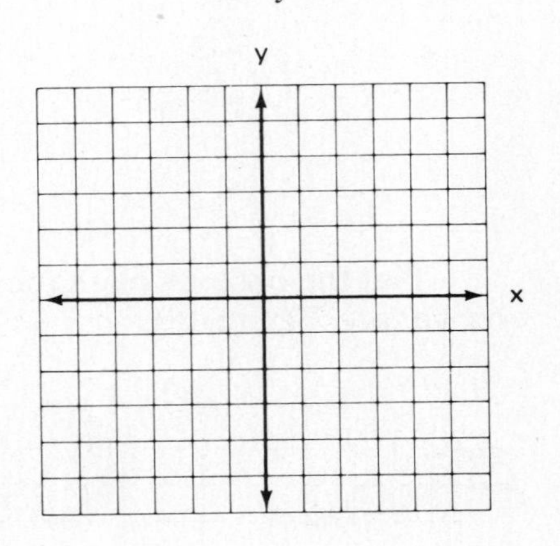

3. $y = x^2 + 3x + 1$

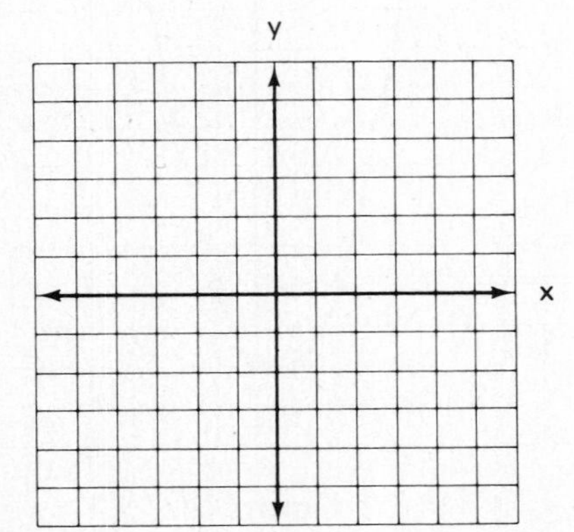

4. $y = -2 + x^2$

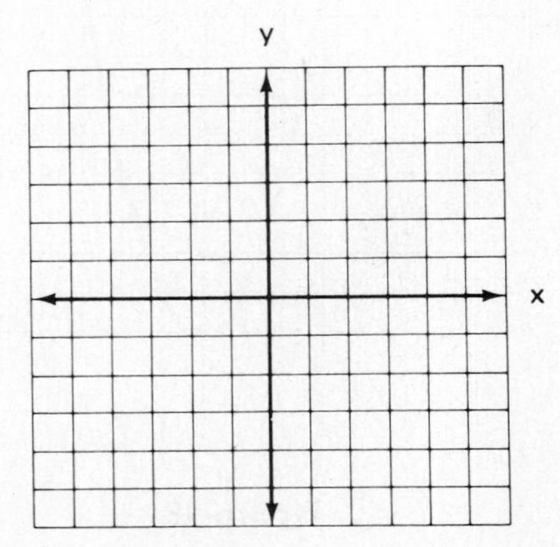

5. $y = x^2 + 2x$

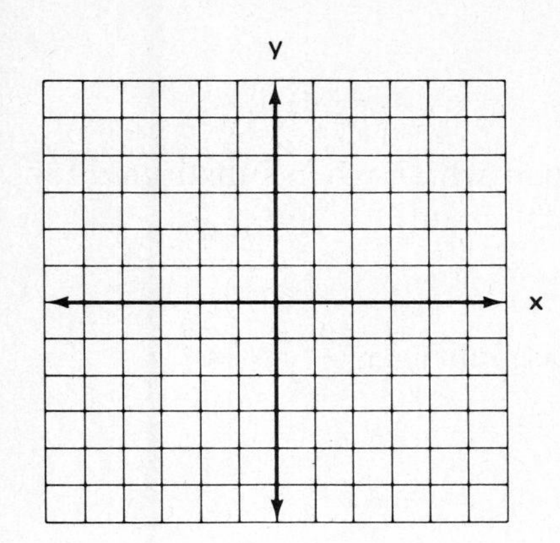

6. $y = -2x^2 + 3$

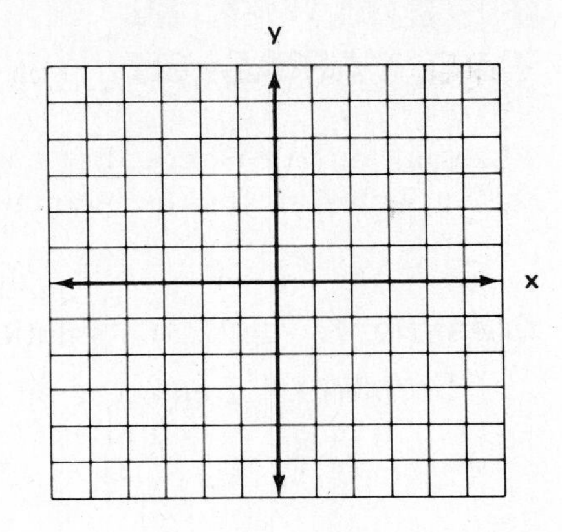

7. $y = x^2 + 2x - 3$

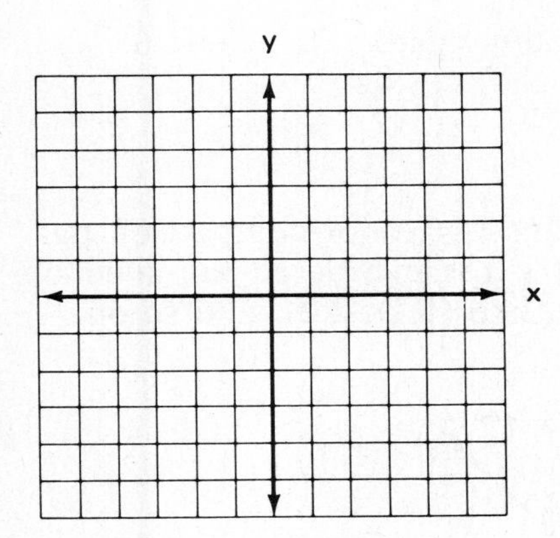

8. $y = x^3$

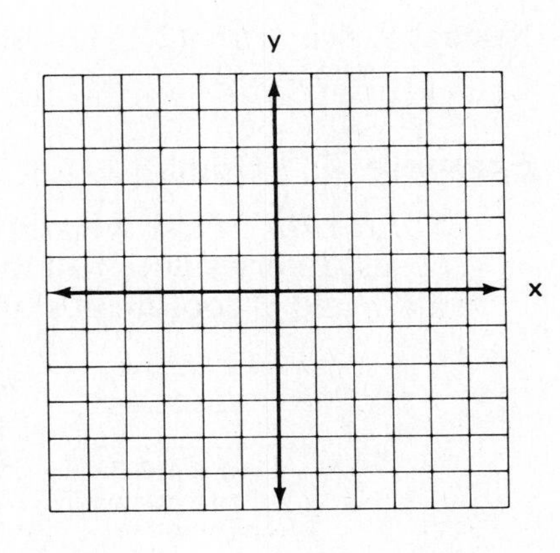

9. $y = -x^3$

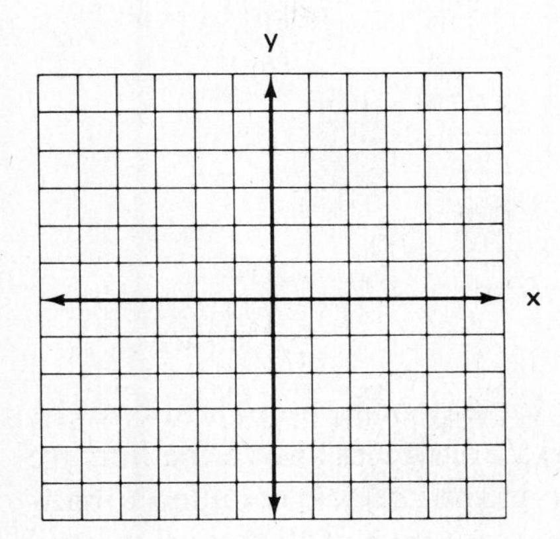

10. $y = x^3 - 2x$

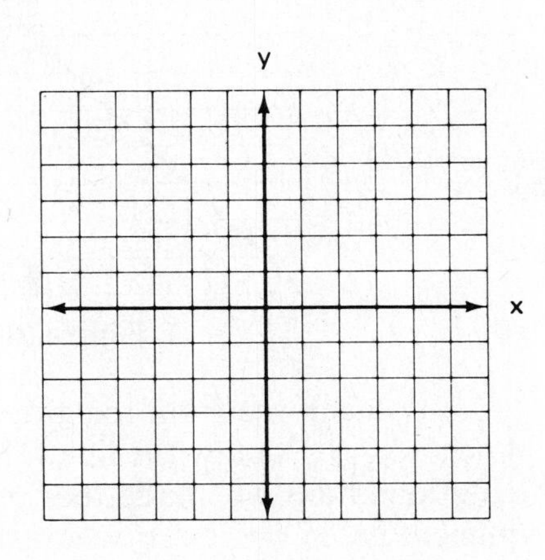

Answers to odd-numbered exercises on page 482.

9.6 GRAPHING LINEAR INEQUALITIES IN TWO VARIABLES

The solution to an inequality is an ordered pair which when substituted for the variables makes the inequality a true statement.

Example 1: Is (2,3) a solution to the inequality $3x - y < 6$?

Solution: Replace x by 2 and y by 3.

$$3x - y < 6$$

$$3 \cdot \boxed{2} - \boxed{3} < 6 \qquad x = 2 \text{ and } y = 3$$

$$6 - 3 < 6$$

$$3 < 6$$

Since $3 < 6$ is true, (2,3) is a solution to the inequality $3x - y < 6$.

Example 2: Graph $y > x$

Solution: First we will graph the line $y = x$ for comparison purposes. Every solution of the equation $y = x$ is an ordered pair in which the x and y coordinates are the same, like (1,1), 2,2), and so on.

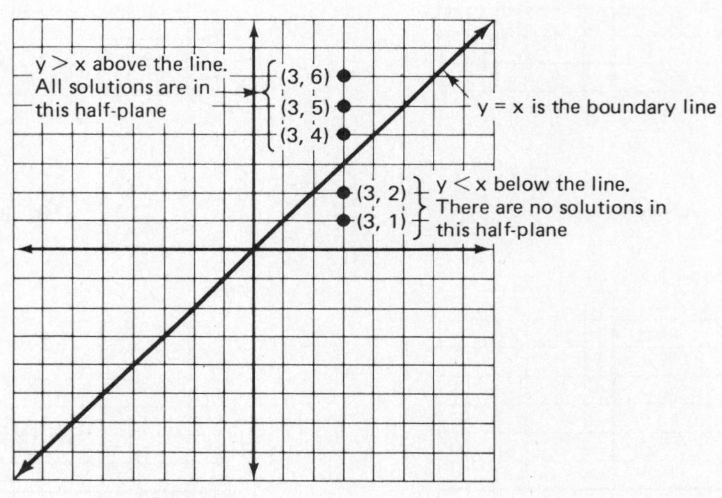

Figure 20.

You can see from the graph that $y > x$ for any ordered pair above the line and $y < x$ below the line. Therefore all ordered pairs above the line are solutions, and we shade that portion of the graph, which is called a **half-plane.** Since $y = x$ everywhere *on* the line, it is not part of the solution, and we draw it as a *broken line*. The graph of $y > x$ is shown in Figure 21.

Note that if we have an inequality containing the symbol $\geq$ or $\leq$, we use a *solid line* to indicate that the line is part of the solution.

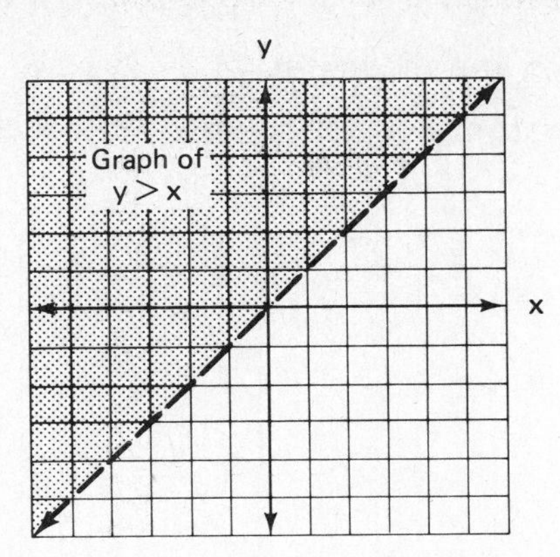

Figure 21.

To graph a linear inequality in two variables:

1. Graph the boundary line (replace the inequality symbol by =). This line is broken if the inequality symbol is > or <, and solid if the inequality symbol is ≥ or ≤.
2. Choose any point not on the boundary line and determine whether it satisfies the inequality.

   If the point satisfies the inequality, shade the side of the line that contains the point.

   If the point does *not* satisfy the inequality, shade the side of the line that does *not* contain the point.

Example 3: Graph $y \leq 2x - 1$

Solution: Replace the inequality symbol ≤ by = and graph the line $y = 2x - 1$. This is the boundary line, and it will be solid since the given inequality symbol is ≤.

x	y
0	−1
1	1
2	3

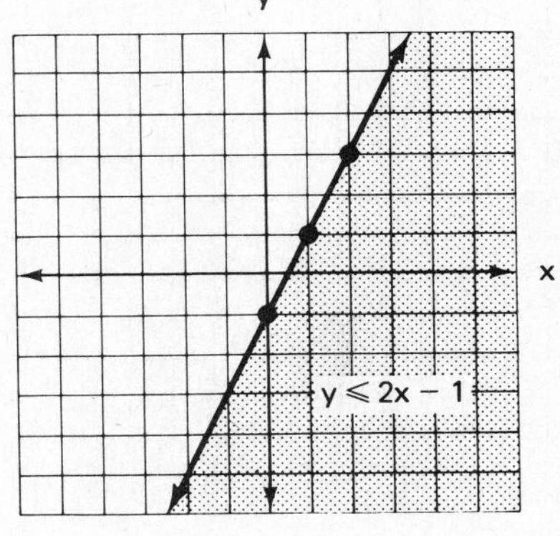

Figure 22.

Now choose any point not on the boundary line [(0,0) is frequently chosen] and substitute it in the original inequality.

$$y \leq 2x - 1$$

$$\boxed{0} \stackrel{?}{\leq} 2 \cdot \boxed{0} - 1 \qquad \text{Substitute } x = 0,\ y = 0.$$

$$0 \leq -1 \text{ is false}$$

Since the point (0,0) does not satisfy the inequality we shade the half-plane that does *not* contain the point (0,0).

Example 4: Graph the inequality $4x - 2y < 8$

Solution: First, graph the line $4x - 2y = 8$ using a broken line.

x	y
0	−4
2	0
3	2

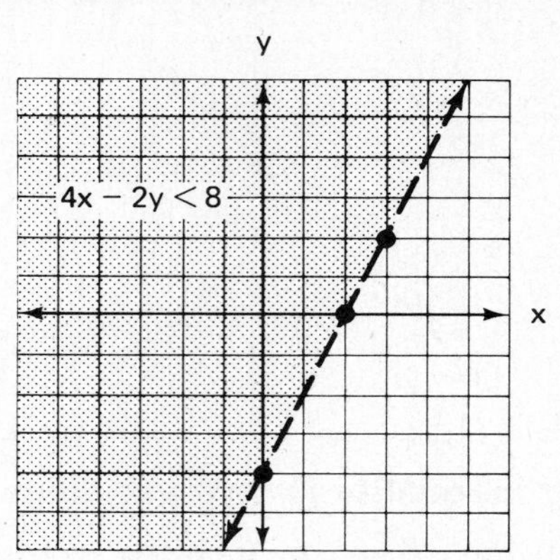

Figure 23.

Try the point (0,0) in the inequality:

$$4x - 2y < 8$$
$$4 \cdot \boxed{0} - 2 \cdot \boxed{0} \overset{?}{<} 8$$

$0 < 8$ is true, so we shade the half-plane containing the point (0,0).

EXERCISE 9.6

Graph the following inequalities:

1. $y > x + 1$

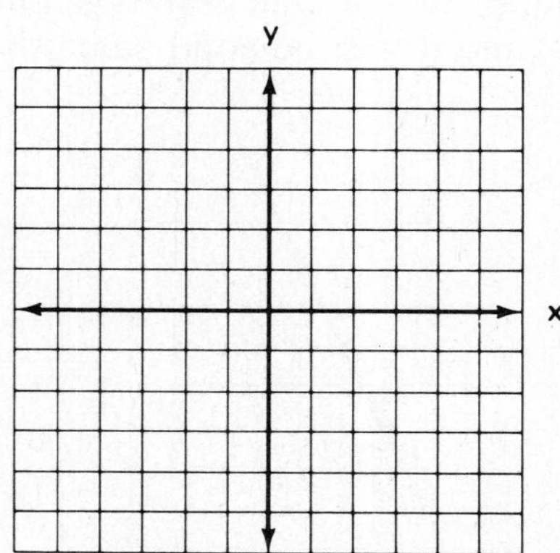

2. $y \leq x - 2$

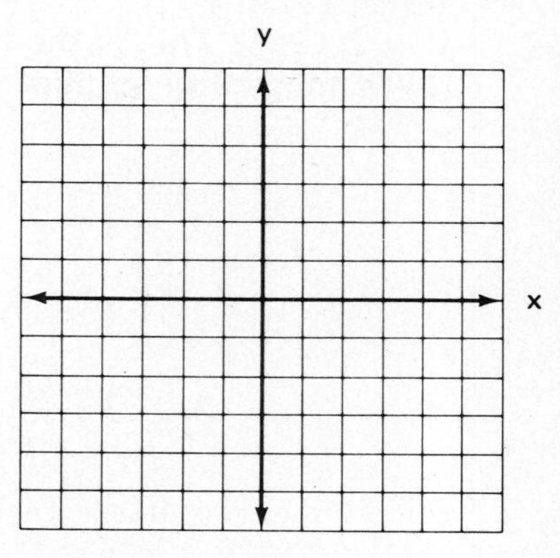

3. $y < 2x + 1$

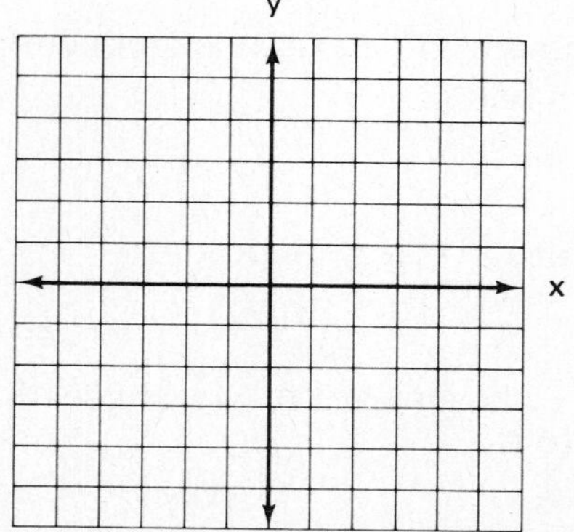

4. $y \geq -x - 2$

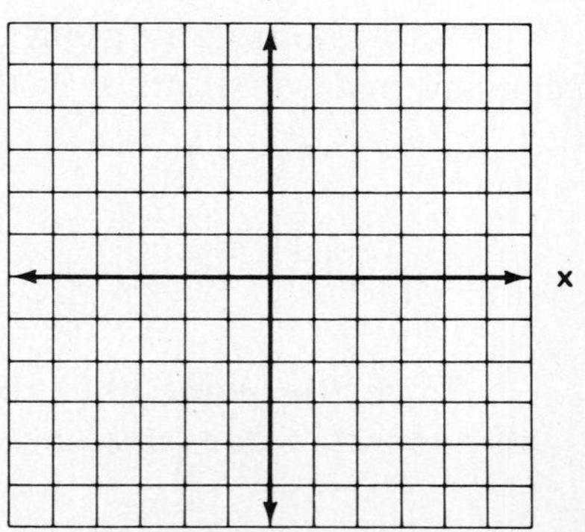

5. $y > 1$

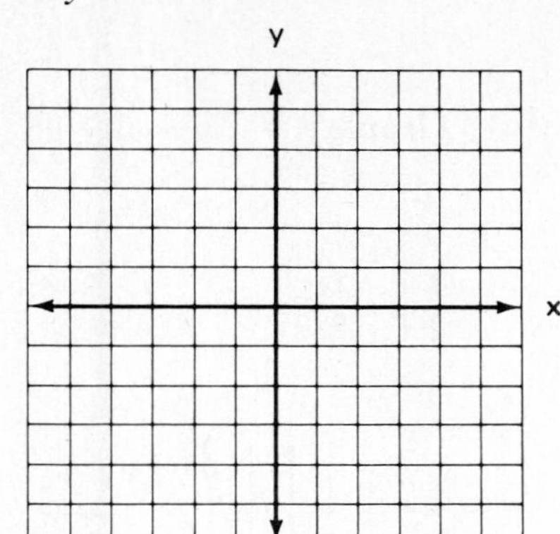

6. $x < -2$

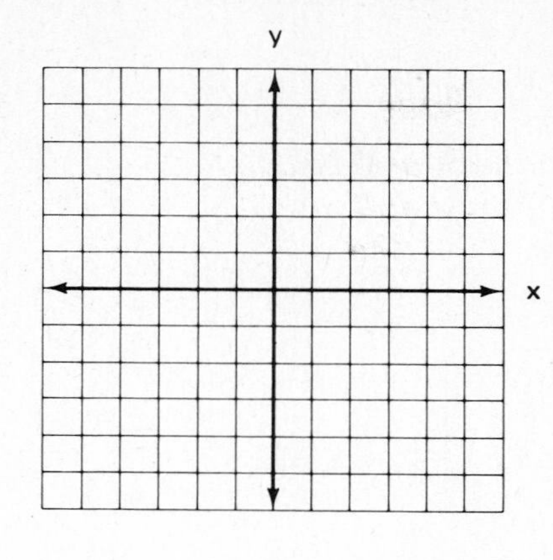

7. $2x - y \leq 4$

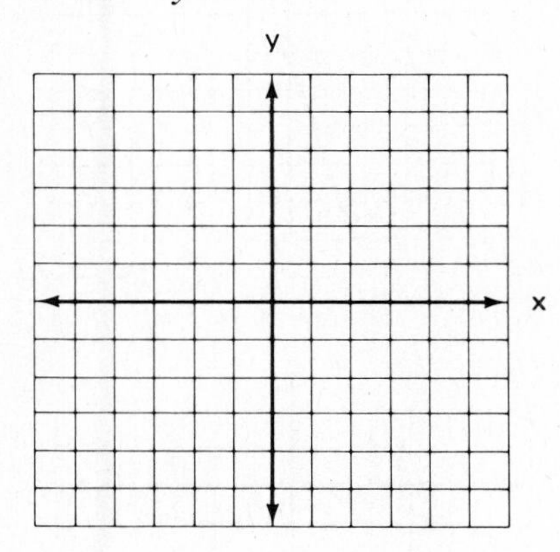

8. $3x \geq 6y + 12$

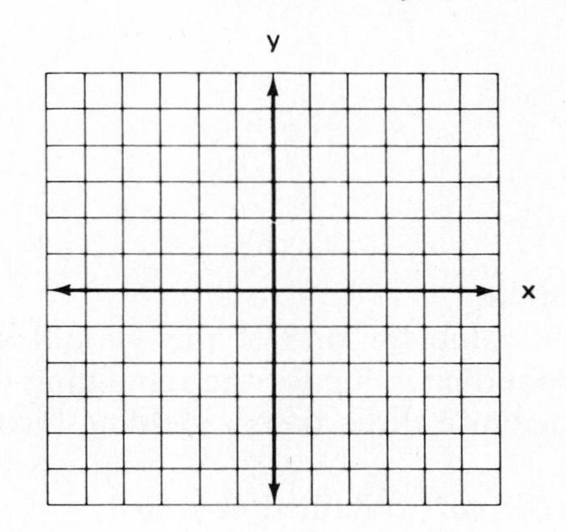

9. $4x + 2y < 10$

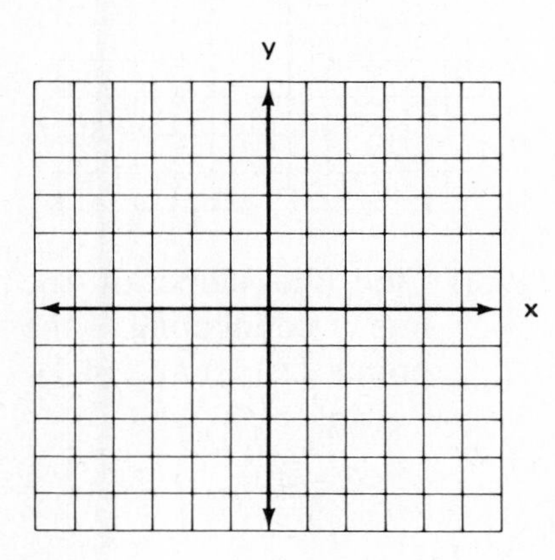

10. $\frac{x}{2} + \frac{y}{3} \leq 1$

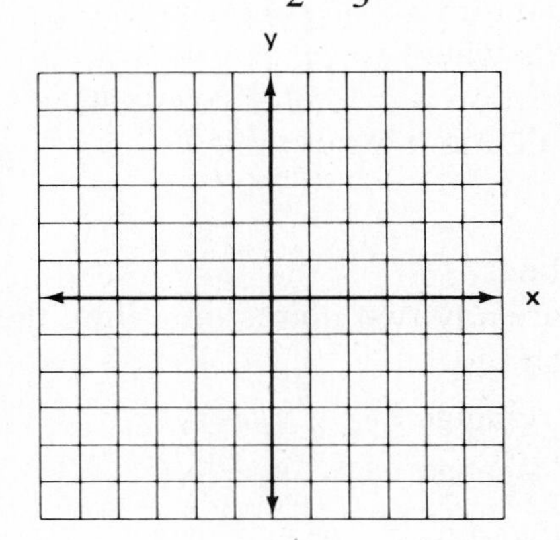

Answers to odd-numbered exercises on page 483.

9.7 Chapter Summary

Examples

(9.1) In an ordered pair (x,y):
x is called the *x-coordinate* or *abscissa*.
y is called the *y-coordinate* or *ordinate*.

(9.1) Graph $y = 3x - 1$.

x	y
0	−1
1	2
2	5

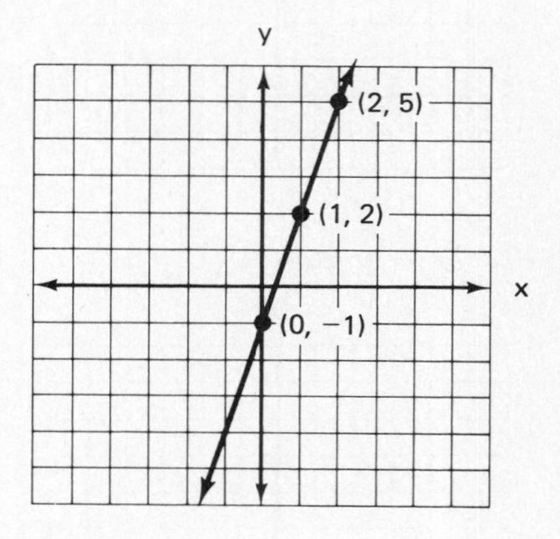

(9.2) To graph a linear equation:
1. Choose an arbitrary value for one of the variables and calculate the corresponding value for the remaining variable. Repeat this procedure three times, yielding three ordered pairs.
2. Plot the three ordered pairs obtained in step 1.
3. Draw a straight line through the points.

(9.2) Vertical and horizontal lines:
1. The graph of the equation $x = c$ (c = constant) is a vertical line through all points where $x = c$.
2. The graph of the equation $y = b$ (b = constant) is a horizontal line through all points where $y = b$.

(9.2)

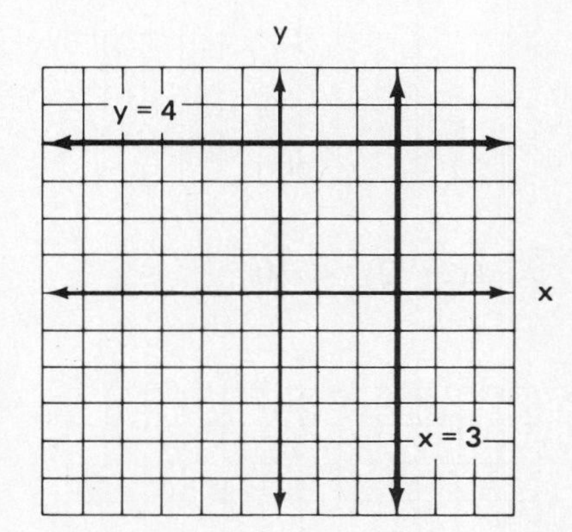

(9.3) To find the slope of a line:
If (x_1, y_1) and (x_2, y_2) are any two points on a line, then the slope of the line is given by

$$m = \text{slope} = \frac{\text{change in } y}{\text{change in } x} = \frac{y_2 - y_1}{x_2 - x_1}$$

(9.3) Find the slope of the line connecting the points $(2,-3)$ and $(4,1)$.

$$m = \frac{-3 - 1}{2 - 4} = \frac{-4}{-2} = 2$$

(9.3) The slope of any horizontal line is 0.
(9.3) The slope of any vertical line is undefined.
(9.3)

SLOPE	LINE
Positive	Goes uphill as it moves from left to right.
Negative	Goes downhill as it moves from left to right.
0	Horizontal line
Undefined	Vertical line

(9.3)

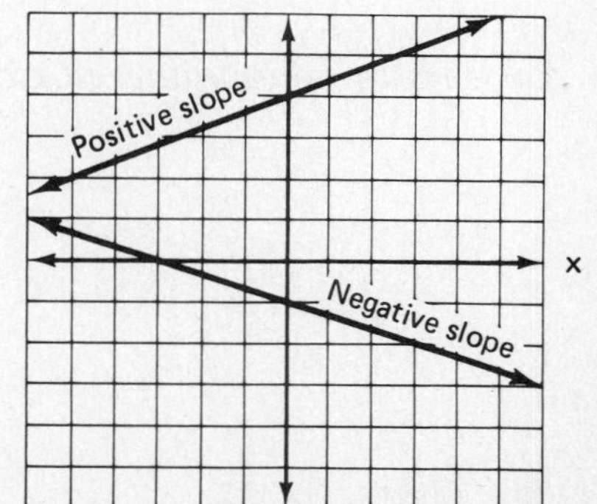

(9.4) The point-slope form of the equation of a line is:

$$y - y_1 = m(x - x_1)$$

where m = slope and (x_1, y_1) is a known point on the line.

(9.4) The standard form of an equation of a line is:

$$Ax + By + C = 0$$

where A, B, and C are integers and A and B are not both zero.

(9.4) The y-intercept is the point where a line crosses the y-axis.

(9.4) The slope-intercept form of the equation of a line is:

$$y = mx + b$$

where m = slope and b = y-intercept.

(9.4) Find the equation of the line through the points (3,−2) and (5,2).

$$m = \frac{y_2 - y_1}{x_2 - x_1}$$

$$= \frac{-2-2}{3-5} = \frac{-4}{-2} = 2$$

$y - y_1 = m(x - x_1)$

$y - 2 = 2(x - 5)$

$y - 2 = 2x - 10$

$y = 2x - 8$ Slope-intercept form.

$2x - y - 8 = 0$ Standard form.

(9.5) To graph a curve:

1. Make a table of ordered pairs by choosing values for one variable and calculating corresponding values for the remaining values.
2. Plot a sufficient number of points to determine the graph.
3. Connect the points with a smooth curve.

(9.5) Graph $y = 2x^2 - 5$.

Choose x	*Calculate* y	*Ordered pairs*
0	$y = 2 \cdot 0^2 - 5 = -5$	(0,−5)
1	$y = 2 \cdot 1^2 - 5 = -3$	(1,−3)
2	$y = 2 \cdot 2^2 - 5 = 3$	(2,3)
−1	$y = 2 \cdot (-1)^2 - 5 = -3$	(−1,−3)
−2	$y = 2 \cdot (-2)^2 - 5 = 3$	(−2,3)

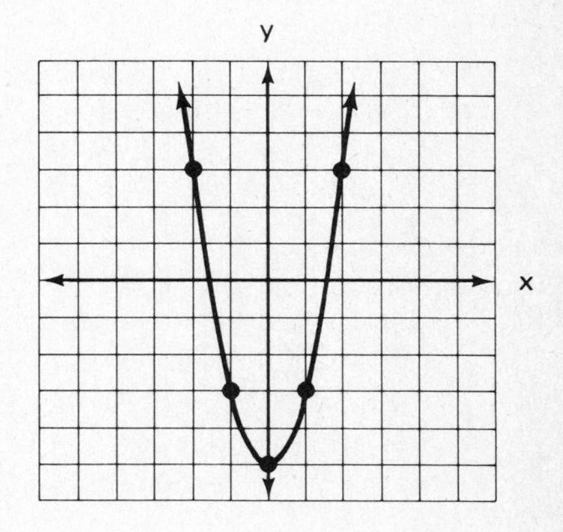

(9.6) To graph a linear inequality in two variables:

1. Graph the boundary line (replace the inequality sign by =). This line is broken if the inequality symbol is > or <, and solid if the inequality symbol is ≥ or ≤.
2. Choose any point not on the boundary line and determine whether it satisfies the inequality.

   If the point satisfies the inequality, shade the side of the line that contains the point.

(9.6) Graph $y < 2x + 3$.
Graph the broken line $y = 2x + 3$.

x	y
0	3
1	5
−1	1

If the point does *not* satisfy the inequality, shade the side of the line that does *not* contain the point.

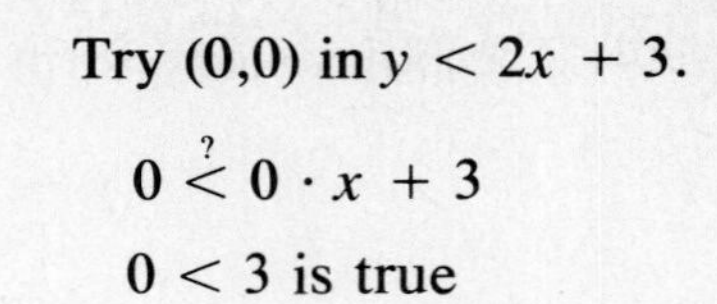

Try (0,0) in $y < 2x + 3$.

$$0 \overset{?}{<} 0 \cdot x + 3$$

$$0 < 3 \text{ is true}$$

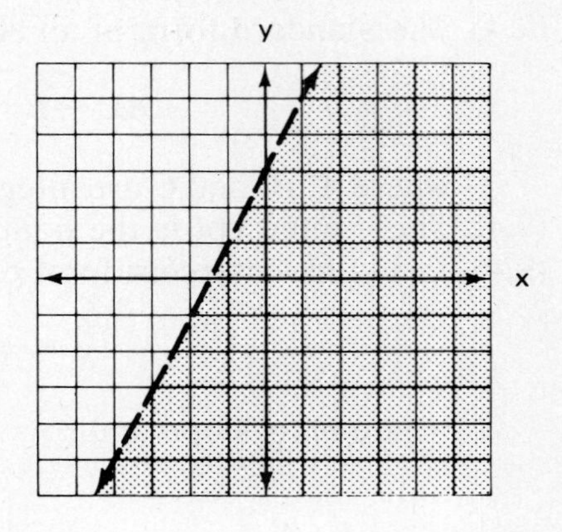

Exercise 9.7 Chapter Review

(9.1) For exercises 1 through 5, draw a set of axes, label them properly, and locate the given points.

1. (3,4)

2. (−2,0)

3. (−3,−1)

4. (0,4)

5. $(2\frac{1}{2},-3\frac{1}{2})$

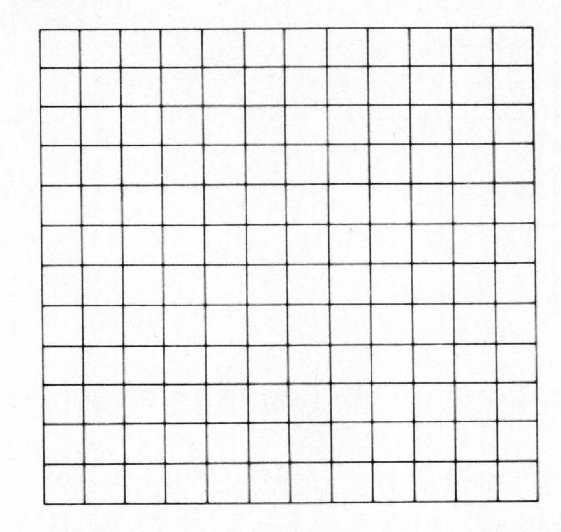

(9.1) In exercises 6 through 10, give the ordered pair which represents each of the points in the figure.

6. A (,)

7. B (,)

8. C (,)

9. D (,)

10. E (,)

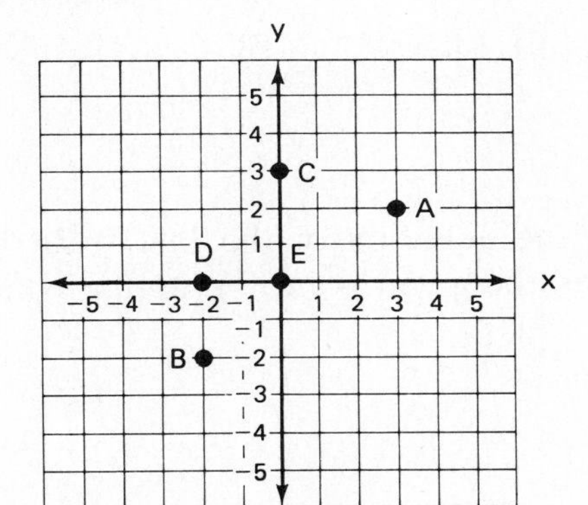

(9.2) Graph each of the given equations:

11. $y = x + 1$

12. $2x + y = -1$

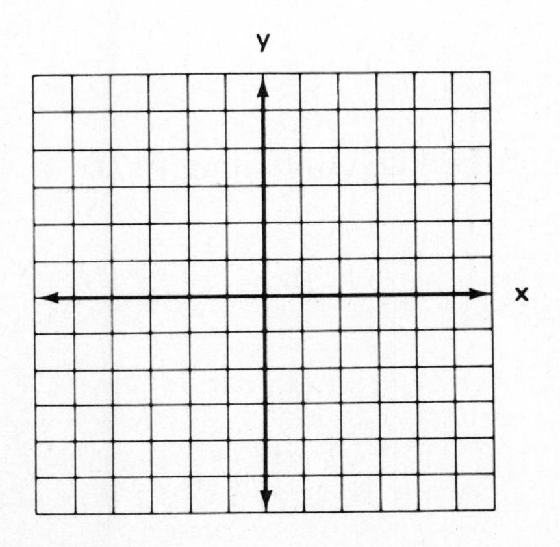

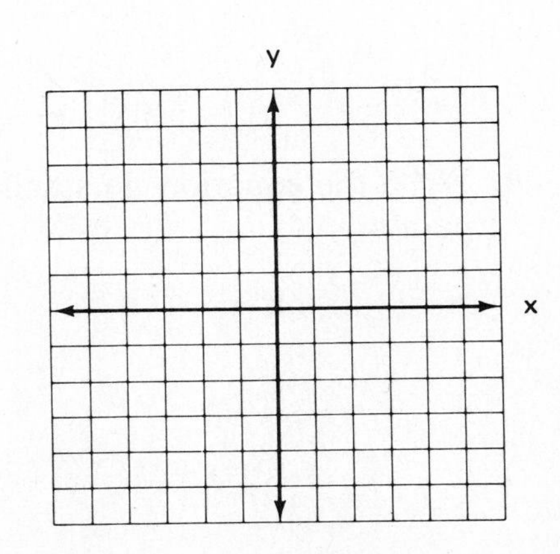

13. $x = -4$

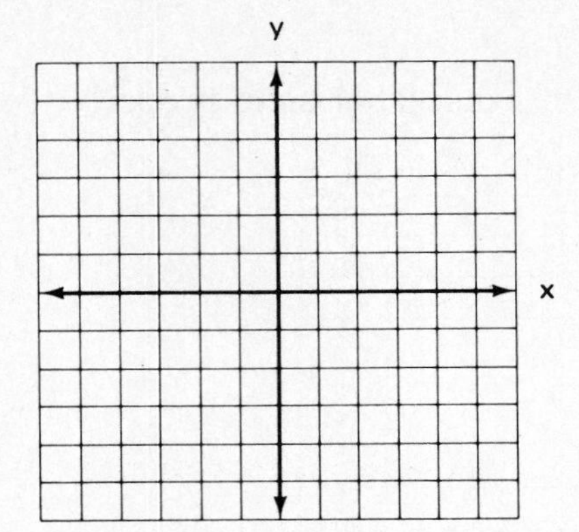

14. $y = -1$

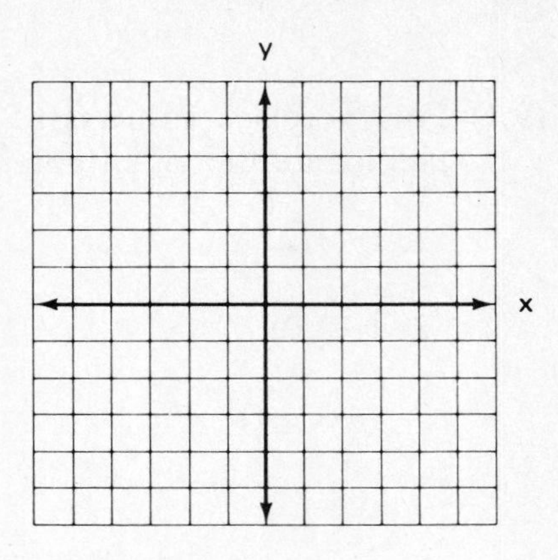

(9.3) Find the slope of the line through the given points.

15. (2,1), (1,5)

16. (4,3), (2,4)

17. (−1,3), (−6,−2)

18. (−4,0), (0,−3)

19. (−2,4), (−2,3)

20. (5,−3), (−3,−3)

(9.4) Write the equation of the line with the given slope which passes through the given point. Write the equation in standard form.

21. (2,1), $m = 4$

22. (3,−3), $m = -3$

23. (−2,4), $m = \frac{1}{2}$

24. (2,−2), $m = -\frac{2}{3}$

25. (−1,−3), $m = 0$

(9.4) Write the equation, in standard form, of the line containing the two points.

26. (3,2), (4,1)

27. (0,6), (2,4)

28. (−3,−1), (4,−2)

29. (−3,4), (−3,6)

30. (2,−2), 5,−2)

(9.4) Find the slope and y-intercept of the lines having the given equation.

31. $y = -3x - 4$

32. $y = \frac{1}{2}x - \frac{2}{3}$

33. $3x - 2y - 4 = 0$

34. $\frac{1}{2}x - 2y = 3$

35. $y = -2$

36. $x = 4$

(9.4) 37. Write the equation of a horizontal line containing the point (−3,1).

(9.4) 38. Write the equation of a vertical line containing the point (−2,−2).

(9.5) Graph the given curves.

39. $y = x^2 - 3$

40. $y = -x^2$

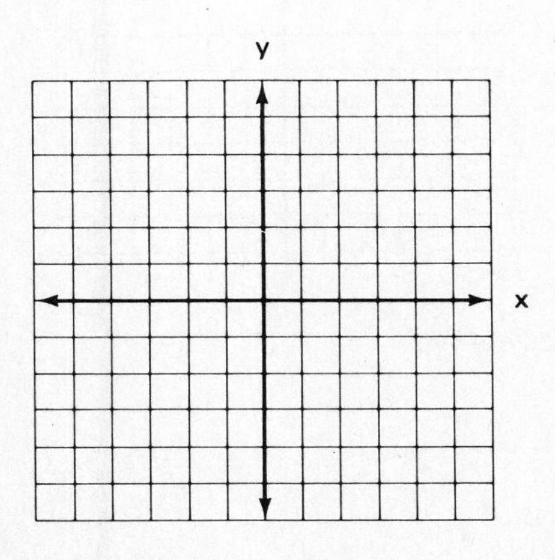

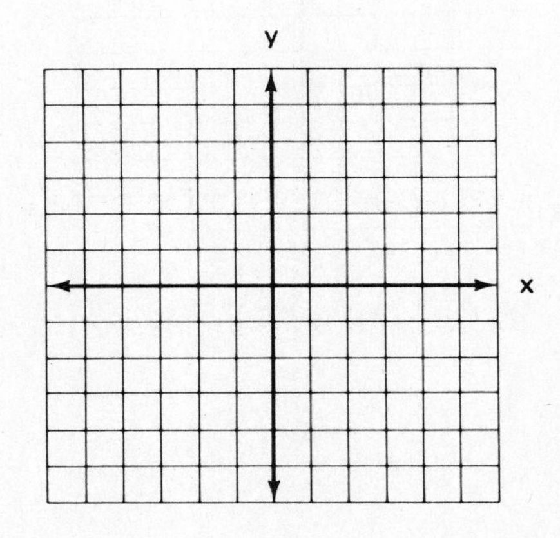

41. $y = x^2 - 2x - 1$

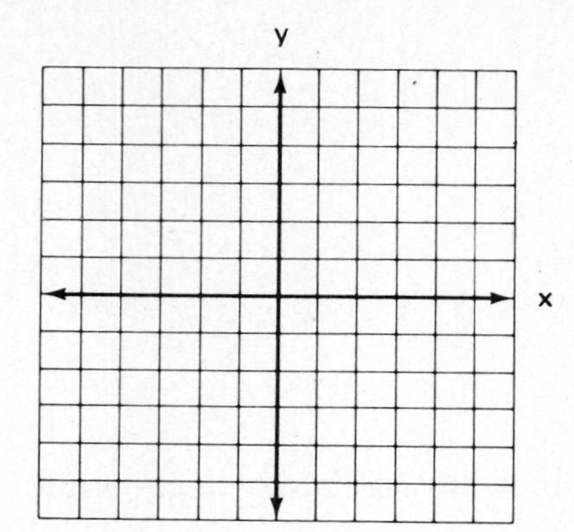

42. $y = x^3 + 1$

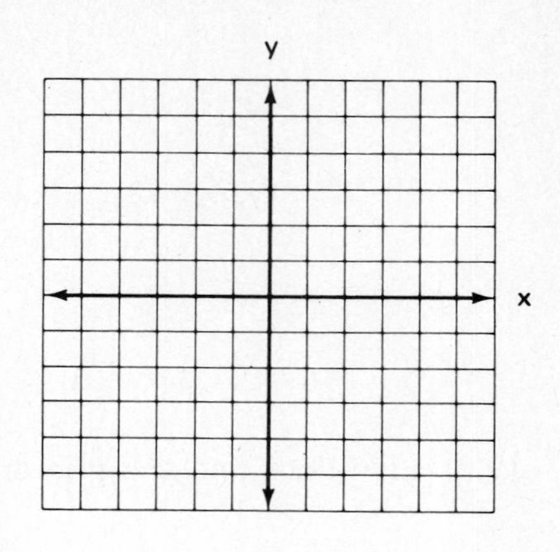

(9.6) Graph the given inequalities.

43. $y < x + 2$

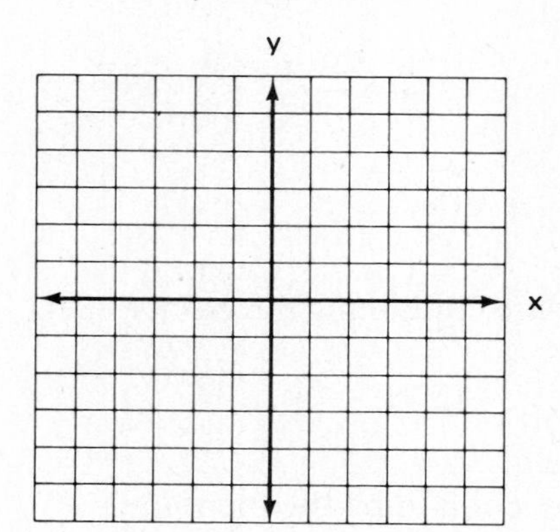

44. $3x - 2y > 6$

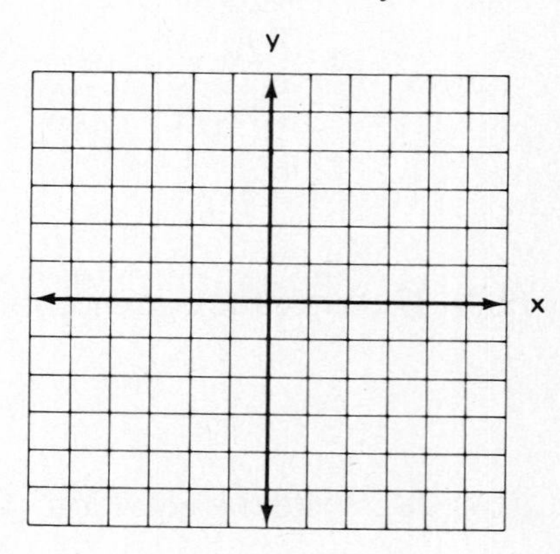

45. $y \leq -3$

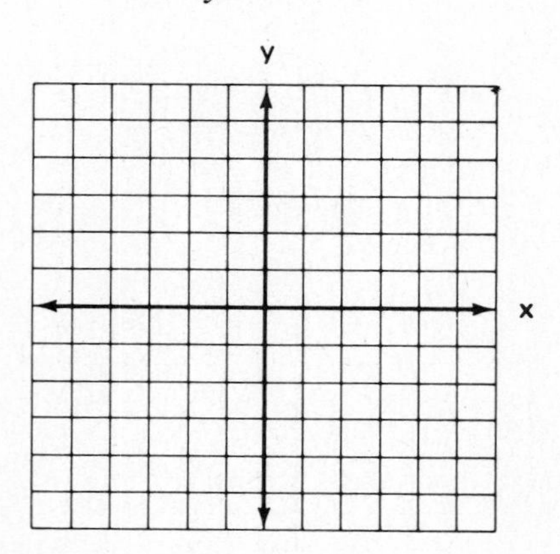

46. $2x \leq 4y + 8$

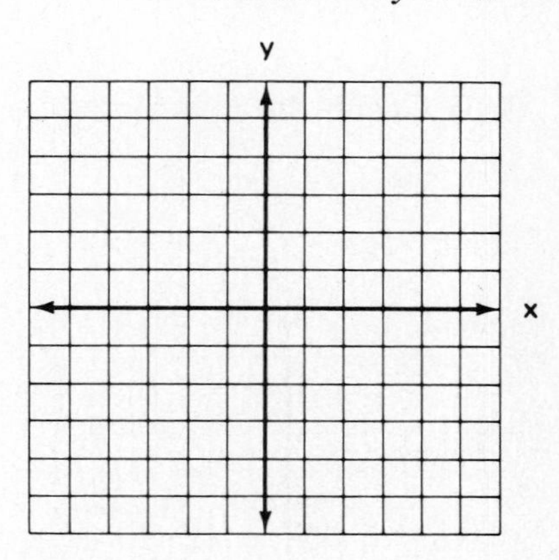

Answers to odd-numbered exercises on page 484.

Chapter 9 Achievement Test

Name ______________________

Class ______________________

This test should be taken before you are tested in class on the material in Chapter 9. Solutions to each problem and the section where the type of problem is found are given on page 485.

Draw a set of axes, label them properly, and locate the given points.

1. $(2,-3)$
2. $(0,-1)$
3. $(-2,1\frac{1}{2})$

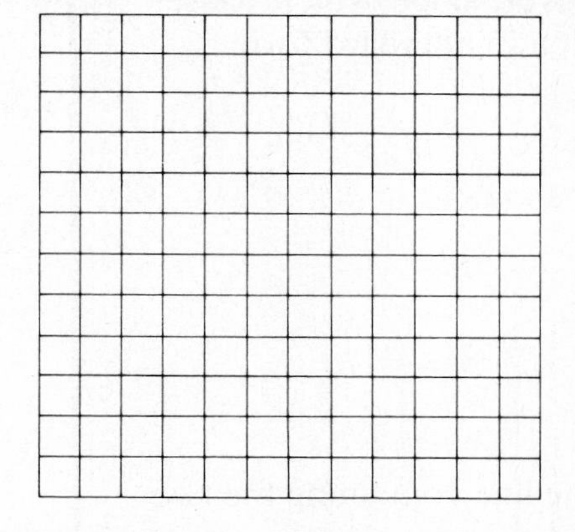

1. ______________
2. ______________
3. ______________

Give the ordered pair which represents each point on the figure.

4. A (,)
5. B (,)
6. C (,)

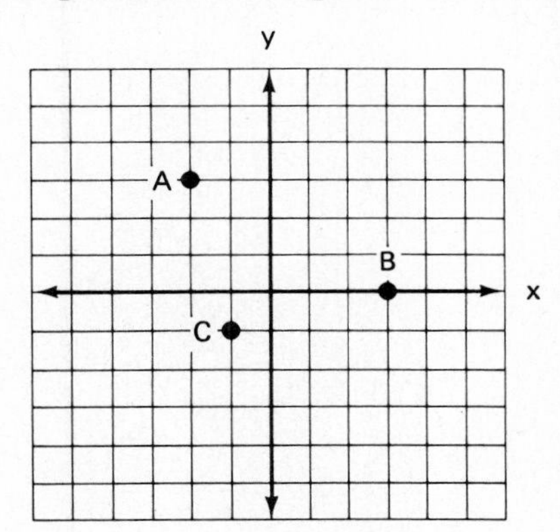

4. ______________
5. ______________
6. ______________

Graph each of the following equations:

7. $y = -x + 3$

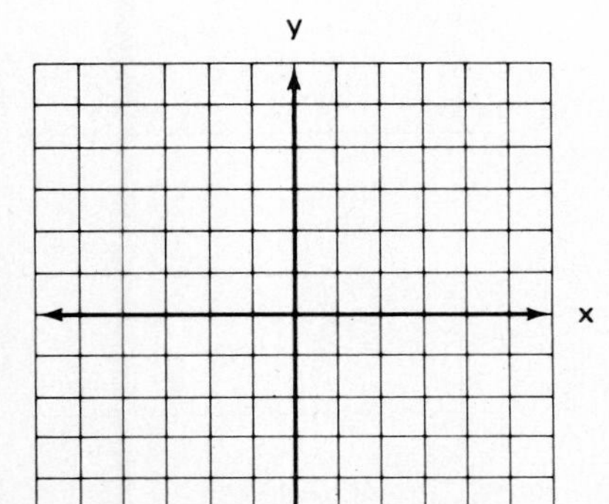

8. $2x - 3y = -6$

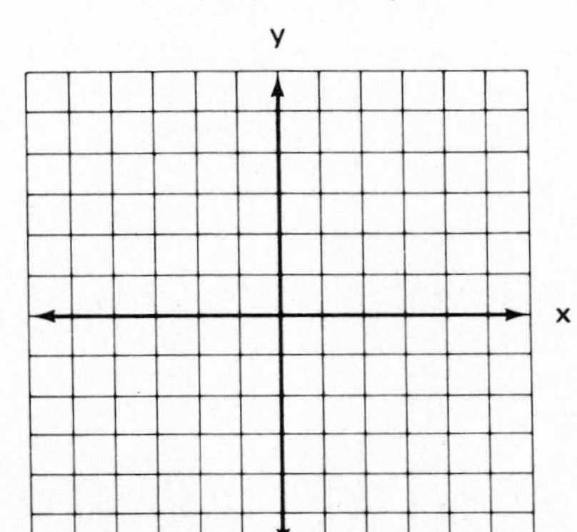

Find the slope of the line containing the given points.

9. $(-2,3), (4,-1)$ 9. ______

10. $(3,-2), (5,-2)$ 10. ______

11. $(-1,4), (-1,-3)$ 11. ______

Write the equation of the line with the given slope which passes through the given point. Write the equation in standard form.

12. $(2,-3)$ $m = 4$ 12. ______

13. $(-1,-4)$ $m = \frac{1}{2}$ 13. ______

14. $(2,1)$ m is undefined 14. ______

Write the equation, in standard form, of the line containing the two points.

15. $(-3,2), (4,-1)$ 15. ______

16. $(-2,-3), (-2,5)$ 16. ______

17. $(4,-1), (-2,-1)$ 17. ______

Find the slope and y-intercept of the line with the following equations:

18. $y = -5x - 3$ 18. ______

19. $y = \frac{3}{4}x + 3$ 19. ______

20. $y = 3$ 20. ______

21. $x = -1$ 21. ______

22. Write the equation of a horizontal line containing the point $(-2,-4)$.

23. Write the equation of a vertical line containing the point $(3,-1)$.

Graph the given curves.

24. $y = x^2 + 2x - 1$

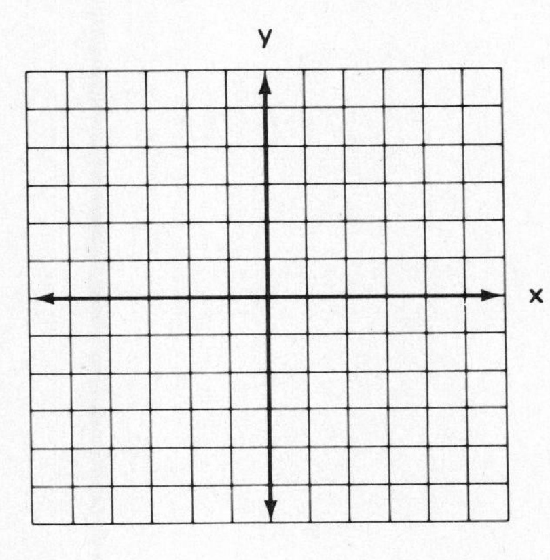

25. $y = x^3 - 3$

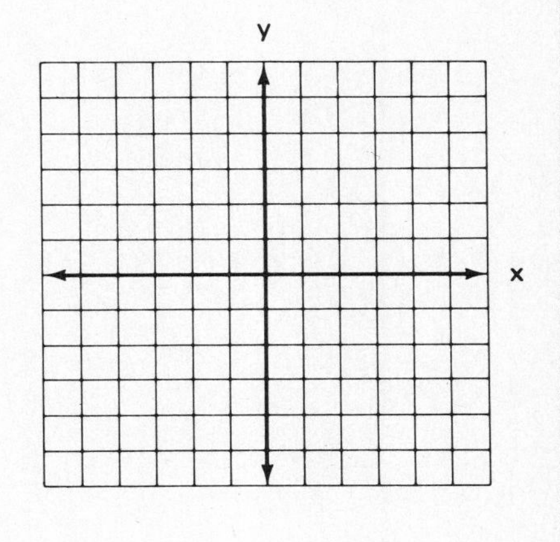

Graph the given inequalities.

26. $y < 2x - 3$

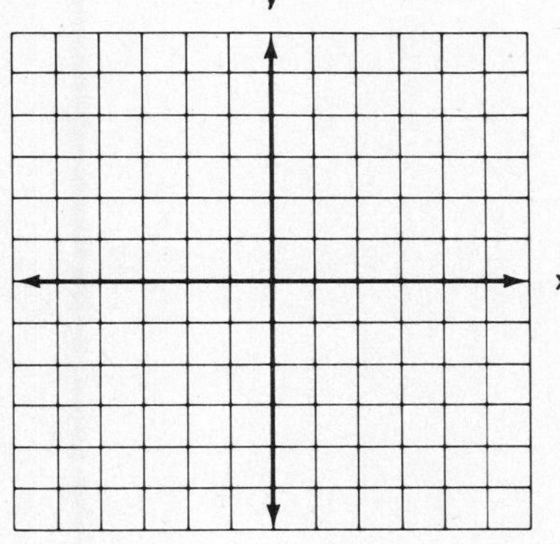

27. $y \geq -2$

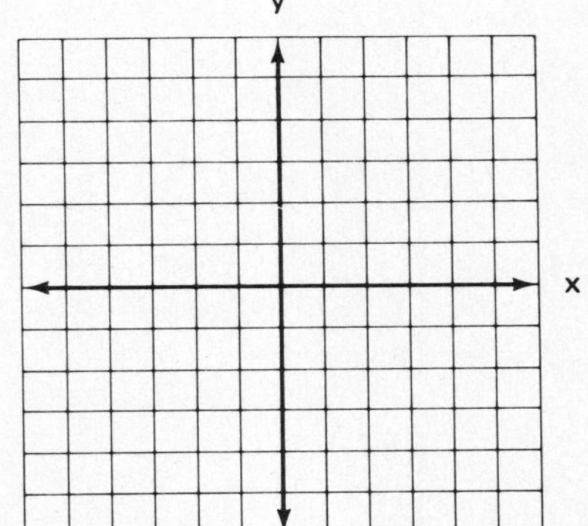

22. ______________________

23. ______________________

10 Systems of Equations

10.1 INTERSECTING, PARALLEL, AND COINCIDING LINES

If we graph two lines on the same set of axes, there are three possible things that can happen:

1. The lines cross each other (they **intersect** in one point).
2. The lines do not intersect at all (they are **parallel**).
3. The two equations represent the same line (they **coincide**).

The following three examples illustrate each of these cases.

(1) $L_1: y = -\frac{3x}{2} + 4$
$L_2: y = x - 1$

(2) $L_1: y = x + 2$
$L_2: y = x - 1$

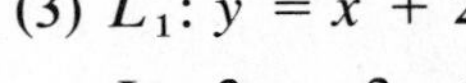

(3) $L_1: y = x + 2$
$L_2: 3x - 3y + 6 = 0$

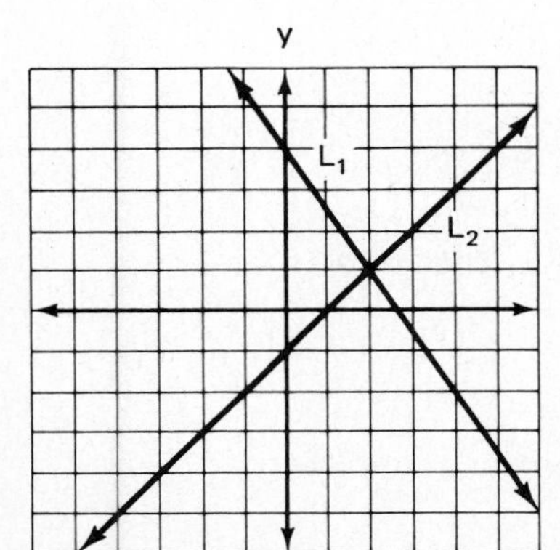

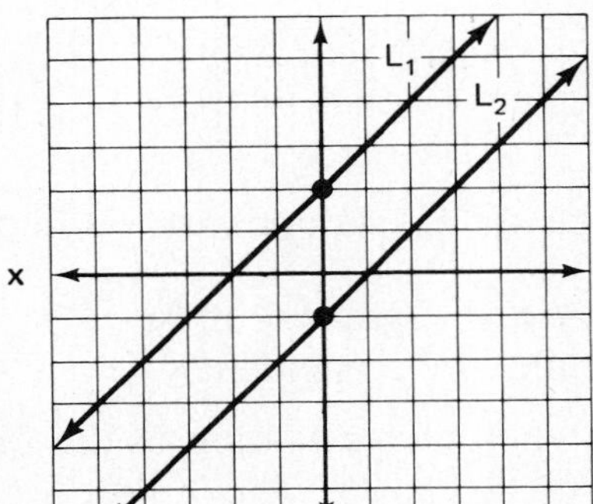

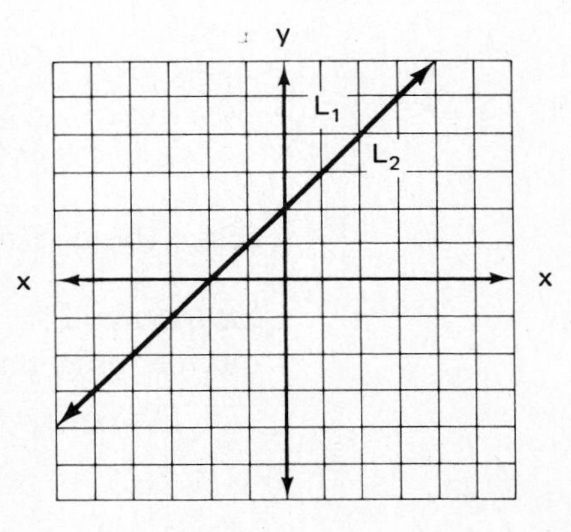

Let's examine each case briefly.

(1) $y = -\frac{3x}{2} + 4$

$y = x - 1$

The equations are in the form $y = mx + b$, and we can see that the coefficient of x (slope) in each of the equations is different.

Since the slopes are different, the lines must **intersect.**

(2) $y = x + 2$

$y = x - 1$

The coefficient of x (slope) in both equations is the same; therefore the lines will never intersect, and they must be **parallel.** Observe that the y-intercepts are different.

(3) $y = x + 2$

$3x - 3y + 6 = 0$

$3y = 3x + 6$

$\frac{3y}{3} = \frac{3x}{3} + \frac{6}{3}$

$y = x + 2$

Putting the second equation in the form $y = mx + b$, we find that it is *identical* to the first equation and we have the same equation twice. Therefore, the second line *coincides* with the first and we call them **coincident** or **coinciding.**

In summary, we have the following:

The graphs of two lines of the form $y = mx + b$:

1. intersect if the coefficients of x (slopes) are different;
2. are parallel if the coefficients of x (slopes) are the same but the y-intercepts are different;
3. are coincident if the two equations are identical.

Example 1: Determine whether the graphs of $3x + y - 4 = 0$ and $3x - 3y = 12$ are intersecting, parallel, or coincident.

Solution: Put each equation in the form $y = mx + b$.

$3x + y - 4 = 0$	$3x - 3y = 12$
$y = \boxed{-3}\, x + 4$	$-3y = -3x + 12$
	$\frac{-3y}{-3} = \frac{-3x}{-3} + \frac{12}{-3}$
	$y = \boxed{1}\, x - 4$

Since the coefficients of x (slope) are -3 and 1, the lines intersect.

Example 2: Are the graphs of $2x + y = -3$ and $-3y = 6x + 9$ intersecting, parallel, or coincident?

Solution: Put each equation in the form $y = mx + b$.

$$2x + y = -3$$
$$\boxed{y = -2x - 3}$$

$$-3y = 6x + 9$$
$$\frac{-3y}{-3} = \frac{6x}{-3} + \frac{9}{-3}$$
$$\boxed{y = -2x - 3}$$

Since the equations are identical, each of the original equations was actually the same, and the lines are coincident.

Example 3: Are the graphs of $3y - 2x = 6$ and $4x - 6y = -1$ intersecting, parallel, or coincident?

Solution: Put each equation in the form $y = mx + b$.

$$3y - 2x = 6$$
$$3y = 2x + 6$$
$$\frac{3y}{3} = \frac{2x}{3} + \frac{6}{3}$$
$$y = \boxed{\frac{2}{3}}x + \boxed{2}$$

$$4x - 6y = -1$$
$$-6y = -4x - 1$$
$$\frac{-6y}{-6} = \frac{-4x}{-6} - \frac{1}{-6}$$
$$y = \boxed{\frac{2}{3}}x + \boxed{\frac{1}{6}}$$

Since the coefficients of x are the same and the y-intercepts are different, the two lines are parallel.

EXERCISE 10.1

Determine whether the following pairs of equations are intersecting, parallel, or coincident.

1. $y = 4x - 1$
   $y = 3x + 5$
2. $y = -2x - 6$
   $y = -2x + 15$
3. $2x - 3y = 6$
   $4x + 3y = 9$
4. $x + y = 1$
   $3y = 3x + 6$
5. $2x - 5 = y$
   $2y - 4x = -10$
6. $3x + 2y - 4 = 0$
   $3x + y - 4 = 0$

7. $-6x + 3 = -21y$
   $2x - 7y = -1$

8. $y - 3x - 5 = 0$
   $2y + 4 = -6x$

9. $y + 1 = 2x$
   $3x - y = -5$

10. $y = \frac{1}{2}x + 4$
    $4y - 2x = 16$

11. $\frac{2}{3}x + \frac{1}{2}y = 2$
    $\frac{1}{2}x - \frac{2}{3}y = 2$

12. $3y + 2 = 0$
    $y + 2x = 3$

Answers to odd-numbered exercises on page 486.

10.2 SOLUTION BY GRAPHING

A set of equations such as

$$2x - y = 5$$
$$x + y = 4$$

is called a **system of linear equations** in two variables. A solution of such a system is an ordered pair (x,y) that satisfies both equations of the system. If we graph both equations on the same set of axes we will discover a hint as to how to find this solution.

$2x - y = 5$

x	y
0	-5
1	-3
2	-1

$x + y = 4$

x	y
0	4
1	3
2	2

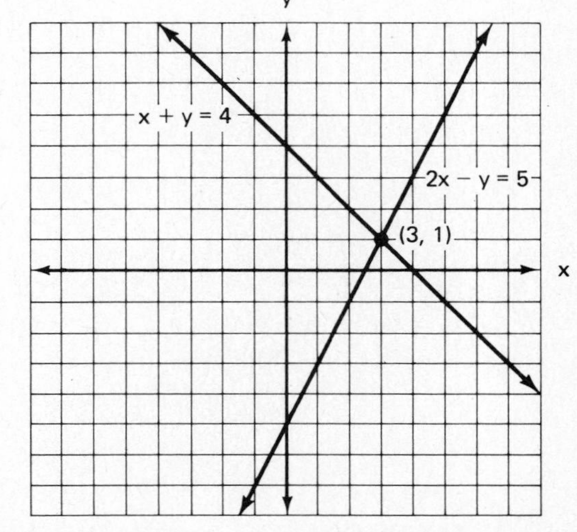

The point of intersection looks as if it is (3,1) and is the only point that lies on both lines. To see if it satisfies each equation, we check.

$$2x - y = 5 \qquad x + y = 4$$

$$2 \cdot \boxed{3} - \boxed{1} \stackrel{?}{=} 5 \qquad \boxed{3} + \boxed{1} = 4 \checkmark$$

$$6 - 1 = 5 \checkmark$$

> To determine whether a point (x,y) is a solution to a system of linear equations, check it in *each* equation.

> **To solve a system of equations by graphing:**
> 1. Graph each equation on the same set of axes.
> 2. One of three possible cases will result:
>   a. The lines *intersect* in one point and the coordinates of the point of intersection are the *only* solution.
>   b. The lines are *parallel* and there is no solution.
>   c. The lines are *coincident* and every ordered pair on the line is a solution to both equations. There are an *infinite* number of solutions.
> 3. Check the solution (if there is one) in the original equations.

Example 1: Solve by graphing:

$$y + x = 5$$
$$y - 2x = -1$$

Solution: Put each equation in $y = mx + b$ form.

$y + x = 5$	$y - 2x = -1$
$y = -x + 5$	$y = 2x - 1$

Different slopes (-1 and $+2$) indicate that the lines must intersect. Graph each equation.

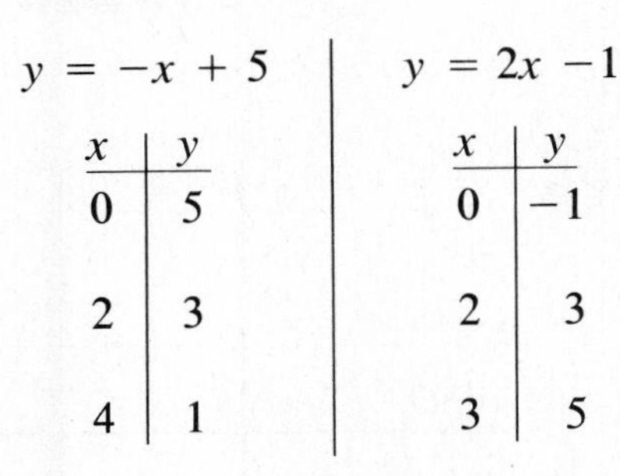

$y = -x + 5$

x	y
0	5
2	3
4	1

$y = 2x - 1$

x	y
0	-1
2	3
3	5

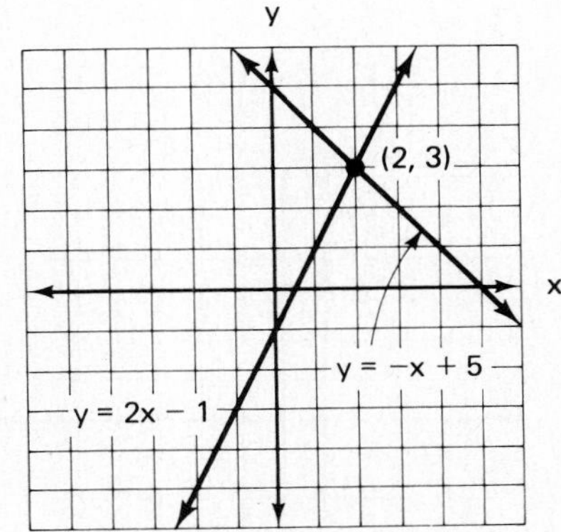

The point of intersection appears to be (2,3).

Check:

$y + x = 5$	$y - 2x = -1$
$\boxed{3} + \boxed{2} = 5\ \checkmark$	$\boxed{3} - 2 \cdot \boxed{2} \stackrel{?}{=} -1$
	$3 - 4 = -1\ \checkmark$

Example 2: Solve by graphing:

$$x + y = 1$$
$$3x = 9 - 3y$$

Solution: Put in $y = mx + b$ form.

$x + y = 1$	$3x = 9 - 3y$
$y = -x + 1$	$3y = -3x + 9$
	$y = -x + 3$

We can see immediately that the slope equals -1 for both equations. Therefore the lines are parallel and there is no solution. We graph the lines for illustration purposes.

$y = -x + 1$

x	y
0	1
2	−1
4	−3

$y = -x + 3$

x	y
0	3
2	1
4	−1

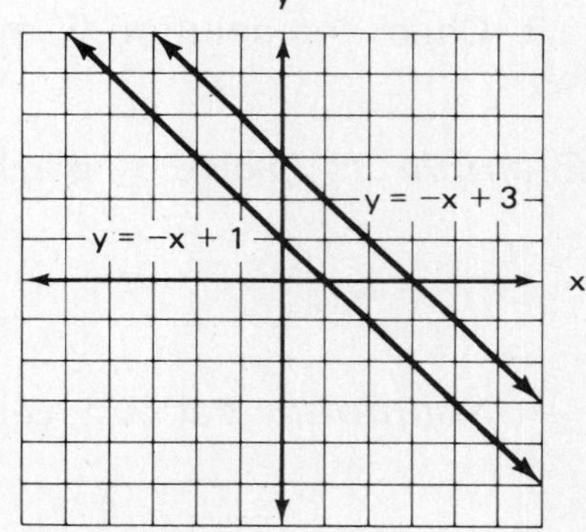

EXERCISE 10.2 Solve the following systems by graphing:

1. $y + x = 4$
   $y - x = 2$

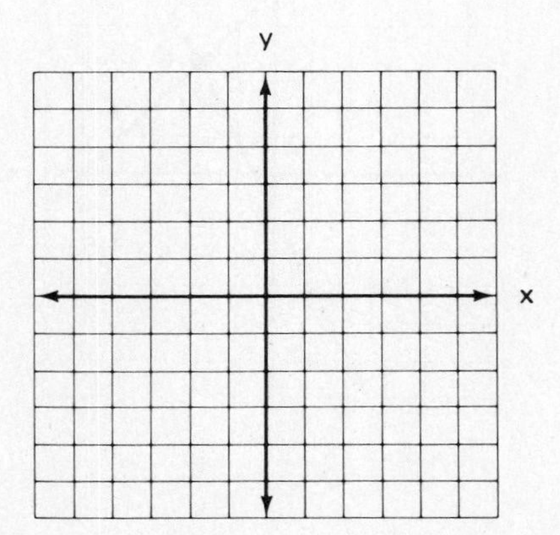

2. $2y - 3x = 4$
   $y + x = 2$

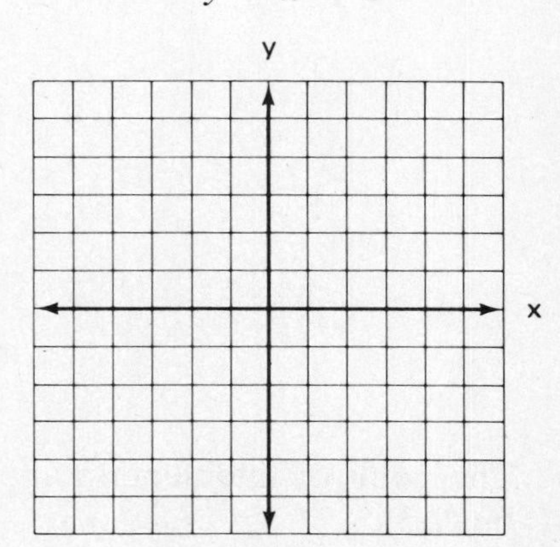

3. $x - 2y = 3$
   $6y - 3x = 9$

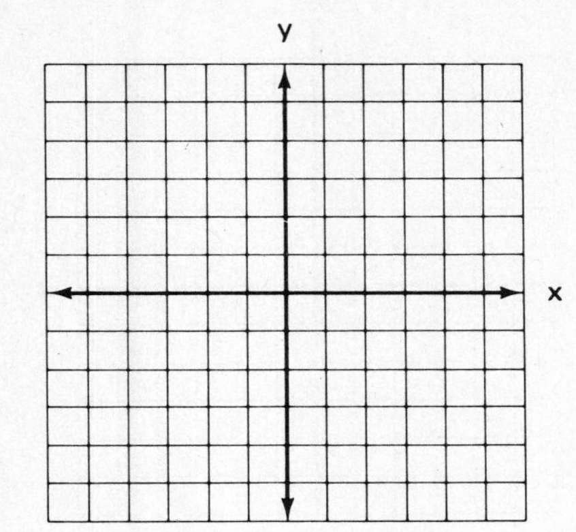

4. $2x + y = 1$
   $2y = -4x + 8$

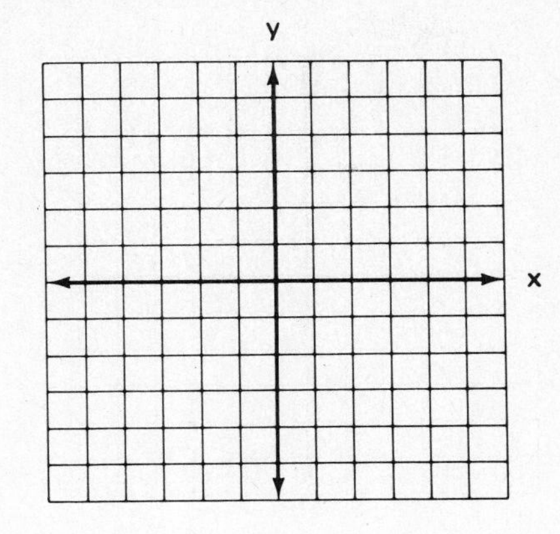

5. $2x - y = 0$
   $x + 4y = 0$

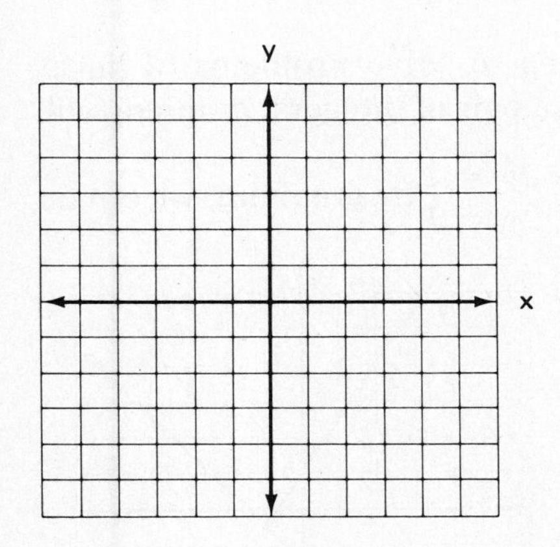

6. $x + y = -2$
   $y = 2x - 8$

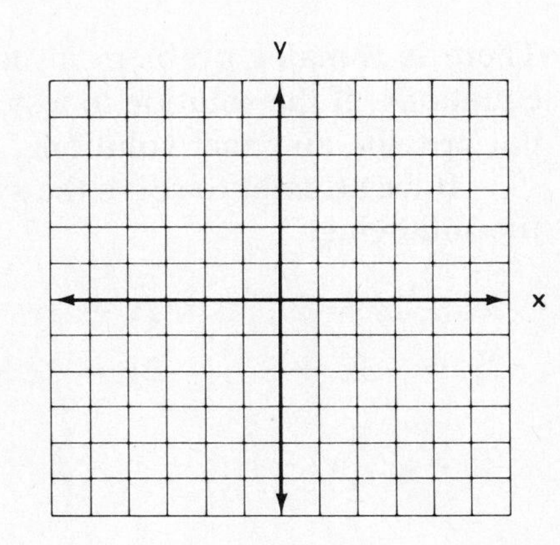

7. $2y - x = 2$
   $2x + y = 6$

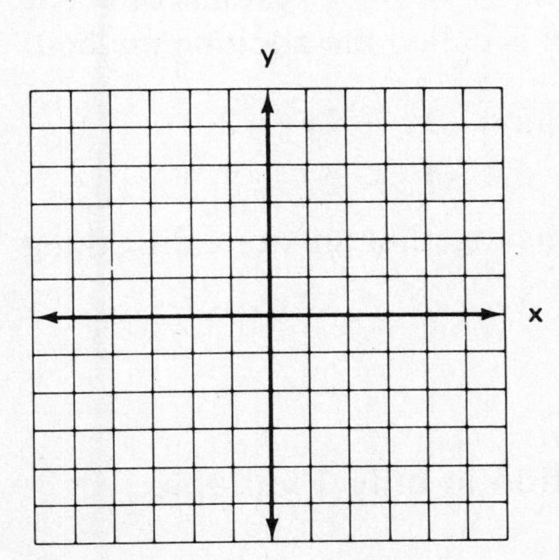

8. $x = 3y + 2$
   $y = 3x + 2$

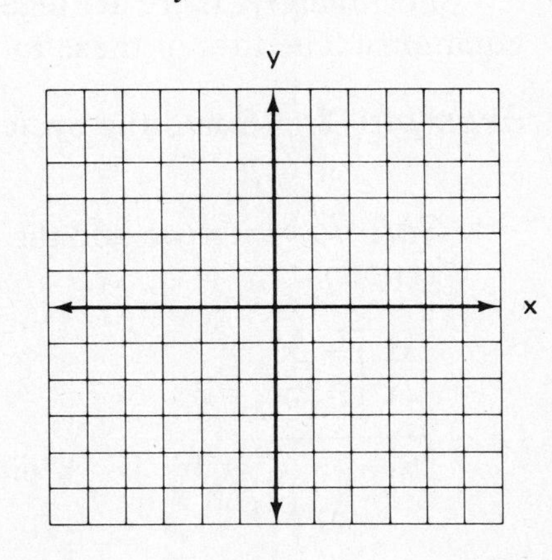

9. $x + 2y = 4$
   $6y = -3x - 12$

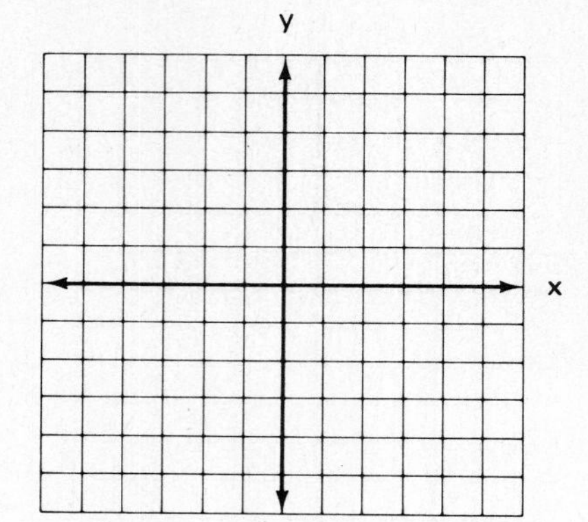

10. $x + 3y = 9$
    $y + \frac{1}{3}x = 1$

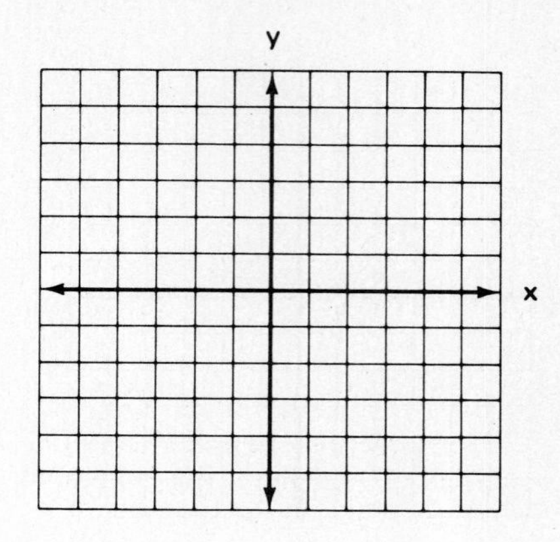

Answers to odd-numbered exercises on page 486.

10.3 SOLUTION BY THE ADDITION METHOD

There is a major problem in using graphing to solve problems of linear equations. If the solution to a system is not a pair of integers, graphing will not provide an exact solution.

If we attempt to solve the system $\begin{matrix} 3x + 2y = 2 \\ 2x - 2y = 3 \end{matrix}$ by graphing, we obtain the following:

$3x + 2y = 2$	$2x - 2y = 3$
$2y = -3x + 2$	$2y = 2x - 3$
$y = -\frac{3}{2}x + 1$	$y = x - \frac{3}{2}$

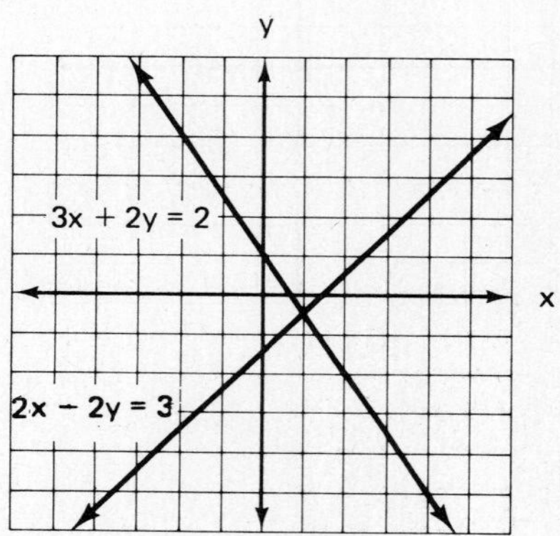

The solution cannot be read exactly.

Fortunately, there are algebraic methods for solving systems of linear equations. The first of these to be discussed is called the **addition method.**

Example 1: Solve the system algebraically: $3x + 2y = 2$
$2x - 2y = 3$

Solution: If we add the two equations together an interesting thing happens:

$$\begin{array}{rl} 3x \boxed{+ 2y} & = 2 \\ 2x \boxed{- 2y} & = 3 \\ \hline 5x \qquad & = 5 \end{array}$$ yields an equation in only 1 variable.

$$x = 1$$

Now substitute the value $x = 1$ in either of the original equations and solve for y:

$$3x + 2y = 2$$
$$3 \cdot \boxed{1} + 2y = 2$$
$$2y = -1$$
$$y = -\frac{1}{2}$$

The solution is $(x,y) = (1,-\frac{1}{2})$.
Check by substituting the solution into both equations.

$3x + 2y = 2$	$2x - 2y = 3$
$3 \cdot \boxed{1} + 2 \cdot \boxed{-\frac{1}{2}} \stackrel{?}{=} 2$	$2 \cdot \boxed{1} - 2 \cdot \boxed{-\frac{1}{2}} \stackrel{?}{=} 3$
$3 - 1 = 2$ ✓	$2 + 1 = 3$ ✓

In solving this system our first step was to add the equations together, and since the coefficients of the y-terms were opposite, the y-term was eliminated. If the coefficients of one of the variables are not opposites of each other, we will make them opposites by multiplying both sides of one or both of the equations by some appropriate number.

Example 2: Solve: $5x - 2y = 14$
$3x + 6y = -6$

Solution: As given, neither pair of variables are opposites. However, if we multiply both sides of the first equation by 3, we accomplish what we want to do.

$15x - 6y = 42$	Multiply both sides of the first equation by 3.
$3x + 6y = -6$	Add the second equation.
$18x = 36$	The y-terms are eliminated.
$x = 2$	

To find the value for y, substitute $x = 2$ in either of the original equations.

$$3x + 6y = -6$$
$$3 \cdot \boxed{2} + 6y = -6$$
$$6y = -12$$
$$y = -2$$

The solution is $(2,-2)$ which we check in both equations.

$5x - 2y = 14$	$3x + 6y = -6$
$5 \cdot \boxed{2} - 2 \cdot \boxed{-2} \stackrel{?}{=} 14$	$3 \cdot \boxed{2} + 6 \cdot \boxed{-2} \stackrel{?}{=} -6$
$10 + 4 = 14$ ✓	$6 - 12 = -6$ ✓

Example 3: Solve:

$$3x + 4y = 7$$
$$2x - 3y = -1$$

Solution: In this case we need to multiply *each* equation by some number in order to make the coefficients of one of the variables opposites. We can make the coefficients of the x-terms opposites by multiplying the first equation by 2 and the second equation by -3.

$6x + 8y = 14$	Multiply both sides of the first equation by 2.
$-6x + 9y = 3$	Multiply both sides of the second equation by -3.
$17y = 17$	Add to eliminate the x-term.
$y = 1$	Solve for y.

Substituting $y = 1$ into the first equation yields:

$$3x + 4 \cdot \boxed{1} = 7$$
$$3x = 3$$
$$x = 1$$

The solution is (1,1).

Check:

$3x + 4y = 7$	$2x - 3y = -1$
$3 \cdot \boxed{1} + 4 \cdot \boxed{1} \stackrel{?}{=} 7$	$2 \cdot \boxed{1} - 3 \cdot \boxed{1} \stackrel{?}{=} -1$
$3 + 4 = 7$ ✓	$2 - 3 = -1$ ✓

We could have solved this system of equations by eliminating the y-term instead of the x-term, as shown in the following example.

Example 4: Solve:

$$3x + 4y = 7$$
$$2x - 3y = -1$$

by eliminating the y-term:

Solution: To make the coefficients of the y-terms opposite we will multiply the first equation by 3 and the second equation by 4.

$9x + 12y = 21$	Multiply both sides of the first equation by 3.
$8x - 12y = -4$	Multiply both sides of the second equation by 4.
$17x = 17$	Add the equations.
$x = 1$	Solve for x.

Put $x = 1$ in equation one and solve for y.

$$3 \cdot \boxed{1} + 4y = 7$$
$$4y = 4$$
$$y = 1$$

As before, the solution is (1,1).

Parallel Lines

A system whose graphs are parallel lines has no solution. The following example shows how to identify such a system when using the addition method.

Example 5: Solve: $2x + 4y = 3$
$3x + 6y = 5$

Solution: Multiply the first equation by 3 and the second equation by -2, and add the equations.

$6x + 12y = 9$	Multiply both sides of equation one by 3.
$-6x - 12y = -10$	Multiply both sides of equation two by -2.
$0 = -1$	Both variables are eliminated, and a false statement results.

Since there is no ordered pair (x,y) for which $0 = -1$, we conclude that there is *no solution* and the lines are *parallel*.

Coinciding Lines

A system whose graphs are coinciding lines has an infinite number of solutions, since every solution to one equation will be a solution to the other equation. The next example shows how coinciding lines are identified when using the addition method.

Example 6: Solve: $6x - 9y = -6$
$-4x + 6y = 4$

Solution: Multiply the first equation by 2 and the second equation by 3.

$12x - 18y = -12$	Multiply both sides of equation one by 2.
$-12x + 18y = 12$	Multiply both sides of equation two by 3.
$0 = 0$	Both variables are eliminated and a true statement results.

This means that each of the given equations is equivalent to the same equation; therefore their graphs are coincident, and every solution to the first equation is also a solution to the second equation. Any point on the line is a solution to the system.

To solve a system of equations by the addition method:

1. Multiply one or both of the equations by numbers that will make the coefficients of either of the variables opposites.
2. Add the equations.

*3. Solve the resulting equation.

4. Substitute the value obtained in step 3 back in either of the original equations and solve for the remaining variable.

* If both variables are eliminated and the resulting statement is *false*, then the lines are parallel and there is no solution.

* If both variables are eliminated and the resulting statement is *true*, then the lines are coincident, and any solution of either equation is a solution to the system.

EXERCISE 10.3

Solve the following systems using the addition method. Some systems may be parallel (no solution), or coinciding (infinite number of solutions).

1. $x + y = 0$
   $2x - y = 6$

2. $x + y = 2$
   $2x - y = -2$

3. $x - y = 7$
   $x + y = 3$

4. $5x + 7y = 18$
   $-5x + y = -2$

5. $-4x + 3y = -3$
   $7x - 3y = -6$

6. $2x - 3y = 4$
   $-4x + 6y = 3$

7. $x + y = 3$
   $-x + 6y = 10$

8. $x + 3y = 10$
   $4x - y = 1$

9. $6x - 4y = 1$
   $x + 5y = 3$

10. $2x - y = 3$
    $-6x + 3y = -9$

11. $2x - 3y = 4$
    $-4x + 6y = 1$

12. $3x + 7y = -4$
    $-x + 2y = -3$

13. $2x + 7y = 12$
    $-4x + 3y = -7$

14. $-3x - 6y = 2$
    $x + 2y = 3$

15. $3x - 2y = 4$
    $3x - 2y = 5$

16. $3x - 2y = 4$
    $2x + 5y = 9$

17. $3x + 7y = 6$
    $-6x - 5y = -12$

18. $2x - 2y = 6$
    $3x - 3y = 9$

19. $5x + 2y = -1$
    $4x + 3y = -12$

20. $3x + 2y = 3$
    $2x + 3y = 7$

21. $2x - 4y = 2$
    $3x - 6y = 5$

22. $x + 2y = 6$
    $2y = 10$

23. $x + 2y = 4$
    $-3x - 6y = -12$

24. $5x + 7y = -11$
    $3x - 2y = 12$

25. $2x + 3y = -4$
    $5x + 7y = 2$

26. $3x - 5y = -4$
    $4x - 2y = -3$

Answers to odd-numbered exercises on page 487.

10.4 SOLUTION BY THE SUBSTITUTION METHOD

A second algebraic method for solving systems of linear equations is called the **substitution method.** We solve one equation for one of the variables and substitute in the other equation.

Example 1: Solve: $x + y = 4$

$$x = y + 2$$

Solution: The second equation says that x and $(y + 2)$ are equal to each other. Therefore, we may substitute $(y + 2)$ for x in the first equation.

$$\boxed{x} + y = 4$$

$$\boxed{(y + 2)} + y = 4 \qquad \text{Substitute } (y + 2) \text{ for } x \text{ in the first equation.}$$

This new equation contains only y and we can solve it.

$$2y + 2 = 4$$
$$2y = 2$$
$$y = 1$$

Now substitute $y = 1$ in either of the original equations and solve for x.

$$x + y = 4 \qquad \text{First equation}$$
$$x + \boxed{1} = 4$$
$$x = 3$$

The solution is (3,1).

Check:

$x + y = 4$	$x = y + 2$
$3 + 1 = 4$ ✓	$3 = 1 + 2$ ✓

Example 2: Solve: $2x + y = 10$
$4x + 3y = 8$

Solution: In this case we must solve one of the equations for one of the variables. It is usually easiest to solve for the variable with the smallest coefficient. We solve the first equation for y, and substitute in the second equation.

$$2x + y = 10 \qquad \text{First equation}$$
$$y = (10 - 2x) \qquad \text{Solved for } y$$

Now substitute $y = (10 - 2x)$ in the second equation and solve for x.

$$4x + 3\boxed{y} = 8$$
$$4x + 3\boxed{(10 - 2x)} = 8$$
$$4x + 30 - 6x = 8$$
$$-2x = -22$$
$$x = 11$$

Now substituting $x = 11$ in $2x + y = 10$ gives us

$$2 \cdot \boxed{11} + y = 10$$

$$22 + y = 10$$

$$y = -12$$

Our solution is $(11, -12)$.

Check:

$2x + y = 10$	$4x + 3y = 8$
$2 \cdot \boxed{11} + (\boxed{-12}) \stackrel{?}{=} 10$	$4 \cdot \boxed{11} + 3(\boxed{-12}) \stackrel{?}{=} 8$
$22 - 12 = 10$ ✓	$44 - 36 = 8$ ✓

Parallel Lines

The following examples illustrate how to recognize a system whose graphs are parallel lines when using the substitution method.

Example 3: Solve:

$$x + y = 4$$
$$3x + 3y = -2$$

Solution: Solve the first equation for x and substitute in the second equation.

$$x + y = 4$$

$$x = (4 - y) \qquad \text{Solved for } x$$

Substitute $x = 4 - y$ in $3x + 3y = -2$.

$$3\boxed{(4 - y)} + 3y = -2$$

$$12 - 3y + 3y = -2$$

$$14 = 0$$

The variable is eliminated and a false statement results.

Since there is no ordered pair (x, y) for which $14 = 0$ we conclude that there is *no solution* and the lines are *parallel.*

Coinciding Lines

The next example illustrates how to recognize a system of coinciding lines when using the substitution method.

Example 4: Solve:

$$x + 2y = -3$$
$$3x + 6y = -9$$

Solution: Solve the first equation for x.

$$x + 2y = -3$$

$$x = -3 - 2y \qquad \text{Solved for } x$$

Now substitute $x = -3-2y$ in the second equation and solve for y.

$$3x + 6y = -9$$

$$3\boxed{(-3-2y)} + 6y = -9$$

$$-9 - 6y + 6y = -9$$

$$0 = 0$$

The variable is eliminated, and a true statement results.

This means that each of the given equations is equivalent to the same equation so their graphs are *coincident*, and any ordered pair that satisfies *either* equation is a solution for the system.

> **To solve a system of equations by the substitution method:**
>
> 1. Solve either equation for one variable in terms of the other variable.
>
> *2. Substitute this result in the other equation and solve.
>
> 3. Substitute the result from step 2 in either of the original equations and solve for the remaining variable.
>
> * If both variables are eliminated and the resulting statement is *false*, then the lines are parallel and there is no solution.
>
> * If both variables are eliminated and the resulting statement is *true*, then the lines coincide and any solution to either equation is a solution to the system.

EXERCISE 10.4

Solve the following systems using the substitution method. Some systems may be parallel (no solution) or coinciding (infinite number of solutions).

1. $2x + y = 4$
   $y = x + 1$

2. $x + y = 6$
   $x = y + 2$

3. $x - y = 1$
   $y = 2x - 3$

4. $x = 2y + 8$
   $2x + y = 8$

5. $x - y = 4$
   $y = 2 - x$

6. $x + y = -3$
   $x = 2y$

7. $x + 2y = 7$
   $x - y = -5$

8. $x - y = 1$
   $2x - y = 3$

9. $y = 3 - 4x$
   $4x - y = 3$

10. $3x - 2y = 8$
    $2y - 3x = 4$

11. $x + y = 10$
    $x + y = 6$

12. $2x + 3y = -6$
    $x - y = -3$

13. $2x + y = -1$
    $3x - 2y = -12$

14. $3x - 8y = -1$
    $5x + 14y = 12$

15. $2x - y = 4$
$-4x + 2y = -8$

16. $2x + 3y = 4$
$-2x - 3y = 2$

17. $2x - 5y = 1$
$4x - 3y = 2$

18. $4x - y = 3$
$-4x - 3y = 1$

19. $2x + 5y = -4$
$8x - 3y = 7$

20. $4x + 15y = 4$
$-3x + 10y = -3$

21. $4x - 6y = 0$
$2x + y = 4$

22. $4x + 3y = 5$
$6x + 5y = 3$

23. $2x - y = 1$
$-4x + 2y = -2$

24. $12x + 8y = 8$
$9x + 6y = 6$

25. $\frac{1}{2}x + y = 2$
$-x + y = 2$

26. $2x + y = 1$
$6x - 3y = 0$

Answers to odd-numbered exercises on page 488.

10.5 SOLVING WORD PROBLEMS USING SYSTEMS OF EQUATIONS

Word problems involving more than one unknown can frequently be simplified by using systems of linear equations to solve them. In each of the examples that follow, there are two unknowns and two equations.

Number Relation Problems

Example 1: A father is twice as old as his son. In six years the father will be three times as old as the son was three years ago. How old is each of them now?

Solution: Let f = the father's age now and s = the son's age now.

The father's age	is	twice the son's age.
f	$=$	$2 \cdot s$

The father's age in 6 years	will be	3 times	the son's age 3 years ago
$f + 6$	$=$	$3 \cdot$	$(s - 3)$

Our system of equations is:

$$f = 2s$$
$$f + 6 = 3(s - 3)$$

Substitute $f = 2s$ in equation two.

$$\boxed{f} + 6 = 3(s - 3)$$
$$\boxed{2s} + 6 = 3s - 9$$
$$-s = -15$$
$$s = 15 = \text{son's age now}$$

Substitute $s = 15$ in $f = 2s$.

$$f = 2 \cdot \boxed{15} = 30 = \text{father's age now}$$

Check in the original problem: The father is 30, which is twice 15, the son's age now. In six years, when the father is 36, he will be three times as old as the son was three years ago when he was 12.

Example 2: The sum of two numbers is 43 and their difference is 13. Find the numbers.

Solution: Let x = the first number and y = the second number. Then

$$\begin{array}{rl} x + y = 43 & \text{The sum is 43.} \\ x - y = 13 & \text{The difference is 13.} \\ \hline 2x \quad = 56 & \text{Add the equations.} \\ x = 28 & \end{array}$$

Substitute $x = 28$ in the equation $x + y = 43$.

$$\boxed{28} + y = 43$$

$$y = 15$$

The two numbers are 28 and 15.

Check:

$x + y = 43$	$x - y = 13$
$\boxed{28} + \boxed{15} = 43$ ✓	$\boxed{28} - \boxed{15} = 13$ ✓

Example 3: There is a famous problem from the ancient Chinese (1st century B.C.) that states that there are a quantity of rabbits and pheasants in a cage. In all, the cage contains 35 heads and 94 feet. How many of each animal are in the cage?

Solution: Let r = the number of rabbits and p = the number of pheasants. Then $r + p = 35$ The number of heads.

$4r + 2p = 94$ The number of feet.

To solve, multiply the first equation by -2 and add to the second equation.

$$\begin{array}{rl} -2r - 2p = & -70 \\ 4r + 2p = & 94 \\ \hline 2r \quad = & 24 \\ r = & 12 \end{array}$$

Now substitute $r = 12$ in $r + p = 35$.

$$\boxed{12} + p = 35$$

$$p = 23$$

The solution is 12 rabbits and 23 pheasants.

Check:

Heads	*Feet*
rabbits + pheasants = 35	rabbits + pheasants = 94
$\boxed{12} + \boxed{23} = 35$ ✓	$\boxed{12} \cdot 4 + \boxed{23} \cdot 2 \stackrel{?}{=} 94$
	$48 + 46 = 94$ ✓

Coin and Mixture Problems

Coin problems and mixture problems are solved using a similar approach. In each we find a *quantity equation* and a *value equation*.

Example 4: A woman has a total of 20 coins, consisting only of dimes and quarters, with a total value of \$3.95. How many does she have of each coin?

Solution: Let d = the number of dimes and q = the number of quarters.

There are two equations to be formulated, a quantity equation and a value equation.

$$\boxed{\text{Quantity equation}}$$

$$(\text{no. of dimes}) + (\text{number of quarters}) = (\text{total no. of coins})$$

$$d + q = 20$$

$$\boxed{\text{Value equation}}$$

$$(\text{value of dimes}) + (\text{value of quarters}) = (\text{total value of coins})$$

$$(10¢)\ (\text{no. of dimes}) + (25¢)\ (\text{no. of quarters}) = 395¢ \quad (\$3.95 = 395¢)$$

$$10d + 25q = 395$$

Solve $d + q = 20$ for d and substitute into $10d + 25q = 395$.

$$d = \boxed{20 - q}$$

$$10(\boxed{20 - q}) + 25q = 395$$

$$200 - 10q + 25q = 395$$

$$15q = 195$$

$$q = 13 \text{ quarters}$$

Substitute $q = 13$ in $d + q = 20$ to find d.

$$d + \boxed{13} = 20$$

$$d = 7 \text{ dimes}$$

Check:

13 quarters	=	13 × \$.25	=	\$3.25
+ 7 dimes	=	7 × \$.10	=	.70
20 coins	=	total value	=	\$3.95 ✓

Example 5: A garden center wants to make a mixture of grass seed from type X seed worth \$1.35 per pound and type Y seed worth \$1.00 per pound. How much of each type will be needed to make 100 pounds of mixture worth \$1.07 per pound?

Solution: Let x = the number of pounds of type X seed and y = the number of pounds of type Y seed.

Quantity Equation

(no. of lbs of X)	+	(no. of lbs of Y)	=	no. of lbs of mixture
x	+	y	=	100

Value Equation

(Value of type X)	+	(Value of type X)	=	(Total value of mixture)
($x \cdot$ \$1.35 per lb)	+	($y \cdot$ \$1.00 per lb)	=	(100 lbs $\cdot$ \$1.07 per lb.)

$135x + 100y = (100)(107)$ Change all values to cents.

Our system of equations is:

$$x + y = 100$$
$$135x + 100y = 10700$$

Solving for x in equation one yields $x = 100 - y$.
Substituting in equation two gives us:

$$135(\boxed{100 - y}) + 100y = 10700$$
$$13500 - 135y + 100y = 10700$$
$$-35y = -2800$$
$$y = 80 \text{ pounds of type } Y \text{ seed}$$
$$x = 100 - y = 100 - \boxed{80} = 20 \text{ pounds of type } X \text{ seed}$$

EXERCISE 10.5

Solve each of the following problems using a system of equations.

1. The sum of two numbers is 46 and their difference is 12. Find the numbers.

$x + y = 46$ — (1)
$x - y = 12$ — (2)
$2x = 58$
$x = \frac{58}{2}$
$x = 29$ — (3)

(3) in (1):
$29 + y = 46$
$y = 46 - 29$
$y = 17$ ✓

2. The sum of two numbers is 50. If the larger number is equal to two more than three times the smaller number, what are the numbers?

$2l + 2w = 58$ — (1)
$l = 4w + 4$ — (2)

3. The length of a rectangle is 4 inches more than four times the width. If the perimeter of the rectangle is 58 inches, find the length and the width.

4. The length of a rectangle is 4 cm more than twice its width. Find the dimensions of the rectangle if the perimeter is 50 cm.

$x + y = 13$ — (1)
$5x + 20y = 115$ — (2)

5. A man has some five-dollar bills and twenty-dollar bills. If he has a total of 13 bills and the total value of the money is $155, how many of each denomination does he have?

6. The total value of 30 coins is $4.50. If the coins consist of quarters and dimes, how many of each coin are there?

$x + y = 52$ — (1)
$x = y + 6$ — (2)

7. Two truck drivers took turns driving on a long trip. If one man drove 6 hours more than the other, and together they drove 52 hours, how long did each man drive?

8. A man's estate of $85,000 is to be divided between his wife and his son. The wife receives $11,000 less than twice what the son receives. How much does each receive?

9. If chocolates cost $4.10 a pound and caramels cost $3.20 a pound, how many pounds of each are there in a 6-pound mixture of chocolates and caramels costing a total of $23.25?

ch + c = 6
4.10ch + 3.20c = 23.25

10. A theater charges $3.50 for adults and $2.00 for children. If there are 109 people attending and the total receipts amount to $341, how many of each type of ticket were sold?

11. A boy has $3.50 in nickels and dimes. How many does he have of each type of coin if he has 45 coins in all?

m + d = 45 — ①
5m + 10d = 3.50 — ②

12. The sum of two angles in a triangle is 90°. Their difference is 34°. Find the angles.

13. The sum of two angles is 180°. Their difference is 72°. Find the angles.

14. A rectangular pasture is enclosed by 1200 meters of fencing. If the length is four times the width, what are the dimensions of the field?

15. Fred is three times as old as his son Freddie. In 10 years, Fred will be 2 years more than twice as old as Freddie will be. How old is each of them now?

16. A father and his three children took a bus ride. An adult ticket cost \$6 more than a child's. If the total cost of the tickets was \$54, how much did each type of ticket cost?

17. The admission charge at a school play is \$1.00 for students and \$3.00 for non-students. If a total of 247 people attended the play and the total money taken in was \$377, how many of each type of ticket were sold?

18. A hat and a pair of gloves cost $56. The hat costs $8 less than three times the cost of the gloves. Find the cost of each.

19. Two dozen cookies and 1 loaf of bread cost $3.85. A half dozen cookies and 2 loaves of bread cost $3.15. What is the cost of a dozen cookies and what is the cost of a loaf of bread?

$2d + 1 = 3.85$ — ①

$\frac{1}{2}d + 2 = 3.15$ — ②

20. How many nickels and quarters would you have if you had 15 coins worth $1.95?

21. The difference in two numbers is 12. The larger number is 8 less than twice the smaller. Find the numbers.

$x - y = 12$ — ①

$x = 2y - 8$ — ②

22. The sum of Max's age and five times David's age is 19. The sum of David's age and five times Max's age is 47. How old are Max and David?

23. The perimeter of a rectangle is 26. The difference between the length and twice the width is 1. Find the dimensions of the rectangle.

24. Tickets to a high school football game are $1.00 for students and $2.50 for non-students. 610 people paid to see the game, and the total gate receipts were $892. How many students and how many non-students paid to see the game?

25. At a dog show, there were a certain number of dogs and their owners. If there were a total of 78 heads and 240 feet, how many dogs and how many owners were in attendance?

26. A grocer mixes navy beans, which sell for $.60 a pound, with kidney beans selling for $.90 a pound. How many pounds of each should he use in a 30-pound mixture of the two kinds of beans that sells for $.80 a pound?

27. The average of two numbers is 80 and one of the numbers is three times the other. Find the numbers.

Answers to odd-numbered exercises on page 488.

10.6 Chapter Summary

(10.1) The graphs of two lines of the form $y = mx + b$:

1. intersect if the coefficients of x are different;
2. are parallel if the coefficients of x are the same but the y-intercepts are different;
3. are coincident if the two equations are identical.

(10.2) To determine whether a point (x,y) is a solution to a system of linear equations, check it in *each* equation.

(10.2) (10.3) (10.4) When solving a system of linear equations, three different possibilities can occur:

1. One solution (x,y).
   Graphing: The lines intersect in one point (x,y), which is the solution.
   Addition or substitution method: Yields one ordered pair (x,y) which is the solution.
2. No solution.
   Graphing: the lines are parallel.
   Addition or substitution method: Both variables are eliminated and the resulting statement is false.
3. An infinite number of solutions (every solution to one equation is also a solution to the other equation).
   Graphing: The lines coincide.
   Addition or substitution method: Both variables are eliminated and the resulting statement is true.

Examples

(10.1)

$y = \boxed{3}x - 4$
$y = \boxed{7}x + 2$
Intersecting

$y = \boxed{-4}x - 1$
$y = \boxed{-4}x + 7$
Parallel

$y = 5x - 1$
$y = 5x - 1$
Coincident

(10.2) Solve: $x - y = 5$
$x + 2y = -1$

Solution by graphing:

$x - y = 5$

x	y
0	−5
3	−2
5	0

$x + 2y = -1$

x	y
1	−1
3	−2
5	−3

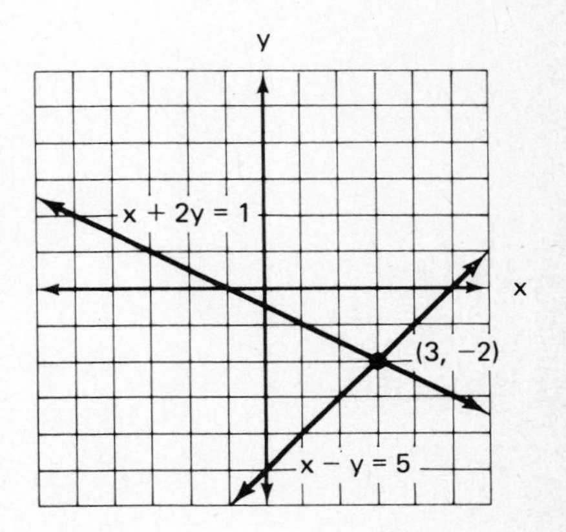

Solution by the addition method:

Multiply the first equation by −1 and add to the second equation.

$$\begin{array}{r} -x + y = -5 \\ x + 2y = -1 \\ \hline 3y = -6 \\ y = -2 \end{array}$$

Substituting $y = -2$ in either original equation yields $x = 3$. The solution is $(3, -2)$.

Solution by the substitution method:

Solve the first equation for x:

$$x = y + 5$$

Substitute for x in the second equation:

$$\begin{aligned} x + 2y &= -1 \\ (y + 5) + 2y &= -1 \\ 3y &= -6 \\ y &= -2 \end{aligned}$$

Now substitute $y = -2$ in either original equation; yielding $x = 3$. Again, the solution is $(3, -2)$.

Exercise 10.6 Chapter Review

(10.1) Determine whether the pairs of equations are intersecting, parallel, or coincident without graphing.

1. $y = 2x + 3$
   $y = 5x - 1$

2. $y = -3x + 2$
   $y = -3x - 5$

3. $3x + 6y = 6$
   $2x + 4y = 4$

4. $2x - y = 3$
   $y + 2 = 3x$

5. $2x + 3y = 4$
   $4x + 2y = -1$

6. $y - \frac{1}{2}x = -3$
   $-4y + 2x = 9$

(10.2) Solve by graphing.

7. $3x - 2y = 1$
   $x + 2y = 3$

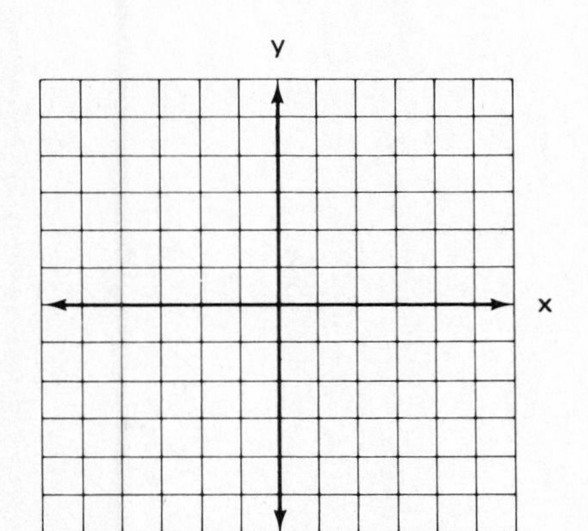

8. $2x - y = -1$
   $2x - y = 2$

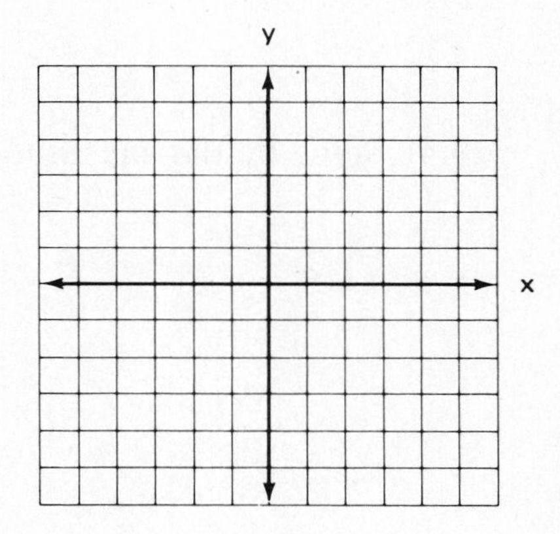

9. $2x - 2y = 4$
   $3y = 3x - 6$

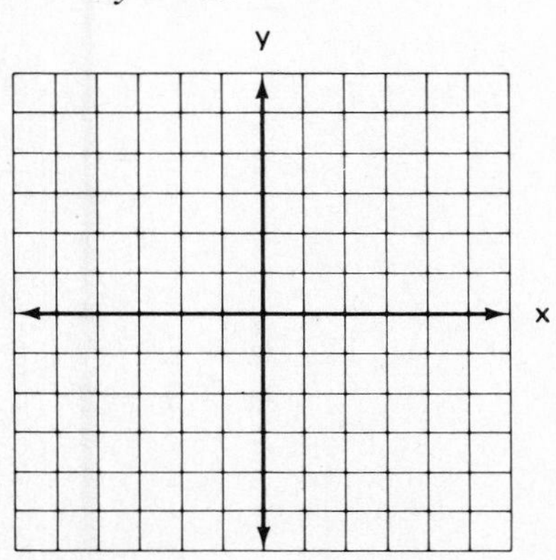

10. $2x + y = 4$
    $3x - y = 1$

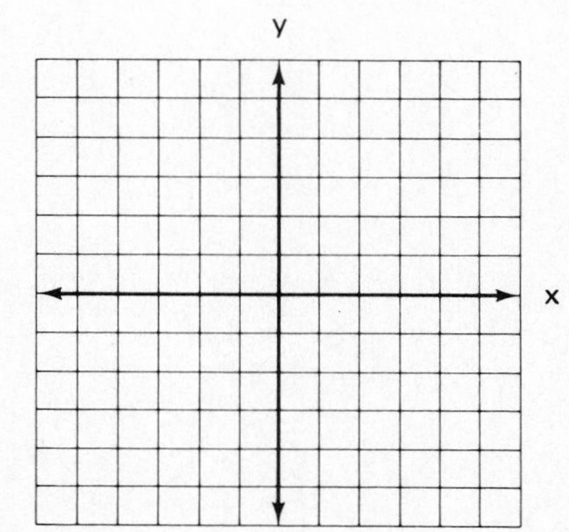

(10.3) Solve by the addition method.

11. $3x - 2y = 1$
    $2x + 2y = 4$

12. $x + 2y = 3$
    $2x + 4y = -2$

13. $3x + 6y = 14$
    $2x - 3y = 7$

14. $3x + 10y = 2$
    $x + 3y = 0$

15. $4x - 6y = -2$
    $3y - 2x = 1$

16. $3x - 9y = -4$
    $-9x + 3y = -4$

(10.4) Solve by the substitution method.

17. $x + y = 5$
    $x - y = 1$

18. $x + y = 3$
    $2x - y = 3$

19. $x + 2y = -3$
    $-2x - 9y = 6$

20. $x + 2y = 4$
    $2x + 3y = 3$

21. $3y - 6x = 3$
    $2x - y = 4$

22. $3x - 2y = -1$
    $2x + y = 3$

(10.5) Solve the word problems using a system of equations.

23. The sum of two numbers is 56 and their difference is 8. Find the numbers.

24. A rectangular aluminum plate has a perimeter of 22 cm. If the length is 2 more than twice the width, find the dimensions of the plate.

25. A child's piggy bank contains only pennies and nickels. If it contains 126 coins having a value of $4.74, how many pennies and how many nickels are in the bank?

26. A father is three times as old as his daughter. In 9 years, the father will be four times as old as his daughter was 2 years ago. How old is each of them now?

27. A total of 28 stamps costs $7.10. If they consist only of 20¢ and 30¢ stamps, how many of each denomination are there?

28. A hardware store mixes 200 pounds of grass seed costing \$1.48 per pound. The mixture consists of seed costing \$1.20 per pound and seed costing \$1.60 per pound. How many pounds of each type are in the mixture?

$x + y = 90°$ — (1)

$x = 4y - 5$ — (2)

29. The sum of two angles is 90°. The larger angle is 5° less than four times the smaller angle. How large are the angles?

$H. + W. = 43$ — (1)

$W = H. + 8$ — (2)

30. A husband and wife take turns driving on a trip. If the wife drives 8 hours more than the husband and together they drive 43 hours, how long does each drive?

Answers to odd-numbered exercises on page 489.

Chapter 10 Achievement Test

Name ______________________

Class ______________________

This test should be taken before you are tested in class on the material in Chapter 10. Solutions to each problem and the section where the type of problem is found are given on page 489.

Determine whether the given pairs of equations are intersecting, parallel, or coincident, without graphing:

1. $6x - 3y = -9$
   $2y = 4x + 6$

2. $3x - 2y = 4$
   $x + 3y = -2$

3. $2x + y = -1$
   $-3y = 6x + 5$

Solve by graphing:

4. $x - 2y = 3$
   $2y = x + 4$

5. $2x - y = 3$
   $x - 2y = 0$

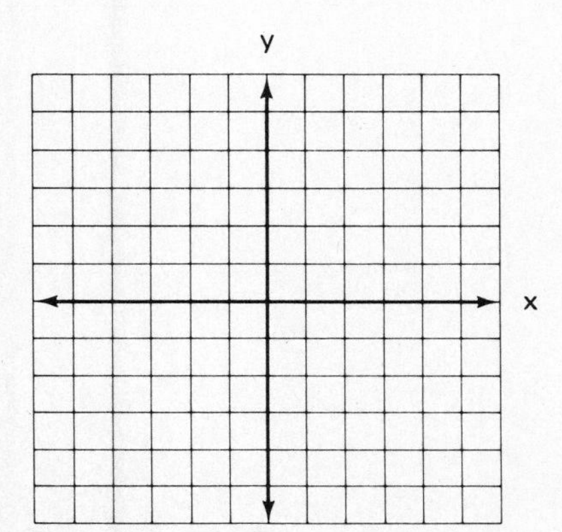

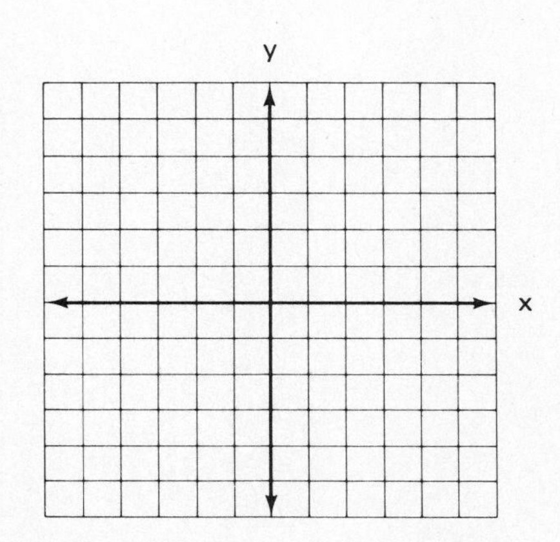

1. ______________________

2. ______________________

3. ______________________

6. $2x - 4y = 2$
$-3x + 6y = -3$

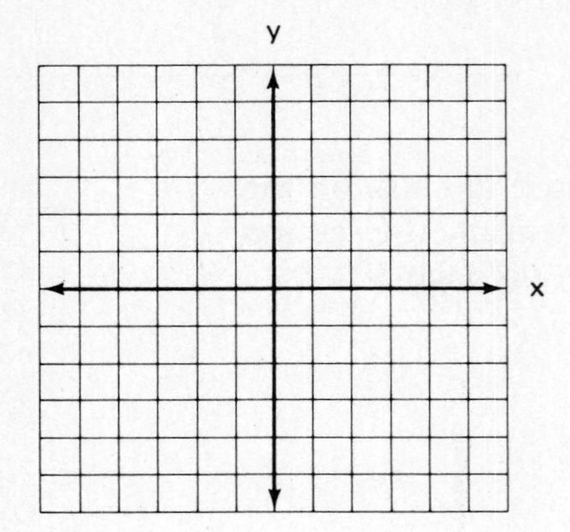

Solve using the addition method:

7. $2x + y = 3$
$4x - y = 3$

7. ____________________

8. $x + 2y = 4$
$3x - 4y = 12$

8. ____________________

9. $x - 2y = -3$
$-3x + 6y = 9$

9. ____________________

10. $2x - 3y = 2$
$8x - 12y = -1$

10. ____________________

Solve using the substitution method:

11. $x - y = -2$
$7x - 3y = 6$

11. ____________________

12. $2x - y = 4$
$-4x + 2y = -8$

12. ____________________

13. $x = 4$
$2x - y = 5$

13. ____________________

14. $4x + 3y = -1$
$4x + 3y = 3$

14. ____________________

Solve each of the following word problems using a system of equations:

15. The sum of two numbers is 70. If the first number is 10 more than twice the second number, what are the numbers?

15. ____________________

16. The perimeter of a rectangle is 130. If the length is 4 times the width, find the dimensions of the rectangle.

16. ____________________

17. A collection of 37 coins has a total value of $5.80. If the coins consist only of quarters and dimes, how many of each coin are there?

17. ____________________

18. One type of candy worth \$2.30 per pound is mixed with a second type of candy worth \$1.80 per pound to form a mixture worth \$2.00 per pound. If the total weight of the mixture is 30 pounds, how many pounds of each type of candy are there in it?

18. ____________________

19. The difference in the ages of a man and his granddaughter is 53 years. Ten years from now the grandfather will be ten times as old as the granddaughter is now. How old is each of them?

19. ____________________

11
Radical Expressions

11.1 SQUARE ROOTS

b is a square root of a if $b^2 = a$

4 is a square root of 16 because $4^2 = 16$.

-4 is also a square root of 16 because $(-4)^2 = 16$.

In fact, every positive number must have two square roots, which have opposite signs of each other.

Zero has only one square root and that is zero.

Negative numbers do not have square roots that are real numbers, because if we square any real number (except zero) we get a *positive* number.

Example 1: What are the square roots of 25?

The square roots of 25 are 5 and -5 since $5^2 = 25$ and $(-5)^2 = 25$.

The symbol $\sqrt{\ }$ is called a **radical sign** and is used to name the *positive square root* of a number. This positive square root is called the **principal square root.** The number under the radical is called the **radicand.**

Example 2:

(a) $\sqrt{9} = 3$ ← principal square root (↙ radical sign; ↖ radicand)

(b) $-\sqrt{9} = -3$

(c) $\sqrt{49} = 7$

(d) $-\sqrt{49} = -7$

(e) $\pm\sqrt{49} = \pm 7$ $\pm$ means "plus or minus" so the result is $+7$ or -7.

(f) $\sqrt{-9}$ is not a real number.

(g) $\sqrt{0} = 0$

Square roots of perfect squares such as $\sqrt{a^2}$ present an interesting situation.

If $a = 4$ we have $\sqrt{4^2} = \sqrt{16} = 4$

If $a = -4$ we have $\sqrt{(-4)^2} = \sqrt{16} = 4$

As you can see the result is positive in both cases.

If we write $\sqrt{a^2} = a$ we will be incorrect whenever a is negative. To make sure that the result is always positive we write:

$$\sqrt{a^2} = |a| \text{ (the absolute value of } a\text{)}$$

Example 3:

(a) $\sqrt{3^2} = |3| = 3$

(b) $\sqrt{(-3)^2} = |-3| = 3$

(c) $\sqrt{x^2} = |x|$ Remember, we don't know whether x represents a positive or negative number.

(d) $\sqrt{(3x)^2} = |3x|$

(e) $\sqrt{x^2 - 2x + 1} = \sqrt{(x-1)^2} = |x - 1|$

EXERCISE 11.1

Find the square roots of the following numbers:

1. 25   2. 4   3. 16   4. 1

5. 100 6. 36 7. 64 8. 81

9. 121 10. 169 11. 144 12. 225

Evaluate the following. If not possible, say so.

13. $\sqrt{4}$ 14. $\sqrt{16}$ 15. $\sqrt{1}$ 16. $\sqrt{16}$

17. $\sqrt{49}$ 18. $-\sqrt{49}$ 19. $\sqrt{-49}$ 20. $\pm\sqrt{49}$

21. $-\sqrt{81}$ 22. $\sqrt{-144}$ 23. $-\sqrt{169}$ 24. $\pm\sqrt{400}$

Simplify:

25. $\sqrt{a^2}$ 26. $\sqrt{h^2}$ 27. $\sqrt{9x^2}$ 28. $\sqrt{(-4)^2}$

29. $\sqrt{(-x)^2}$ 30. $\sqrt{0}$ 31. $\sqrt{(x-2)^2}$ 32. $\sqrt{x^2-4x+4}$

33. $-\sqrt{x^2-4x+4}$

Answers to odd-numbered exercises on page 490.

11.2 RATIONAL, IRRATIONAL, AND REAL NUMBERS

> A **rational number** is a number that can be expressed as a fraction, $\frac{a}{b}$, where a and b are integers and $b \neq 0$. (The integers are the positive and negative counting numbers and zero.)

Rational numbers have the property that they may always be expressed as either **terminating** or **repeating decimals.**

Example 1: The following are rational numbers:

(a) $\frac{3}{4}$ Fraction

(b) 6 6 can be expressed as the fraction $\frac{6}{1}$.

(c) .75 is a *terminating* decimal.

(d) .353535 . . . . is a *repeating* decimal.

(e) $\frac{8}{5}$ Fraction

An **irrational number** is a number that *cannot* be expressed as a fraction $\frac{a}{b}$, where a and b are integers.

One source of irrational numbers is square roots of numbers that are not perfect squares, like $\sqrt{2}$ and $\sqrt{3}$. There are many other irrational numbers that do not involve square roots; for instance, π.

When written in decimal form, irrational numbers never terminate, nor do they have a block of repeating digits.

Example 2: The following are irrational numbers:

(a) $\sqrt{5}$

(b) $\sqrt{12}$

(c) $-\sqrt{6}$

(d) $\pi = 3.14159265\ldots.$ Neither repeating nor terminating.

(e) $\frac{3}{\sqrt{3}}$

(f) .463821392714 . . . . Neither repeating nor terminating.

The **real numbers** consist of all the rational numbers plus all the irrational numbers. Each real number is represented by a point on the *real number line*.

Example 3: The following are real numbers:

(a) $\frac{9}{16}$

(b) $\sqrt{21}$

(c) $-\sqrt{5}$

(d) .231231231 . . . . Repeating decimal (rational).

(e) .025173846129 . . . . Non-repeating decimal (irrational).

(f) .341 Terminating decimal (rational).

Note: $\sqrt{-16}$ is *not* a real number since there is no rational or irrational number we could square that would equal -16.

EXERCISE 11.2

Put an x in the columns that apply to the number on the left.

	NUMBER	RATIONAL	IRRATIONAL	REAL
1.	3			
2.	$\frac{0}{4}$			
3.	$\sqrt{8}$			
4.	$\frac{22}{7}$			
5.	π			
6.	6.11111. . .			
7.	.1262626. . .			
8.	$\sqrt{9}$			
9.	$-\sqrt{9}$			
10.	$\sqrt{-9}$			
11.	$-\sqrt{-9}$			
12.	6.301425187. . .			
13.	0			
14.	-6			
15.	$17\frac{4}{5}$			

Answers to odd-numbered exercises on page 491.

11.3 MULTIPLYING AND FACTORING RADICALS

Compare the following statements:

1. $\sqrt{4} \cdot \sqrt{9} = 2 \cdot 3 = 6$
2. $\sqrt{4 \cdot 9} = \sqrt{36} = 6$

This suggests that $\sqrt{4} \cdot \sqrt{9} = \sqrt{4 \cdot 9}$, and indeed this is true in general.

$\sqrt{a} \cdot \sqrt{b} = \sqrt{a \cdot b}$; $a, b \geq 0$

Note that $\sqrt{-4} \cdot \sqrt{-9} \neq \sqrt{36}$, since $\sqrt{36} = 6$ but $\sqrt{-4}$ and $\sqrt{-9}$ are not real numbers.

This rule can be used for both multiplication and factoring of radicals.

Example 1: Multiply the following:

(a) $\sqrt{2} \cdot \sqrt{3} = \sqrt{2 \cdot 3} = \sqrt{6}$

(b) $\sqrt{2} \cdot \sqrt{32} = \sqrt{2 \cdot 32} = \sqrt{64} = 8$

(c) $\sqrt{7} \cdot \sqrt{7} = \sqrt{7 \cdot 7} = \sqrt{49} = 7$

When factoring, it is desirable to factor a product into two factors, one of which is a perfect square. In this way, the radical can be simplified.

Example 2: Simplify $\sqrt{28}$ by factoring.

$\sqrt{28} = \sqrt{\boxed{4} \cdot 7}$ 4 is a perfect square.

$= \sqrt{4} \cdot \sqrt{7}$ $\sqrt{a \cdot b} = \sqrt{a} \cdot \sqrt{b}$

$= 2\sqrt{7}$ $\sqrt{4} = 2$

We could also have factored $\sqrt{28}$ as $\sqrt{2 \cdot 14}$, but neither 2 nor 14 is a perfect square, so nothing is gained.

Example 3: Simplify $\sqrt{75}$

$\sqrt{75} = \sqrt{\boxed{25}} \cdot \sqrt{3}$ 25 is a perfect square.

$= 5\sqrt{3}$

Example 4: Simplify $\sqrt{48}$

$\sqrt{48} = \sqrt{\boxed{16}} \cdot \sqrt{3}$ 16 is a perfect square.

$= 4\sqrt{3}$

You may have noticed that $\sqrt{48}$ could have been factored as:

$\sqrt{48} = \sqrt{\boxed{4}} \cdot \sqrt{12}$ 4 is a perfect square.

$= 2\sqrt{12}$

$= 2\sqrt{\boxed{4}} \cdot \sqrt{3}$ 4 is a perfect square.

$= 2 \cdot 2 \cdot \sqrt{3}$

$= 4\sqrt{3}$

which is the same answer as before. However, it is always easier to factor out the *largest* perfect square immediately.

When the radicand involves letters, we must restrict them to positive values, since square roots of negative numbers are not real numbers. For the remainder of this chapter, we will assume that all letters represent positive numbers, and therefore we will not need to use absolute values.

Example 5: Simplify the following:

(a) $\sqrt{x^3} = \sqrt{x^2} \cdot \sqrt{x} = x\sqrt{x}$ x^2 is a perfect square.

(b) $\sqrt{x^5} = \sqrt{x^4} \cdot \sqrt{x} = x^2\sqrt{x}$ $x^4 = (x^2)^2$ is a perfect square.

(c) $\sqrt{x^8} = x^4$

$x^8 = (x^4)^2$ and is a perfect square already. In fact, every even power of a variable is a perfect square, and its square root is the variable raised to half that power.

Example 6: Simplify $\sqrt{12x^3}$

$\sqrt{12x^3} = \sqrt{\boxed{4} \cdot 3 \cdot \boxed{x^2} \cdot x}$ 4 and x^2 are perfect squares.

$= \sqrt{4x^2} \cdot \sqrt{3x}$ Separate the perfect squares.

$= 2x\sqrt{3x}$

Example 7: Simplify $\sqrt{50x^5y^6}$

$\sqrt{50x^5y^6} = \sqrt{\boxed{25} \cdot 2 \cdot \boxed{x^4} \cdot x \cdot \boxed{y^6}}$ 25, x^4, and y^6 are perfect squares.

$= \sqrt{25x^4y^6} \cdot \sqrt{2x}$ Separate the perfect squares.

$= 5x^2y^3\sqrt{2x}$

Example 8: Multiply and simplify $\sqrt{2x} \cdot \sqrt{14xy}$

$\sqrt{2x} \cdot \sqrt{14xy} = \sqrt{28x^2y}$ Multiply.

$= \sqrt{\boxed{4} \cdot 7\,\boxed{x^2} \cdot y}$ Factor and find the perfect squares.

$= \sqrt{4x^2} \cdot \sqrt{7y}$ Separate the perfect squares.

$= 2x\sqrt{7y}$

Example 9: Multiply and simplify $\sqrt{2a^3b} \cdot \sqrt{12a^3}$

$\sqrt{2a^3b} \cdot \sqrt{12a^3} = \sqrt{24a^6b}$ Multiply.

$= \sqrt{\boxed{4} \cdot 6 \cdot \boxed{a^6} \cdot b}$ Factor and find the perfect squares.

$= 2a^3\sqrt{6b}$

EXERCISE 11.3

Simplify:

1. $\sqrt{3} \cdot \sqrt{2}$
2. $\sqrt{5} \cdot \sqrt{7}$
3. $\sqrt{6} \cdot \sqrt{6}$
4. $\sqrt{18}$
5. $\sqrt{40}$
6. $\sqrt{20}$
7. $\sqrt{72}$
8. $\sqrt{50}$
9. $\sqrt{12}$
10. $\sqrt{32}$
11. $\sqrt{y^4}$
12. $\sqrt{x^6}$
13. $\sqrt{16x^6}$
14. $\sqrt{a^9b^7}$
15. $\sqrt{54z^5}$
16. $\sqrt{t^{30}}$
17. $\sqrt{t^{25}}$
18. $\sqrt{28x^7y^4}$
19. $\sqrt{(x-1)^2}$
20. $\sqrt{(x-1)^4}$

Multiply and simplify:

21. $\sqrt{10}\sqrt{5}$
22. $\sqrt{6}\sqrt{8}$
23. $\sqrt{5}\sqrt{15}$
24. $\sqrt{2}\sqrt{14}$
25. $\sqrt{6}\sqrt{3}$
26. $\sqrt{x^5}\sqrt{x^3}$

27. $\sqrt{a^3}\sqrt{a}$ 28. $\sqrt{3x}\cdot\sqrt{4x}$ 29. $\sqrt{6xy}\sqrt{8x}$

30. $\sqrt{3x^3y}\sqrt{8xy^5}$ 31. $\sqrt{5x^5}\sqrt{10x^{10}}$ 32. $\sqrt{2x^2y}\sqrt{18x^3y}$

33. $\sqrt{8x^3}\cdot\sqrt{2x}\sqrt{3x^5}$ 34. $\sqrt{10ab}\sqrt{20a^3b}\sqrt{50a^4b^4}$

35. $\sqrt{3xy}\sqrt{2x^3}\sqrt{6y^5}$

Answers to odd-numbered exercises on page 492.

11.4 ADDING AND SUBTRACTING RADICALS

Like or **similar** square roots are square roots that have the same radicand (number under the radical). Like radicals are added or subtracted by applying the distributive rule and adding or subtracting the coefficients.

Example 1: Add $5\sqrt{3}+2\sqrt{3}$

$$5\sqrt{3}+2\sqrt{3} = (5+2)\sqrt{3}$$ Add the coefficients using the distributive rule.

$$= 7\sqrt{3}$$

Example 2: Combine $2\sqrt{5}-8\sqrt{5}+3\sqrt{5}$

$$2\sqrt{5}-8\sqrt{5}+3\sqrt{5} = (2-8+3)\sqrt{5}$$ Combine the coefficients.

$$= -3\sqrt{5}$$

In order to add or subtract, the radicals must be *like*. If they are not, we apply the methods of the last section and attempt to make them *like*.

Example 3: Add $\sqrt{12}+\sqrt{27}$

$$\sqrt{12}+\sqrt{27} = \sqrt{4\cdot 3}+\sqrt{9\cdot 3}$$ Simplify each radical.

$$= 2\sqrt{3}+3\sqrt{3}$$ Combine the like radicals.

$$= 5\sqrt{3}$$

Example 4: Combine $\sqrt{50}+\sqrt{32}-\sqrt{18}$

$$\sqrt{50}+\sqrt{32}-\sqrt{18} = \sqrt{25\cdot 2}+\sqrt{16\cdot 2}-\sqrt{9\cdot 2} \qquad \text{Simplify.}$$
$$= 5\sqrt{2}+4\sqrt{2}-3\sqrt{2} \qquad \text{Combine.}$$
$$= 6\sqrt{2}$$

Example 5: Combine $3\sqrt{12}-2\sqrt{75}$

$$3\sqrt{12}-2\sqrt{75} = 3\sqrt{4\cdot 3}-2\sqrt{25\cdot 3} \qquad \text{Simplify.}$$
$$= 3\cdot 2\sqrt{3}-2\cdot 5\sqrt{3}$$
$$= 6\sqrt{3}-10\sqrt{3} \qquad \text{Combine.}$$
$$= -4\sqrt{3}$$

Example 6: Combine $\sqrt{72}-\sqrt{48}$

$$\sqrt{72}-\sqrt{48} = \sqrt{36\cdot 2}-\sqrt{16\cdot 3}$$
$$= 6\sqrt{2}-4\sqrt{3}$$

Since the radicals are not *like* radicals, they cannot be combined further.

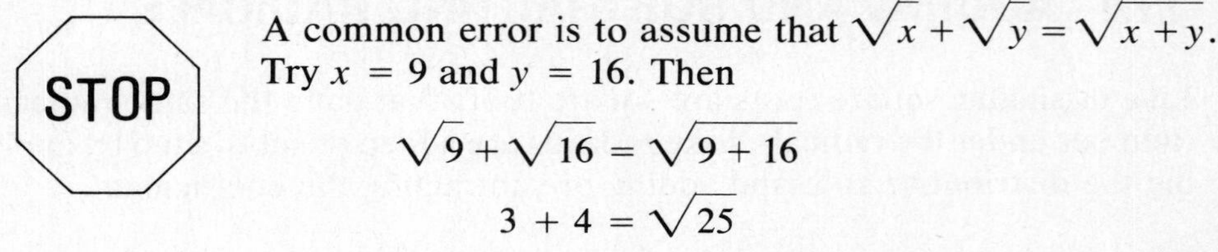

A common error is to assume that $\sqrt{x}+\sqrt{y}=\sqrt{x+y}$. Try $x = 9$ and $y = 16$. Then

$$\sqrt{9}+\sqrt{16} = \sqrt{9+16}$$
$$3 + 4 = \sqrt{25}$$

But $7 \neq 5$!

The sum of square roots is *not* equal to the square root of the sum.

EXERCISE 11.4

Combine where possible:

1. $3\sqrt{5}+2\sqrt{5}$
2. $5\sqrt{2}-7\sqrt{2}$
3. $3\sqrt{x}+7\sqrt{x}$
4. $6\sqrt{3}+3\sqrt{x}-8\sqrt{x}$
5. $2\sqrt{10}-3\sqrt{3}+4\sqrt{5}$
6. $5\sqrt{6}+2\sqrt{3}-4\sqrt{6}+\sqrt{3}$

7. $\sqrt{18} - \sqrt{8}$

8. $\sqrt{50} - \sqrt{8}$

9. $2\sqrt{12} + 3\sqrt{48}$

10. $\sqrt{32} - 3\sqrt{27} + 2\sqrt{8}$

11. $2\sqrt{20} + 2\sqrt{80}$

12. $2\sqrt{44} - \sqrt{99}$

13. $\sqrt{98} - 2\sqrt{50} + 3\sqrt{8}$

14. $\sqrt{45} - 2\sqrt{20}$

15. $3\sqrt{8} + 2\sqrt{18} - 4\sqrt{72}$

16. $2\sqrt{40} + 4\sqrt{20} + 4\sqrt{90} - 3\sqrt{125}$

17. $\sqrt{48x} + \sqrt{27x}$

18. $6\sqrt{45ab} - 3\sqrt{5ab} - 2\sqrt{20ab}$

19. $3\sqrt{28a} + 4\sqrt{63a} - \sqrt{7a}$

20. $2\sqrt{8a} + 2\sqrt{32b} - 4\sqrt{18c}$

Answers to odd-numbered exercises on page 492.

11.5 DIVISION OF RADICALS AND RATIONALIZING DENOMINATORS

Division of Radicals

Compare the following statements:

1. $\sqrt{\dfrac{36}{9}} = \sqrt{4} = 2$

2. $\dfrac{\sqrt{36}}{\sqrt{9}} = \dfrac{6}{3} = 2$

Therefore $\sqrt{\frac{36}{9}} = \frac{\sqrt{36}}{\sqrt{9}}$, and this is true in general.

$$\sqrt{\frac{a}{b}} = \frac{\sqrt{a}}{\sqrt{b}}$$

Example 1: $\frac{\sqrt{24}}{\sqrt{6}} = \sqrt{\frac{24}{6}} = \sqrt{4} = 2$

Example 2: $\frac{\sqrt{26x^5}}{\sqrt{13x^3}} = \sqrt{\frac{26x^5}{13x^3}} = \sqrt{2x^2} = x\sqrt{2}$

Example 3: $\sqrt{\frac{36}{25}} = \frac{\sqrt{36}}{\sqrt{25}} = \frac{6}{5}$

Rationalizing Denominators.

If the denominator of a fraction is not a *rational* number, we attempt to make it *rational* by applying the following procedure.

To rationalize the denominator of a fraction, multiply the numerator and denominator of the fraction by a quantity that makes the radicand of the denominator a perfect square.

Example 4: Rationalize the denominator of $\frac{\sqrt{5}}{\sqrt{2}}$

Solution:

$\frac{\sqrt{5}}{\sqrt{2}} = \frac{\sqrt{5}}{\sqrt{2}} \cdot \boxed{\frac{\sqrt{2}}{\sqrt{2}}}$ Multiply the numerator and denominator by $\sqrt{2}$.

$= \frac{\sqrt{5 \cdot 2}}{\sqrt{2 \cdot 2}}$

$= \frac{\sqrt{10}}{\sqrt{4}}$

$= \frac{\sqrt{10}}{2}$ The denominator is now *rational.*

Example 5: Rationalize the denominator of $\sqrt{\frac{2}{3}}$

Solution:

$$\sqrt{\frac{2}{3}} = \frac{\sqrt{2}}{\sqrt{3}}$$ Apply the division rule.

$$= \frac{\sqrt{2}}{\sqrt{3}} \cdot \boxed{\frac{\sqrt{3}}{\sqrt{3}}}$$ Rationalize the denominator.

$$= \frac{\sqrt{6}}{3}$$ The denominator is now a rational number.

Example 6: Rationalize the denominator of $\frac{3\sqrt{2}}{\sqrt{3}}$

Solution:

$$\frac{3\sqrt{2}}{\sqrt{3}} = \frac{3\sqrt{2}}{\sqrt{3}} \cdot \boxed{\frac{\sqrt{3}}{\sqrt{3}}}$$

$$= \frac{\cancel{3}\sqrt{6}}{\cancel{3}}$$

$$= \sqrt{6}$$

Example 7: Rationalize the denominator of $\sqrt{\frac{5}{12x}}$

Solution:

$$\sqrt{\frac{5}{12x}} = \frac{\sqrt{5}}{\sqrt{12x}}$$ Apply the division rule.

$$= \frac{\sqrt{5}}{\sqrt{12x}} \cdot \boxed{\frac{\sqrt{3x}}{\sqrt{3x}}}$$ Rationalize the denominator.

$$\frac{\sqrt{15x}}{\sqrt{36x^2}}$$

$$\frac{\sqrt{15x}}{6x}$$ The denominator is rational.

EXERCISE 11.5

Divide and simplify:

1. $\frac{\sqrt{45}}{\sqrt{15}}$
2. $\frac{\sqrt{27}}{\sqrt{3}}$
3. $\frac{\sqrt{20}}{\sqrt{5}}$

4. $\dfrac{\sqrt{75}}{\sqrt{15}}$

5. $\dfrac{\sqrt{18}}{\sqrt{3}}$

6. $\dfrac{\sqrt{100}}{\sqrt{25}}$

7. $\dfrac{\sqrt{48}}{\sqrt{3}}$

8. $\dfrac{\sqrt{64}}{\sqrt{16}}$

9. $\sqrt{\dfrac{49}{4}}$

10. $\sqrt{\dfrac{25}{16}}$

11. $\sqrt{\dfrac{100}{36}}$

12. $\sqrt{\dfrac{49}{81}}$

13. $\dfrac{\sqrt{12}}{\sqrt{3x}}$

14. $\dfrac{\sqrt{20x^3}}{\sqrt{5x}}$

15. $\dfrac{\sqrt{75x^3y}}{\sqrt{15xy}}$

16. $\dfrac{\sqrt{60x^5}}{\sqrt{15x^3}}$

17. $\dfrac{\sqrt{18x^5y^3}}{\sqrt{3xy^3}}$

18. $\dfrac{\sqrt{15a^7b^3}}{\sqrt{3a^3b}}$

Rationalize the denominator:

19. $\dfrac{\sqrt{3}}{\sqrt{2}}$

20. $\dfrac{\sqrt{5}}{\sqrt{3}}$

21. $\dfrac{2}{\sqrt{3}}$

22. $\dfrac{9}{\sqrt{7}}$

23. $\dfrac{2\sqrt{5}}{\sqrt{6}}$

24. $\dfrac{12\sqrt{2}}{\sqrt{3}}$

25. $\dfrac{4\sqrt{5}}{\sqrt{2}}$

26. $\sqrt{\dfrac{3}{5}}$

27. $\sqrt{\dfrac{1}{7}}$

28. $\dfrac{\sqrt{5}}{\sqrt{x}}$

29. $\sqrt{\dfrac{4}{y}}$

30. $\sqrt{\dfrac{x}{y}}$

31. $\dfrac{\sqrt{5}}{\sqrt{12}}$

32. $\dfrac{\sqrt{3}}{\sqrt{8}}$

33. $\dfrac{\sqrt{2}}{\sqrt{27x}}$

Answers to odd-numbered exercises on page 492.

11.6 OTHER RADICAL EXPRESSIONS

The following examples illustrate still other expressions containing radicals.

Products Containing Radicals

Example 1: Multiply $\sqrt{2}\,(3\sqrt{2}+\sqrt{7})$

$$\sqrt{2}\,(3\sqrt{2}+\sqrt{7}) = \sqrt{2}\cdot 3\sqrt{2}+\sqrt{2}\cdot\sqrt{7}$$ Apply the distributive rule.

$$= 3\sqrt{2\cdot 2}+\sqrt{2\cdot 7}$$
$$= 3\cdot 2 + \sqrt{14}$$
$$= 6 + \sqrt{14}$$

Example 2: Multiply $\sqrt{5}\,(3\sqrt{5}-2)$

$$\sqrt{5}\,(3\sqrt{5}-2) = 3\sqrt{5}\cdot\sqrt{5}-2\sqrt{5}$$ Apply the distributive rule.

$$= 3\cdot 5 - 2\sqrt{5}$$
$$= 15 - 2\sqrt{5}$$

Binomials containing radicals are multiplied using the **FOIL** system in the same way as a binomial containing variables.

Example 3: Multiply $(\sqrt{2}-3)(\sqrt{2}+4)$

$$(\sqrt{2}-3)(\sqrt{2}+4) = \underbrace{\sqrt{2}\cdot\sqrt{2}}_{\text{First}} + \underbrace{4\cdot\sqrt{2}}_{\text{Outer}} - \underbrace{3\cdot\sqrt{2}}_{\text{Inner}} - \underbrace{3\cdot 4}_{\text{Last}}$$
$$= 2 + 4\sqrt{2}-3\sqrt{2}-12$$
$$= \sqrt{2}-10$$

Example 4: Multiply $(2\sqrt{3}-5)(4\sqrt{3}+2)$

$$(2\sqrt{3}-5)(4\sqrt{3}+2) = \underbrace{2\sqrt{3}\cdot 4\sqrt{3}}_{\text{F}} + \underbrace{2\cdot 2\sqrt{3}}_{\text{O}} - \underbrace{5\cdot 4\sqrt{3}}_{\text{I}} - \underbrace{(5)(2)}_{\text{L}}$$
$$= 24 + 4\sqrt{3} - 20\sqrt{3} - 10$$
$$= 14 - 16\sqrt{3}$$

Example 5: Multiply $(\sqrt{3}-4)(\sqrt{3}+4)$

$$(\sqrt{3}-4)(\sqrt{3}+4) = \sqrt{3}\cdot\sqrt{3}\ \underbrace{+4\sqrt{3}-4\sqrt{3}}\ - 16$$

↖ Middle term drops out.

$$= 3 - 16$$
$$= -13$$

The binomials in this problem are exactly alike except that the signs of the second terms are different. Such binomials are called **conjugates** of each other. Whenever conjugates are multiplied together, the middle term of the result drops out. This property is quite useful in simplifying a fraction with a binomial denominator containing square roots.

Example 6:

(a) The conjugate of $\sqrt{3}-4$ is $\sqrt{3}+4$

(b) The conjugate of $2\sqrt{3}+\sqrt{5}$ is $2\sqrt{3}-\sqrt{5}$

(c) The conjugate of $4-\sqrt{x}$ is $4+\sqrt{x}$

> To rationalize a binomial denominator containing radicals multiply the numerator and denominator by the **conjugate** of the denominator.

Example 7: Rationalize the denominator of $\dfrac{4}{\sqrt{3}-2}$

Solution:

$$\frac{4}{\sqrt{3}-2}=\frac{4}{\sqrt{3}-2}\cdot\boxed{\frac{\sqrt{3}+2}{\sqrt{3}+2}}$$

Multiply the numerator and denominator by the conjugate of the denominator.

$$=\frac{4\sqrt{3}+4\cdot 2}{\sqrt{3}\cdot\sqrt{3}\ \boxed{+2\sqrt{3}-2\sqrt{3}}\ -4}$$

The middle term equals zero.

$$=\frac{4\sqrt{3}+8}{3-4}$$

$$=\frac{4\sqrt{3}+8}{-1}$$

$$=-4\sqrt{3}-8$$

Example 8: Rationalize the denominator of $\dfrac{\sqrt{5}-2}{\sqrt{5}+2}$

Solution:

$$\frac{\sqrt{5}-2}{\sqrt{5}+2}=\frac{\sqrt{5}-2}{\sqrt{5}+2}\cdot\boxed{\frac{\sqrt{5}-2}{\sqrt{5}-2}}$$

Multiply by the conjugate.

$$=\frac{\sqrt{5}\cdot\sqrt{5}-2\sqrt{5}-2\sqrt{5}+(-2)(-2)}{\sqrt{5}\cdot\sqrt{5}-2\sqrt{5}+2\sqrt{5}+(+2)(-2)}$$

$$=\frac{5-4\sqrt{5}+4}{5-4}$$

$$= \frac{9 - 4\sqrt{5}}{1}$$

$$= 9 - 4\sqrt{5}$$

Example 9: Rationalize the denominator of $\dfrac{6}{\sqrt{5} - \sqrt{2}}$

Solution:

$$\frac{6}{\sqrt{5} - \sqrt{2}} \cdot \boxed{\frac{\sqrt{5} + \sqrt{2}}{\sqrt{5} + \sqrt{2}}}$$ Multiply by the conjugate.

$$= \frac{6(\sqrt{5} + \sqrt{2})}{5 - 2}$$

$$= \frac{\overset{2}{\cancel{6}}(\sqrt{5} + \sqrt{2})}{\underset{1}{\cancel{3}}}$$

$$= 2\sqrt{5} + 2\sqrt{2}$$

EXERCISE 11.6

Multiply and simplify:

1. $\sqrt{3}\,(4\sqrt{3} + 2)$
2. $\sqrt{5}\,(3 - \sqrt{5})$
3. $2\sqrt{7}\,(\sqrt{3} - \sqrt{7})$
4. $\sqrt{10}\,(\sqrt{5} - \sqrt{2})$
5. $3\sqrt{2}\,(\sqrt{2} - \sqrt{6})$
6. $\sqrt{x}\,(\sqrt{x} - \sqrt{2})$
7. $\sqrt{y}\,(3 + 2\sqrt{y})$
8. $3\sqrt{5}\,(\sqrt{15} + 2\sqrt{5})$
9. $4\sqrt{2}\,(2\sqrt{2} + \sqrt{6})$
10. $2\sqrt{3x}\,(\sqrt{3x} - \sqrt{x})$

11. $(\sqrt{2}+3)(\sqrt{2}-1)$

12. $(\sqrt{5}+3)(\sqrt{5}-2)$

13. $(\sqrt{3}-\sqrt{2})(\sqrt{3}-2\sqrt{2})$

14. $(2\sqrt{7}-3)(3\sqrt{7}+3)$

15. $(\sqrt{6}+2)(\sqrt{6}+5)$

16. $(\sqrt{5}-3)(\sqrt{5}+3)$

17. $(\sqrt{5}-2\sqrt{7})(\sqrt{5}+2\sqrt{7})$

18. $(3-\sqrt{6})(3+\sqrt{6})$

19. $(\sqrt{3}+\sqrt{2})(\sqrt{3}-\sqrt{6})$

20. $(\sqrt{5}-7)(\sqrt{2}+\sqrt{3})$

21. $(\sqrt{3}+5)^2$

22. $(2\sqrt{2}-3)^2$

23. $(2+\sqrt{3})^2$

24. $(3\sqrt{3}-\sqrt{5})^2$

Write the conjugate of each of the following:

25. $3-\sqrt{7}$

26. $3\sqrt{2}+5$

27. $\sqrt{5}+3\sqrt{6}$

28. $\sqrt{7}+2\sqrt{2}$

29. $x-\sqrt{5}$

30. $3\sqrt{x}+y$

Rationalize the denominators and simplify:

31. $\dfrac{2}{\sqrt{2}+1}$

32. $\dfrac{6}{\sqrt{2}+2}$

33. $\dfrac{6}{\sqrt{3}+\sqrt{5}}$

34. $\dfrac{8}{\sqrt{5}-2}$

35. $\dfrac{-2}{1-\sqrt{3}}$

36. $\dfrac{3}{3+\sqrt{3}}$

37. $\dfrac{6}{4-2\sqrt{3}}$

38. $\dfrac{-5}{\sqrt{5}+2\sqrt{2}}$

39. $\dfrac{\sqrt{3}-1}{\sqrt{3}+1}$

40. $\dfrac{3-\sqrt{5}}{2+\sqrt{5}}$

41. $\dfrac{3+2\sqrt{2}}{3-\sqrt{2}}$

42. $\dfrac{\sqrt{3}+5}{\sqrt{3}+3}$

43. $\dfrac{2-3\sqrt{5}}{4+\sqrt{5}}$

44. $\dfrac{\sqrt{5}-1}{\sqrt{5}-2}$

Answers to odd-numbered exercises on page 492.

11.7 RADICAL EQUATIONS

The easiest way to solve an equation containing a radical is to change it to an equation *without* any radicals. We accomplish this by applying the rule that says if two numbers are equal, then their squares are equal.

> If $a = b$, then $a^2 = b^2$

> **To solve a radical equation:**
>
> 1. Isolate the radical on one side of the equation.
> 2. Square both sides of the equation.
> 3. Solve the resulting equation.
> 4. Check the solution in the original equation.

Example 1: Solve $\sqrt{x} = 5$

Solution:

$\sqrt{x} = 5$	The radical is isolated on the left.
$(\sqrt{x})^2 = 5^2$	Square both sides of the equation.
$x = 25$	

Check:

$$\sqrt{x} = 5$$
$$\sqrt{25} = 5 \checkmark$$

Example 2: Solve $\sqrt{2x-1} - 3 = 0$

Solution:

$\sqrt{2x-1} - 3 = 0$	
$\sqrt{2x-1} = 3$	Isolate the radical on the left.
$2x - 1 = 9$	Square both sides of the equation.
$2x = 10$	Solve for x.
$x = 5$	

Check:

$$\sqrt{2x-1} - 3 = 0$$
$$\sqrt{2 \cdot 5 - 1} - 3 \stackrel{?}{=} 0$$
$$\sqrt{9} - 3 \stackrel{?}{=} 0$$
$$3 - 3 = 0 \checkmark$$

It is absolutely necessary to check every solution of a radical equation. This is so because it is possible to obtain an apparent solution that does not satisfy the original equation.

The next example illustrates this point.

Example 3: Solve $\sqrt{3x+1} + 4 = 0$

Solution:

$\sqrt{3x+1} + 4 = 0$	
$\sqrt{3x+1} = -4$	Isolate the radical.
$3x + 1 = 16$	Square both sides.
$3x = 15$	Solve for x.
$x = 5$	

Check:

$$\sqrt{3x+1}+4=0$$
$$\sqrt{3(5)+1}+4 \stackrel{?}{=} 0$$
$$\sqrt{16}+4 \stackrel{?}{=} 0$$
$$4+4 \stackrel{?}{=} 0$$
$$8 \neq 0 \ \checkmark$$

Therefore $x = 5$ *is not a solution* since it does not satisfy the original equation. We conclude that there is *no solution*.

EXERCISE 11.7

Solve and check:

1. $\sqrt{x}=4$
2. $\sqrt{y}=3$
3. $\sqrt{x}=5$
4. $\sqrt{2y}=4$
5. $\sqrt{x-2}=3$
6. $\sqrt{x+4}=5$
7. $\sqrt{2x+3}=9$
8. $6+\sqrt{x-1}=8$
9. $5-\sqrt{y+2}=3$
10. $\sqrt{x}=-2$
11. $\sqrt{x-2}-5=3$
12. $\sqrt{y-1}+2=0$
13. $\sqrt{6x}=6$
14. $2\sqrt{2y}=4$
15. $2\sqrt{5x-1}=6$
16. $\sqrt{2x-2}=-4$
17. $4-\sqrt{2x-1}=5$
18. $\sqrt{4x-5}=\sqrt{x+7}$

19. $\sqrt{4x-7}=\sqrt{3x+3}$

20. $\sqrt{3x-2}=\sqrt{3x+5}$

Answers to odd-numbered exercises on page 493.

11.8 THE PYTHAGOREAN THEOREM

A triangle that contains a right angle (90°) is called a **right triangle.** The longest side of the right triangle is called the **hypotenuse** and the other two (shorter) sides are called the **legs** of the triangle. The hypotenuse is labeled c and the legs are labeled a and b.

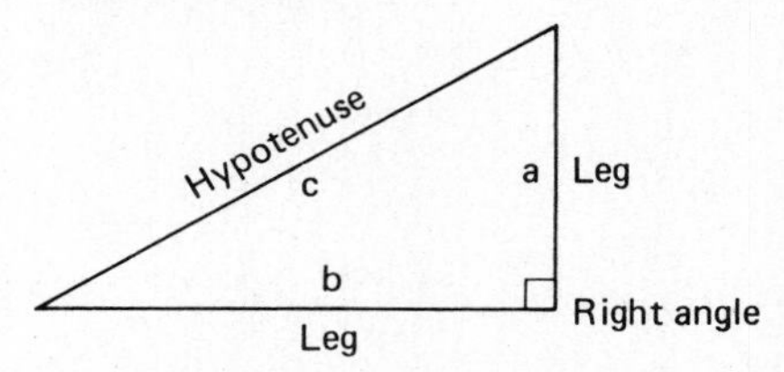

The three sides of any right triangle are related by the following theorem, named after the Greek mathematician, Pythagoras.

Pythagorean Theorem:

In any right triangle the sum of the squares of the two legs, ($a^2 + b^2$), equals the square of the hypotenuse (c^2).

$$a^2 + b^2 = c^2$$

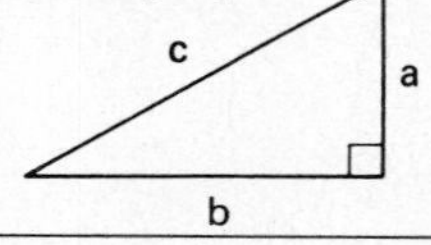

With this theorem, if we know any two sides of a right triangle, we can find the remaining side.

Example 1: Find the hypotenuse of the right triangle.

Solution:

$$c^2 = a^2 + b^2$$
$$c^2 = 5^2 + 12^2$$
$$c^2 = 25 + 144$$
$$c^2 = 169$$
$$c = \sqrt{169}$$
$$c = 13$$

c
5
12

Example 2: Find the unknown side in the triangle.

Solution:

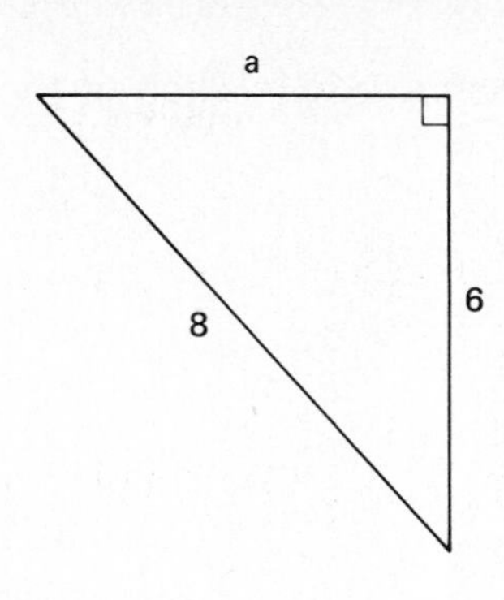

$$a^2 + b^2 = c^2$$
$$a^2 + 6^2 = 8^2$$
$$a^2 + 36 = 64$$
$$a^2 = 28$$
$$a = \sqrt{28} = \sqrt{4 \cdot 7}$$
$$a = 2\sqrt{7}$$

Example 3: Find the diagonal of the given rectangle.

Solution: The diagonal of the rectangle forms the hypotenuse of a right triangle.

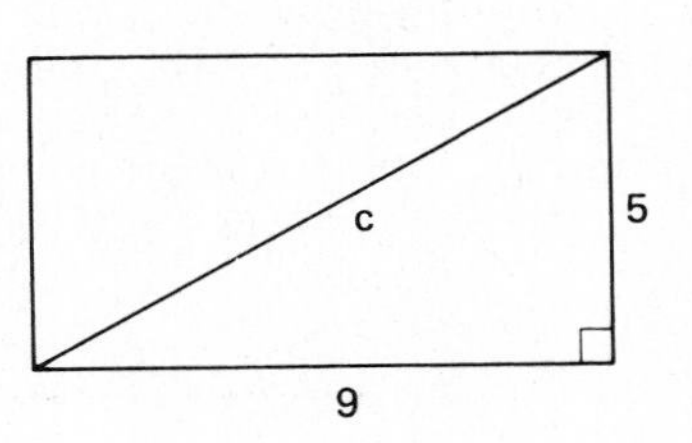

$$c^2 = a^2 + b^2$$
$$c^2 = 5^2 + 9^2$$
$$c^2 = 25 + 81$$
$$c^2 = 106$$
$$c = \sqrt{106}$$

Since the Pythagorean Theorem holds only for right triangles, we can use it to determine whether a given triangle is a right triangle or not.

Example 4: Is the given triangle a right triangle or not?

Solution:

$c^2 = a^2 + b^2$ only in a right triangle.

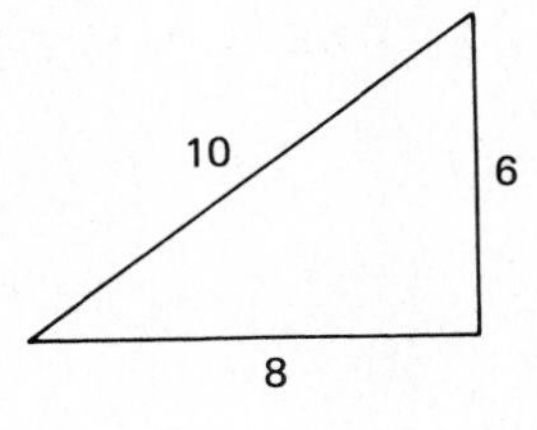

$$10^2 \stackrel{?}{=} 6^2 + 8^2$$
$$100 \stackrel{?}{=} 36 + 64$$
$$100 = 100$$

The triangle is a right triangle.

Example 5: Determine whether the given triangle is a right triangle or not.

Solution:

$c^2 = a^2 + b^2$ only in a right triangle.

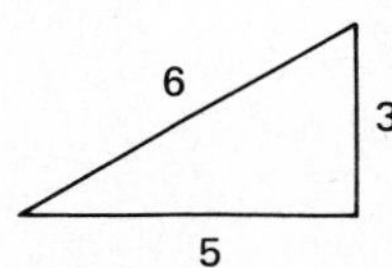

$$6^2 \stackrel{?}{=} 3^2 + 5^2$$
$$36 \stackrel{?}{=} 9 + 25$$
$$36 \neq 34$$

Therefore the triangle is *not* a right triangle.

EXERCISE 11.8

Find the missing side in each of the right triangles.

1. 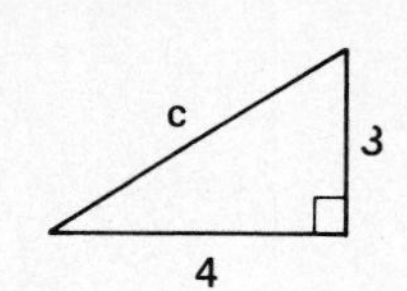

2. 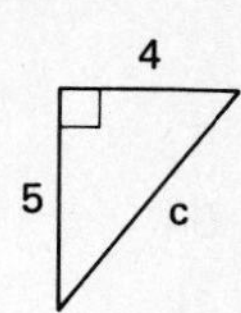

3. 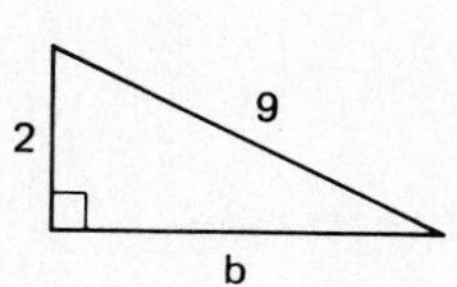

4. 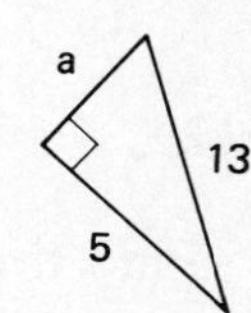

5. 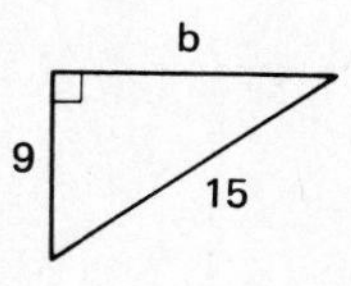

6. 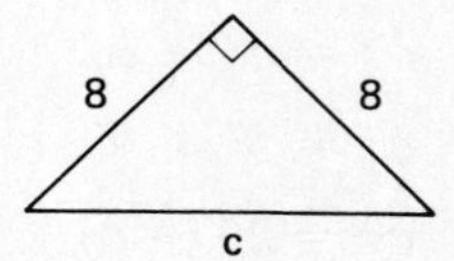

In a right triangle, if a and b represent the legs and c is the hypotenuse, find the length of the missing side.

7. $a = 6, b = 10$

8. $a = 8, b = 10$

9. $a = 4, c = 5$

10. $c = \sqrt{2}, a = 1$

11. $a = 3, b = 3$

12. $b = 12, c = 15$

13. $b = 5, c = 10$

14. $a = 2\sqrt{3}, b = 3\sqrt{2}$

15. $b = 2\sqrt{5}$, $c = \sqrt{47}$

16. $a = 2\sqrt{2}$, $c = 2\sqrt{26}$

Determine whether or not the given triangles are right triangles.

17. $a = 1$, $b = 1$, $c = \sqrt{2}$

18. $a = 5$, $b = 6$, $c = 30$

19. $a = 3$, $b = 5$, $c = \sqrt{34}$

20. $a = 2$, $b = 2\sqrt{3}$, $c = 4$

21. $a = 2\sqrt{2}, b = 3\sqrt{2}, c = 5\sqrt{2}$

22. $a = \sqrt{2} + 1, b = \sqrt{2} - 1, c = 1$

23. Find the diagonal of a rectangle with a length of 12 cm and a width of 5 cm.

24. A baseball diamond is a square 90 feet on each side. How far is it from home plate to second base?

25. Find the width of a rectangle that has a length of 24 inches and a diagonal of 26 inches.

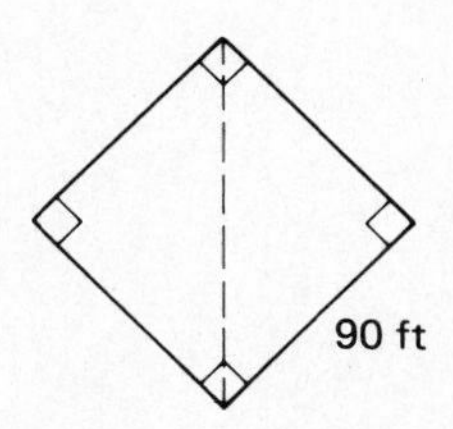

26. How long would a ladder have to be to reach 20 feet up the side of a building if the base of the ladder is 6 feet away from the building?

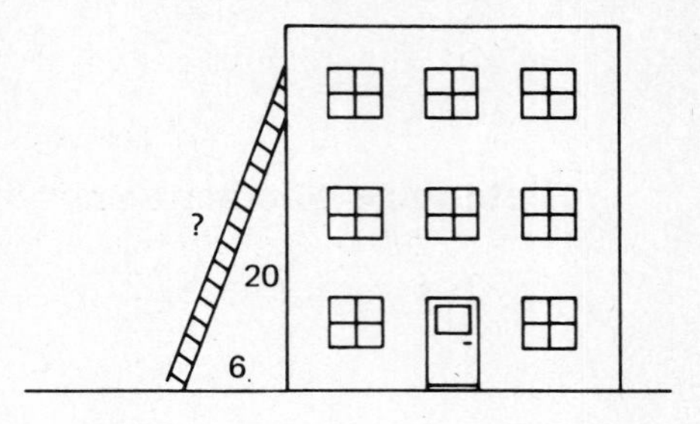

27. Two cars leave the same point. One travels south for 24 miles and the other travels west a distance of 18 miles. How far apart are the cars?

28. The diagonal of a square is 10 cm. How long is the side of the square?

29. How long does a guy wire have to be to reach the top of a mast 48 feet high, if the wire is fastened to the ground 20 feet from the base of the mast?

30. A plane flies due north at 400 miles per hour. A second plane leaves the same place at the same time flying due west at 300 miles per hour. How far apart are the two planes after 2 hours?

Answers to odd-numbered exercises on page 493.

11.9 Chapter Summary

Examples

(11.1) b is a **square root** of a if $b^2 = a$.

$\sqrt{\ }$ is called a **radical sign.**

The number under a radical sign is called the **radicand.** The symbol $\sqrt{\ }$ indicates a positive square root called the **principal square root.**

$$\sqrt{a^2} = |a|$$

(11.1) 4 and -4 are square roots of 16 since $4^2 = 16$ and $(-4)^2 = 16$.

$$\sqrt{16} = 4$$

(11.2) A **rational number** can always be expressed as a fraction $\frac{a}{b}$, where a and b are integers and $b \neq 0$.

A **rational number** can always be written as a terminating or repeating decimal.

An **irrational number** can never be expressed as a fraction $\frac{a}{b}$, where a and b are integers, and $b \neq 0$.

An **irrational number** can never be written as a terminating or repeating decimal.

The **real numbers** consist of all the rational numbers plus all the irrational numbers.

The square root of any negative number is not a real number.

(11.2)

Number	*Rational*	*Irrational*	*Real*
$\frac{2}{3}$	✓		✓
5.1242424... repeating	✓		✓
7.6 terminating	✓		✓
7.3145927.... non-repeating		✓	✓
$-\sqrt{7}$		✓	✓
$\sqrt{-7}$	no	no	no

(11.3) $\sqrt{a} \cdot \sqrt{b} = \sqrt{a \cdot b}, \quad a$ and $b \geqslant 0$

(11.3)

(a) $\sqrt{3} \cdot \sqrt{12}$

$= \sqrt{3 \cdot 12} = \sqrt{36} = 6$

(b) $\sqrt{32} = \sqrt{16 \cdot 2}$

$= \sqrt{16} \cdot \sqrt{2} = 4\sqrt{2}$

(c) $\sqrt{27x^5}$

$= \sqrt{9 \cdot 3 \cdot x^4 \cdot x}$

$= \sqrt{9x^4} \cdot \sqrt{3x}$

$= 3x^2\sqrt{3x}$

(11.4) **Like** or **similar** square roots have the same radicand.

To add like square roots, add their coefficients and keep the same radicand.

(11.4)

(a) $3\sqrt{7} + 5\sqrt{7}$

$= 8\sqrt{7}$

(b) $\sqrt{18} - \sqrt{32}$

$= \sqrt{9 \cdot 2} - \sqrt{16 \cdot 2}$

$= 3\sqrt{2} - 4\sqrt{2}$

$= -\sqrt{2}$

(c) $\sqrt{18} + \sqrt{75}$

$= \sqrt{9 \cdot 2} + \sqrt{25 \cdot 3}$

$= 3\sqrt{2} + 5\sqrt{3}$

They cannot be combined further.

(11.5) $\sqrt{\frac{a}{b}} = \frac{\sqrt{a}}{\sqrt{b}}$

(11.5)

(a) $\sqrt{\frac{49}{36}} = \frac{\sqrt{49}}{\sqrt{36}} = \frac{7}{6}$

(b) $\frac{\sqrt{75}}{\sqrt{3}} = \sqrt{\frac{75}{3}} = \sqrt{25} = 5$

(11.5) To rationalize the denominator of a fraction, multiply the numerator and the denominator of the fraction by a quantity that makes the denominator a perfect square.

(11.5) Rationalize the denominator in $\sqrt{\frac{3}{8}}$.

$$\sqrt{\frac{3}{8}} = \frac{\sqrt{3}}{\sqrt{8}} \cdot \boxed{\frac{\sqrt{2}}{\sqrt{2}}}$$

$$= \frac{\sqrt{6}}{\sqrt{16}} = \frac{\sqrt{6}}{4}$$

(11.6) The **conjugate** of a binomial containing radicals is the same binomial with the sign of the second term changed.

(11.6) The conjugate of $\sqrt{3} - 2$ is $\sqrt{3} + 2$.

(11.6) To rationalize a binomial containing radicals, multiply the numerator and denominator by the conjugate of the denominator.

(11.6) Rationalize the denominator of $\frac{4}{\sqrt{3} - 2}$.

$$\frac{4}{\sqrt{3} - 2}$$

$$= \frac{4}{(\sqrt{3} - 2)} \cdot \frac{(\sqrt{3} + 2)}{(\sqrt{3} + 2)}$$

$$= \frac{4(\sqrt{3} + 2)}{3 - 4}$$

$$= \frac{4\sqrt{3} + 8}{-1}$$

$$= -4\sqrt{3} - 8$$

(11.7) To solve a radical equation:

1. Isolate the radical on one side of the equation.
2. Square both sides of the equation.
3. Solve the resulting equation.
4. Check the solution in the original equation.

(11.7) Solve for x: $5 - \sqrt{2x + 1} = 2$.

$3 = \sqrt{2x + 1}$ Isolate the radical.

$9 = 2x + 1$ Square both sides.

$8 = 2x$ Solve for x.

$4 = x$

Check:

$$5 - \sqrt{2 \cdot 4 + 1} \stackrel{?}{=} 2$$
$$5 - \sqrt{9} \stackrel{?}{=} 2$$
$$5 - 3 = 2 \ \checkmark$$

$x = 4$ is the correct solution.

(11.8) The **Pythagorean Theorem:** In any right triangle, the sum of the squares of the two legs equals the square of the hypotenuse.

$$a^2 + b^2 = c^2$$

c a b

(11.8) Find the hypotenuse of the given triangle.

$$a^2 + b^2 = c^2$$
$$a^2 + 4^2 = 5^2$$
$$a^2 + 16 = 25$$
$$a^2 = 9$$
$$a = 3$$

a 4 5

Exercise 11.9 Chapter Review

(11.1) Find the square roots of the following:

1. 49 2. 81 3. 121

(11.1) Evaluate the following if possible:

4. $\sqrt{64}$ 5. $\sqrt{25}$ 6. $-\sqrt{25}$

7. $\pm\sqrt{25}$ 8. $\sqrt{-25}$ 9. $\sqrt{0}$

(11.1) Simplify:

10. $\sqrt{y^2}$ 11. $\sqrt{(-y)^2}$ 12. $-\sqrt{y^2}$

13. $\sqrt{(h-3)^2}$ 14. $\sqrt{x^2+2x+1}$

(11.2) Indicate which of the following are (a) rational, (b) irrational, (c) real, or (d) not a real number.

15. -6 16. $5\frac{3}{8}$ 17. $\sqrt{43}$

18. .456456456. . . 19. $\sqrt{-8}$ 20. 6.21384092318. . .

21. $-\sqrt{100}$

(11.3) Simplify:

22. $\sqrt{32}$ 23. $\sqrt{80}$ 24. $\sqrt{72}$

25. $\sqrt{y^8}$ 26. $\sqrt{18t^3}$ 27. $\sqrt{(y-3)^6}$

(11.3) Multiply and simplify:

28. $\sqrt{3}\sqrt{12}$ 29. $\sqrt{3}\sqrt{15}$ 30. $\sqrt{6}\sqrt{18}$

31. $\sqrt{y^5}\sqrt{y^7}$ 32. $\sqrt{3x^3}\sqrt{3x^8}$ 33. $\sqrt{3xy^3}\sqrt{2x}\sqrt{8y}$

(11.4) Combine where possible:

34. $5\sqrt{7}+2\sqrt{7}$ 35. $4\sqrt{x}-11\sqrt{x}$ 36. $\sqrt{12}+\sqrt{27}$

37. $\sqrt{66}-\sqrt{55}$ 38. $\sqrt{45}+\sqrt{18}$ 39. $\sqrt{98}+2\sqrt{50}-3\sqrt{8}$

40. $2\sqrt{28y}-3\sqrt{7y}$ 41. $3\sqrt{27x}-5\sqrt{12y}$

(11.5) Divide and simplify:

42. $\dfrac{\sqrt{45}}{\sqrt{20}}$ 43. $\dfrac{\sqrt{60}}{\sqrt{15}}$

44. $\sqrt{\dfrac{25}{64}}$ 45. $\sqrt{\dfrac{90x^3}{15x}}$

(11.5) Rationalize the denominator:

46. $\dfrac{\sqrt{5}}{\sqrt{7}}$ 47. $\sqrt{\dfrac{3}{5}}$ 48. $\dfrac{2\sqrt{3}}{\sqrt{7}}$

49. $\dfrac{\sqrt{6}}{\sqrt{y}}$ 50. $\dfrac{\sqrt{7}}{\sqrt{8}}$

(11.6) Multiply and simplify:

51. $\sqrt{5}\,(2\sqrt{3} - 1)$

52. $\sqrt{6}\,(2\sqrt{3} + \sqrt{2})$

53. $2\sqrt{x}\,(\sqrt{3x} - \sqrt{2})$

54. $(2 - \sqrt{3})(4 - 3\sqrt{3})$

55. $(\sqrt{7} + 2\sqrt{3})(2\sqrt{7} - \sqrt{3})$

56. $(2\sqrt{3} + \sqrt{5})(2\sqrt{3} - \sqrt{5})$

(11.6) Write the conjugate of the following:

57. $4 - \sqrt{5}$

58. $\sqrt{6} - 3\sqrt{y}$

(11.6) Rationalize the denominator and simplify:

59. $\dfrac{4}{\sqrt{2} - 2}$

60. $\dfrac{3 - \sqrt{7}}{3 + \sqrt{7}}$

61. $\dfrac{\sqrt{2} - \sqrt{3}}{2\sqrt{2} - \sqrt{3}}$

(11.7) Solve and check:

62. $\sqrt{3y + 1} = 5$

63. $3 + \sqrt{x - 1} = 2$

64. $\sqrt{4x - 1} = \sqrt{3x + 5}$

65. $5 - \sqrt{2y + 2} = -7$

(11.8) In a right triangle a, b, c, find the length of the missing side.

66. $a = 3, b = 8$

67. $a = 6, c = 9$

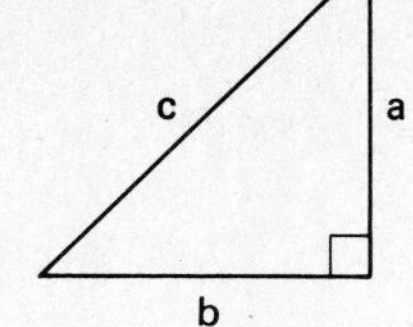

(11.8) Determine whether or not the given triangles are right triangles:

68. $a = \sqrt{3}, b = 2\sqrt{2}, c = \sqrt{11}$

69. $a = 4, b = 5, c = 9$

(11.8)

70. Find the width of a rectangle having a length of 16 and a diagonal of 20.

71. How far up the building will a 32-foot ladder reach if the base of the ladder is 5 feet away from the building?

Answers to odd-numbered exercises on page 493.

Chapter 11 Achievement Test

Name ______________________

Class ______________________

This test should be taken before you are tested in class on the material in Chapter 11. Solutions to each problem and the section where the type of problem is found are given on page 494.

1. Find the square roots of 64. 1. ________

Evaluate if possible:

2. $\sqrt{36}$ 2. ________
3. $-\sqrt{36}$ 3. ________
4. $\sqrt{-36}$ 4. ________

Indicate which of the following are rational or irrational:

5. .343434. . . . . . . 5. ________
6. .341286541. . . 6. ________
7. $\frac{5}{16}$ 7. ________
8. $\sqrt{21}$ 8. ________

Simplify the following:

9. $\sqrt{45}$ 9. ________
10. $\sqrt{18x^5}$ 10. ________

Multiply and simplify:

11. $\sqrt{5}\sqrt{15}$ 11. ________
12. $\sqrt{y^7}\sqrt{y^5}$ 12. ________

Combine if possible:

13. $6\sqrt{5} - 9\sqrt{5}$

14. $\sqrt{48} + 3\sqrt{27} - 3\sqrt{12}$

15. $2\sqrt{3} - 4\sqrt{2} + 3\sqrt{7}$

Divide and simplify:

16. $\dfrac{\sqrt{75}}{\sqrt{25}}$

17. $\sqrt{\dfrac{36}{64}}$

18. $\dfrac{\sqrt{30x^3}}{\sqrt{15x}}$

Rationalize the denominator:

19. $\dfrac{\sqrt{3}}{\sqrt{10}}$

20. $\dfrac{3\sqrt{5}}{\sqrt{3}}$

Multiply and simplify:

21. $\sqrt{3}\,(2\sqrt{5} - \sqrt{6})$

22. $(\sqrt{6} - 2\sqrt{3})(\sqrt{6} + 2\sqrt{3})$

23. $(\sqrt{7} - 5)(2\sqrt{7} + 3)$

Write the conjugate:

24. $\sqrt{2} - 3\sqrt{5}$

Rationalize the denominator and simplify:

25. $\dfrac{3}{\sqrt{2} - 1}$

13. ______________

14. ______________

15. ______________

16. ______________

17. ______________

18. ______________

19. ______________

20. ______________

21. ______________

22. ______________

23. ______________

24. ______________

25. ______________

26. $\dfrac{4 - 2\sqrt{7}}{5 + \sqrt{7}}$

26. 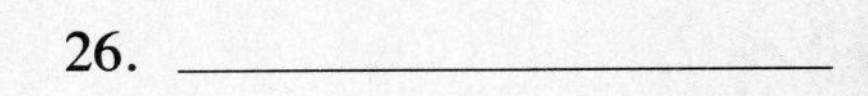

Solve and check:

27. $\sqrt{4y + 1} = 5$

27. ____________________

28. $7 + \sqrt{2x + 1} = 4$

28. ____________________

In the right triangle, find the length of the missing side.

29. $a = 7, c = 9$

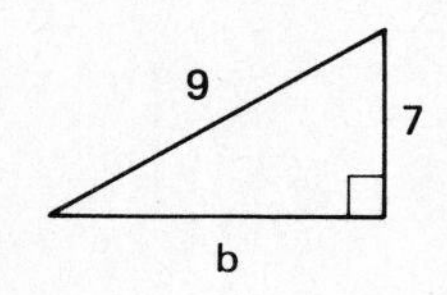

29. ____________________

Determine whether or not the given triangle is a right triangle.

30. $a = \sqrt{5}, b = 2\sqrt{7}, c = \sqrt{33}$

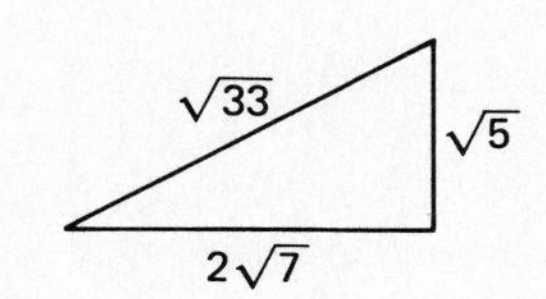

30. ____________________

31. Find the length of a wire attached to the top of a 60-foot tower that is fixed to the ground 45 feet from the base of the tower.

31. ____________________

12
Quadratic Equations

12.1 STANDARD FORM OF A QUADRATIC EQUATION

Any equation that can be written in the form

$$ax^2 + bx + c = 0, \text{ where } a, b \text{ and } c \text{ are real numbers and } a \neq 0$$

is called a **quadratic equation.** The above form is called the **standard form** of a quadratic equation. In this chapter we will require that $a > 0$ and a, b, and c are integers.

Example 1: Write $-2x = 3 - 9x^2$ in standard form.

Solution:

$$-2x = 3 - 9x^2$$

$9x^2 - 2x - 3 = 0$ Add $9x^2$ and -3 to both sides.

$a = 9, b = -2, c = -3$ in $ax^2 + bx + c = 0$.

Example 2: Write $\frac{1}{3}x + \frac{1}{2} = \frac{3}{4}x^2 - \frac{2}{3}$ in standard form.

Solution:

$$\frac{1}{3}x + \frac{1}{2} = \frac{3}{4}x^2 - \frac{2}{3}$$ Multiply by the LCD, 12.

$$\boxed{\overset{4}{\cancel{12}}} \cdot \frac{1}{\underset{1}{\cancel{3}}}x + \boxed{\overset{6}{\cancel{12}}} \cdot \frac{1}{\underset{1}{\cancel{2}}} = \boxed{\overset{3}{\cancel{12}}} \cdot \frac{3}{\underset{1}{\cancel{4}}}x^2 - \boxed{\overset{4}{\cancel{12}}} \cdot \frac{2}{\underset{1}{\cancel{3}}}$$

$$4x + 6 = 9x^2 - 8$$ Add $-9x^2$ and 8 to both sides.

$$-9x^2 + 4x + 14 = 0$$ We want positive coefficients of x^2.

$$\boxed{-1}(-9x^2 + 4x + 14) = \boxed{-1} \cdot 0$$ Multiply both sides by -1.

$$9x^2 - 4x - 14 = 0$$ in standard form.

$$a = 9, b = -4, c = -14$$

EXERCISE 12.1

Write each equation in standard form.

1. $4x^2 = -2 + 3x$
2. $5 - 4x + 7x^2 = 0$
3. $x = 3x^2 + 1$
4. $3 = 4x^2$
5. $3x^2 + 2x - 1 = 4x^2 + 5$
6. $-6x^2 + 2x - 1 = 3x^2 + 4x - 5$
7. $-2 - x^2 = 4x^2 + x$
8. $2x^2 - 3x - 1 = -3(x^2 + 4)$

9. $\frac{2x^2}{5} + 3x - \frac{1}{2} = 0$

10. $\frac{1}{2}x + \frac{2}{3} = \frac{x^2}{4} - \frac{3}{4}$

Answers to odd-numbered exercises on page 495.

12.2 SOLVING QUADRATIC EQUATIONS BY FACTORING

What do we know about two numbers whose product is zero? A little thought should convince you that at least one of them must be zero. This is known as the **principle of zero products** and is stated as follows:

> **The principle of zero products:**
> If $a \cdot b = 0$, then either $a = 0$, $b = 0$ or both a and $b = 0$.

We can use this principle to solve any quadratic equation that is factorable by use of the following procedure.

> **To solve a quadratic equation by factoring:**
> 1. Put the equation in standard form $ax^2 + bx + c = 0$.
> 2. Factor the left side of the equation.
> 3. Set each factor equal to zero and solve for x.
> 4. Check each solution in the original equation.

Example 1: Solve $x^2 - 2x - 8 = 0$ by factoring.

Solution:

$x^2 - 2x - 8 = 0$	Already in standard form.
$(x - 4)(x + 2) = 0$	Factor the left side.
$x - 4 = 0 \quad \vert \quad x + 2 = 0$	Set each factor equal to 0.
$x = 4 \quad \vert \quad x = -2$	Solve each equation for x.

There are *two* solutions, $x = 4$ or $x = -2$.

Check:

$x^2 - 2x - 8 = 0$	$x^2 - 2x - 8 = 0$
$(4)^2 - 2(4) - 8 \stackrel{?}{=} 0$	$(-2)^2 - 2(-2) - 8 \stackrel{?}{=} 0$
$16 - 8 - 8 = 0$ ✓	$4 + 4 - 8 = 0$ ✓

Example 2: Solve $3y^2 = 2 - 5y$ by factoring.

Solution:

$$3y^2 = 2 - 5y$$

$3y^2 + 5y - 2 = 0$	Put in standard form.
$(3y-1)(y+2) = 0$	Factor.
$3y - 1 = 0 \quad \vert \quad y + 2 = 0$	Set each factor equal to 0.
$y = \frac{1}{3} \quad \vert \quad y = -2$	Solve each equation for y.

Check:

$3y^2 = 2 - 5y$	$3y^2 = 2 - 5y$
$3\left(\frac{1}{3}\right)^2 \overset{?}{=} 2 - 5\left(\frac{1}{3}\right)$	$3(-2)^2 \overset{?}{=} 2 - 5(-2)$
$3 \cdot \frac{1}{9} \overset{?}{=} 2 - \frac{5}{3}$	$3 \cdot 4 \overset{?}{=} 2 + 10$
$\frac{1}{3} \overset{?}{=} \frac{6}{3} - \frac{5}{3}$	$12 = 12$ ✓
$\frac{1}{3} = \frac{1}{3}$ ✓	

Example 3: Solve $3x^2 - 6x = 0$ by factoring.

Solution:

$3x^2 - 6x = 0$	Already in standard form.
$3x(x - 2) = 0$	Factor.
$3x = 0 \quad \vert \quad x - 2 = 0$	Set each factor equal to 0.
$x = 0 \quad \vert \quad x = 2$	Solve for x in each equation.

Check:

$3x^2 - 6x = 0$	$3x^2 - 6x = 0$
$3 \cdot 0^2 - 6 \cdot 0 \overset{?}{=} 0$	$3 \cdot 2^2 - 6 \cdot 2 \overset{?}{=} 0$
$0 = 0$	$12 - 12 = 0$

Never divide both sides of a quadratic equation by an expression containing the variable. This can cause one of the solutions to be missed. This is illustrated in the following example.

Example 4: Solve $2x^2 = 6x$.

Correct solution:

$$2x^2 = 6x$$

$2x^2 - 6x = 0$ Put in standard form.

$2x(x - 3) = 0$ Factor.

$2x = 0 \mid x - 3 = 0$ Set each equation equal to 0.

$x = 0 \mid x = 3$

Incorrect solution:

$$2x^2 = 6x$$

$\frac{2x^2}{2x} = \frac{6x}{2x}$ Divide both sides by $2x$, which you're not supposed to do!

$x = 3$ Only one solution results.

Dividing by $2x$ causes us to lose the solution $x = 0$.

EXERCISE 12.2

Solve by factoring and check:

1. $y^2 - 16 = 0$
2. $x^2 - 5x + 4 = 0$
3. $x^2 + 10x + 16 = 0$
4. $y^2 - 7y = 0$
5. $t^2 + t - 12 = 0$
6. $x^2 - 64 = 0$
7. $x^2 = 14x$
8. $a^2 + a - 20 = 0$

9. $x^2 + 7x + 6 = 0$

10. $3x^2 = 12x$

11. $4x^2 + 4x - 3 = 0$

12. $5x^2 - 14x - 3 = 0$

13. $3a^2 + 5a - 2 = 0$

14. $12x^2 - 11x + 2 = 0$

15. $2x^2 = 5 - 3x$

16. $11x = 4 - 3x^2$

17. $x^2 - 3x + 3 = 2x - 3$

18. $2x^2 + 3x - 18 = 3x$

19. $2x^2 - 5x + 8 = x^2 + x$

20. $-11a + 6a^2 - 1 = 2a^2 - 8$

Answers to odd-numbered exercises on page 495.

12.3 SOLVING QUADRATIC EQUATIONS OF THE FORM $ax^2 - c = 0$

If $b = 0$, a quadratic equation of the form $ax^2 - c = 0$ results.

To solve a quadratic equation of the form $ax^2 - c = 0$:

1. Add c to both sides to obtain $ax^2 = c$.
2. Divide both sides by a, yielding $x^2 = \frac{c}{a}$.
3. Take the square root of both sides, obtaining $x = \pm\sqrt{\frac{c}{a}}$.
4. Check the solution in the original equation.

Example 1: Solve $3x^2 - 12 = 0$

Solution:

$3x^2 - 12 = 0$	
$3x^2 = 12$	Add 12 to both sides.
$x^2 = 4$	Divide both sides by 3.
$x = \pm\sqrt{4} = \pm 2$	Take the square root of both sides.

This means that $x = +2$ or $x = -2$.

Check:

$3x^2 - 12 = 0$	$3x^2 - 12 = 0$
$3(\boxed{+2})^2 - 12 \stackrel{?}{=} 0$	$3(\boxed{-2})^2 - 12 \stackrel{?}{=} 0$
$12 - 12 = 0$ ✓	$12 - 12 = 0$ ✓

Example 2: Solve $3x^2 - 24 = 0$

Solution:

$3x^2 - 24 = 0$	
$3x^2 = 24$	Add 24 to both sides.
$x^2 = 8$	Divide both sides by 3.
$x = \pm\sqrt{8}$	Take the square root of both sides.
$x = \pm\sqrt{4 \cdot 2} = \pm 2\sqrt{2}$	Note there are two solutions.

Check:

$3x^2 - 24 = 0$	$3x^2 - 24 = 0$
$3(\boxed{+2\sqrt{2}})^2 - 24 \stackrel{?}{=} 0$	$3(\boxed{-2\sqrt{2}})^2 - 24 \stackrel{?}{=} 0$
$3 \cdot 8 - 24 \stackrel{?}{=} 0$	$3 \cdot 8 - 24 \stackrel{?}{=} 0$
$24 - 24 = 0$ ✓	$24 - 24 = 0$ ✓

EXERCISE 12.3

Solve and check:

1. $x^2 - 16 = 0$
2. $y^2 - 36 = 0$
3. $2x^2 - 50 = 0$

4. $a^2 = 64$

5. $3x^2 - 27 = 0$

6. $3x^2 - 48 = 0$

7. $6m^2 - 54 = 0$

8. $y^2 = 144$

9. $3x^2 - 36 = 0$

10. $36x^2 - 25 = 0$

11. $20x^2 = 45$

12. $5y^2 - 10 = 0$

13. $6a^2 = 150$

14. $3x^2 - 10 = 0$

15. $3t^2 - 15 = 0$

16. $49t^2 = 24$

17. $5x^2 - 36 = 0$

18. $t^2 - \frac{4}{9} = 0$

19. $\frac{x^2}{3} - 3 = 0$

20. $\frac{2}{3}y^2 - \frac{27}{2} = 0$

Answers to odd-numbered exercises on page 495.

12.4 COMPLETING THE SQUARE

While solving quadratic equations by factoring, you may well have wondered what happens if the equation isn't factorable. Obviously, we will need another method for solving quadratic equations like $x^2 - 4x - 7 = 0$, which are not factorable. The method is called **completing the square** and involves construction of **perfect square trinomials.**

Examples of perfect square trinomials are:

$$x^2 + 2x + 1 = (x + 1)^2$$

$$x^2 + \boxed{6}\,x + \boxed{9} = (x + 3)^2$$

If we take $\frac{1}{2}$ of 6 and square it, we get $3^2 = 9$.

$$x^2 \boxed{-8}\,x + \boxed{16} = (x - 4)^2$$

Take $\frac{1}{2}$ of -8 and square it, obtaining $+ 16$.

You should notice that if we take one half of the coefficient of the middle term and square it, we get the last term. This process is called *completing the square* and is illustrated in the following example.

Example 1: What must we add to complete the square?

(a) $x^2 + 6x + \_\_\_\_\_\_ = (x + \_\_\_\_\_\_)^2$

$$\left(\frac{1}{2} \cdot 6\right)^2 = 3^2 = 9$$

or $\quad x^2 + 6x + 9 = (x + 3)^2$

(b) $x^2 - 10x + \_\_\_\_\_\_ = (x + \_\_\_\_\_\_)^2$

$$\left(\frac{1}{2} \cdot (-10)\right)^2 = (-5)^2 = 25$$

$$x^2 - 10x + 25 = (x - 5)^2$$

(c) $x^2 + 3x + \_\_\_\_\_\_ = (x + \_\_\_\_\_\_)^2$

$$\left(\frac{1}{2} \cdot 3\right)^2 = \left(\frac{3}{2}\right)^2 = \frac{9}{4}$$

$$x^2 + 3x + \frac{9}{4} = \left(x + \frac{3}{2}\right)^2$$

We will now make use of the process of completing the square to solve quadratic equations.

Example 2: Solve $x^2 - 4x - 7 = 0$ by completing the square.

Solution: Put the constant term on the right side.

$x^2 - 4x - 7 = 0$ Add 7 to both sides.

$x^2 - 4x + \_\_\_\_\_\_ = 7$ Notice the space that is left.

Now complete the square by adding the square of half the coefficient of x to *both* sides of the equation.

$$\left(\frac{-4}{2}\right)^2 = (-2)^2 = 4$$

$x^2 - 4x + \boxed{4} = 7 + \boxed{4}$	Factor the left side, which is a perfect square trinomial.
$(x - 2)^2 = 11$	
$x - 2 = \pm\sqrt{11}$	Take the square root of both sides.
$x = 2 \pm \sqrt{11}$	Add 2 to both sides.

The solutions are $2 + \sqrt{11}$ or $2 - \sqrt{11}$.

Check:

$x^2 - 4x - 7 = 0$	$x^2 - 4x - 7 = 0$
$(\boxed{2+\sqrt{11}})^2 - 4(\boxed{2+\sqrt{11}}) - 7 \stackrel{?}{=} 0$	$(\boxed{2-\sqrt{11}})^2 - 4(\boxed{2-\sqrt{11}}) - 7 \stackrel{?}{=} 0$
$4 + 4\sqrt{11} + 11 - 8 - 4\sqrt{11} - 7 \stackrel{?}{=} 0$	$4 - 4\sqrt{11} + 11 - 8 + 4\sqrt{11} - 7 \stackrel{?}{=} 0$
$0 = 0$ ✓	$0 = 0$ ✓

To complete the square, the coefficient of the x^2-term must be 1. If it is not, each term must be divided by the coefficient of the x^2-term before applying the procedure.

To solve a quadratic equation $ax^2 + bx + c = 0$ by completing the square:

1. Subtract the constant term, c, from both sides of the equation.
2. If $a \neq 1$, divide each term on both sides by a.
3. Complete the square by adding the square of half the coefficient of x, $(\frac{b}{2})^2$, to both sides.
4. Factor the left side of the equation (it will be a perfect square trinomial).
5. Take the square root of both sides of the equation and solve for x.

Example 3: Solve $2x^2 + 3x - 2 = 0$ by completing the square.

Solution:

$2x^2 + 3x - 2 = 0$	
$2x^2 + 3x + \underline{\hspace{2cm}} = 2$	Add 2 to both sides.
$x^2 + \frac{3}{2}x + \underline{\hspace{2cm}} = 1$	Divide each term by 2.
$\left(\frac{1}{2} \cdot \frac{3}{2}\right)^2 = \left(\frac{3}{4}\right)^2 = \frac{9}{16}$	Take half of $\frac{3}{2}$ and square it.

$$x^2 + \frac{3}{2}x + \boxed{\frac{9}{16}} = 1 + \boxed{\frac{9}{16}}$$ Add this result to both sides.

$$\left(x + \frac{3}{4}\right)^2 = \frac{25}{16}$$ $1 + \frac{9}{16} = \frac{16}{16} + \frac{9}{16} = \frac{25}{16}$

$$x + \frac{3}{4} = \pm\sqrt{\frac{25}{16}} = \pm\frac{5}{4}$$ Take the square root of both sides.

$$x = -\frac{3}{4} \pm \frac{5}{4}$$ Subtract $\frac{3}{4}$ from both sides.

The solutions are $-\frac{8}{4} = -2$ or $\frac{2}{4} = \frac{1}{2}$

Check:

$2x^2 + 3x - 2 = 0$	$2x^2 + 3x - 2 = 0$
$2(-2)^2 + 3(-2) - 2 \stackrel{?}{=} 0$	$2\left(\frac{1}{2}\right)^2 + 3 \cdot \frac{1}{2} - 2 \stackrel{?}{=} 0$
$2 \cdot 4 - 6 - 2 \stackrel{?}{=} 0$	$2 \cdot \frac{1}{4} + \frac{3}{2} - 2 \stackrel{?}{=} 0$
$8 - 6 - 2 = 0$ ✓	$\frac{1}{2} + \frac{3}{2} - \frac{4}{2} = 0$ ✓

EXERCISE 12.4

Complete the square:

1. $x^2 + 4x + \_\_\_\_\_\_ = (x + \_\_\_\_\_\_)^2$
2. $x^2 - 10x + \_\_\_\_\_\_ = (x - \_\_\_\_\_\_)^2$
3. $y^2 - 8y + \_\_\_\_\_\_ = (y - \_\_\_\_\_\_)^2$
4. $y^2 + 3y + \_\_\_\_\_\_ = (y + \_\_\_\_\_\_)^2$
5. $x^2 - 7x + \_\_\_\_\_\_ = (x - \_\_\_\_\_\_)^2$
6. $x^2 - \frac{2}{3}x + \_\_\_\_\_\_ = (x - \_\_\_\_\_\_)^2$
7. $x^2 + \frac{1}{4}x + \_\_\_\_\_\_ = (x + \_\_\_\_\_\_)^2$
8. $x^2 - x + \_\_\_\_\_\_ = (x - \_\_\_\_\_\_)^2$

Solve by completing the square

9. $x^2 + 6x + 8 = 0$

10. $y^2 - 2y - 3 = 0$

11. $t^2 - 8t + 15 = 0$

12. $x^2 - 4x - 12 = 0$

13. $x^2 + 10x + 21 = 0$

14. $x^2 - 4x - 1 = 0$

15. $y^2 - 3y - 10 = 0$

16. $z^2 + 6z - 4 = 0$

17. $y^2 + y - 12 = 0$

18. $x^2 + 3x - 5 = 0$

19. $m^2 + 2m - 5 = 0$

20. $t^2 + 3t + 1 = 0$

21. $2x^2 + 6x - 4 = 0$

22. $3y^2 + 2y - 6 = 0$

23. $2x^2 - 5x - 5 = 0$

24. $2y^2 - y - 7 = 0$

Answers to odd-numbered exercises on page 496.

12.5 THE QUADRATIC FORMULA

Since we can use the process of completing the square on any quadratic equation whether it is factorable or not, let us now apply this procedure on the standard form of the quadratic equation $ax^2 + bx + c = 0$. In doing so, we will derive a formula known as the **quadratic formula.**

Derivation of the Quadratic Formula

Start with the standard form and complete the square.

$$ax^2 + bx + c = 0$$

$a \neq 0$, standard form.

$$ax^2 + bx = \boxed{-c}$$

Add $-c$ to both sides.

$$\frac{\cancel{a}x^2}{\boxed{\cancel{a}}} + \frac{bx}{\boxed{a}} = \frac{-c}{\boxed{a}}$$

Divide by a.

$$x^2 + \frac{b}{a}x \boxed{+\frac{b^2}{4a^2}} = \frac{-c}{a} \boxed{+\frac{b^2}{4a^2}}$$

Add $\left(\frac{1}{2} \cdot \frac{b}{a}\right)^2 = \frac{b^2}{4a^2}$ to both sides.

$$\left(x + \frac{b}{2a}\right)^2 = \frac{-c}{a} + \frac{b^2}{4a^2}$$

Factor the left side.

$$\left(x + \frac{b}{2a}\right)^2 = \frac{-4ac + b^2}{4a^2} = \frac{b^2 - 4ac}{4a^2}$$

Add fractions on the right.

$$x + \frac{b}{2a} = \pm\sqrt{\frac{b^2 - 4ac}{4a^2}} = \frac{\pm\sqrt{b^2 - 4ac}}{2a}$$

Take the square root of both sides.

$$x = -\frac{b}{2a} \pm \frac{\sqrt{b^2 - 4ac}}{2a}$$

Add $\frac{-b}{2a}$ to both sides.

$$x = \frac{-b \pm \sqrt{b^2 - 4ac}}{2a}$$

is called the **quadratic equation.**

(*Note:* $\boxed{\pm}$ gives two solutions.)

The formula must be memorized, since it provides solutions to *any* quadratic equation. In solving a quadratic equation, try to factor it first and if it is not factorable, apply the quadratic formula. The quadratic formula replaces the tedious method of completing the square.

To solve a quadratic equation $ax^2 + bx + c = 0$ using the quadratic formula

$$x = \frac{-b \pm \sqrt{b^2 - 4ac}}{2a}$$

1. Write the given equation in standard form $ax^2 + bx + c = 0$.
2. Identify the constants a, b, and c.
3. Substitute the values for a, b, and c into the quadratic formula.
4. Evaluate to give the solutions.

Example 1: Solve $x^2 - 5x + 3 = 0$

Solution: $x^2 - 5x + 3 = 0$ is in standard form as given, and $a = 1, b = -5, c = 3$.

Substitute these values in the quadratic formula.

$$x = \frac{-b \pm \sqrt{b^2 - 4ac}}{2a}$$

$$x = \frac{-(\boxed{-5}) \pm \sqrt{(\boxed{-5})^2 - 4 \cdot \boxed{1} \cdot \boxed{3}}}{2(\boxed{1})}$$ Substitute for a, b, and c.

$$= \frac{5 \pm \sqrt{25 - 12}}{2}$$

$$= \frac{5 \pm \sqrt{13}}{2}$$

Example 2: Solve $2x^2 - 5x - 3 = 0$

(a) by factoring
(b) using the quadratic formula

Solution: (a)

$$2x^2 - 5x - 3 = 0$$

$$(2x + 1)(x - 3) = 0$$ Factor.

$2x + 1 = 0$ | $x - 3 = 0$ Set each factor equal to 0.

$2x = -1$ | $x = 3$

$x = -\frac{1}{2}$

The solutions are $x = -\frac{1}{2}$ or $x = 3$.

Solution: (b)

$2x^2 - 5x - 3 = 0$ is in standard form.

$a = 2, b = -5, c = -3$ Identify *a, b,* and *c*.

$$x = \frac{-b \pm \sqrt{b^2 - 4ac}}{2a}$$ Use the quadratic formula.

$$x = \frac{-(\boxed{-5}) \pm \sqrt{(\boxed{-5})^2 - 4(\boxed{2}) \cdot (\boxed{-3})}}{2(2)}$$ Substitute for *a, b,* and *c*.

$$x = \frac{5 \pm \sqrt{25 + 24}}{4} = \frac{5 \pm \sqrt{49}}{4} = \frac{5 \pm 7}{4}$$

$$x = \frac{5 \boxed{+} 7}{4} = \frac{12}{4} = 3 \quad \text{or} \quad x = \frac{5 \boxed{-} 7}{4} = \frac{-2}{4} = -\frac{1}{2}$$

The solutions are $x = 3$ or $x = -\frac{1}{2}$ as before.

EXERCISE 12.5

Solve using the quadratic formula:

1. $x^2 + 5x + 4 = 0$
2. $y^2 + 6y - 8 = 0$
3. $2y^2 - 8y + 3 = 0$
4. $t^2 + t - 20 = 0$
5. $x^2 + 6x + 3 = 0$
6. $x^2 + 5x - 2 = 0$
7. $x^2 + x - 1 = 0$
8. $2y^2 - 7y + 3 = 0$
9. $5t^2 - 8t + 3 = 0$

10. $3x^2 - 2x - 8 = 0$

11. $x^2 - 9 = 0$

12. $y^2 - 16 = 0$

13. $12x^2 - x = 6$

14. $y^2 + 2y = 30$

15. $t^2 + 4t = 1$

16. $4m^2 - 4m = 7$

17. $3x^2 - 6x = 2x^2 + 3$

18. $x(x - 5) = -3$

19. $(3y - 1)^2 = 6y$

20. $x^2 + 2(3x - 2) = 0$

21. $x(x + 2) + 3(x - 2) = -1$

22. $(x + 1)^2 = 3(1 - x)$

23. $(x + 3)(x + 4) = 15$

24. $(2x + 1)(x - 1) = -6(x - 1)$

Answers to odd-numbered exercises on page 496.

12.6 FRACTIONAL EQUATIONS RESULTING IN QUADRATIC EQUATIONS

All fractional equations that we have considered so far have resulted in *linear equations*. Now we will discuss fractional equations which lead to *quadratic equations*. This procedure may result in equations whose solutions are not solutions to the original equations. Therefore, *all* apparent solutions must be checked in the *original* equation.

Example 1: Solve $\frac{3}{x+1} - \frac{2}{x-1} = -1$

Solution: Multiply by the LCD $= (x + 1)(x - 1)$ to clear of fractions.

$$\boxed{(x+1)(x-1)}\left[\frac{3}{x+1} - \frac{2}{x-1}\right] = \boxed{(x+1)(x-1)}\,(-1)$$

$$\frac{\cancel{(x+1)}(x-1)}{1} \cdot \frac{3}{\cancel{x+1}} - \frac{(x+1)\cancel{(x-1)}}{1} \cdot \frac{2}{\cancel{x-1}} = (x+1)(x-1)(-1)$$

$$3x - 3 - 2x - 2 = -x^2 + 1$$

$$x - 5 = -x^2 + 1$$

$$x^2 + x - 6 = 0 \qquad \text{Write in standard form.}$$

$$(x + 3)(x - 2) = 0 \qquad \text{Factor.}$$

$$x + 3 = 0 \quad \Big| \quad x - 2 = 0$$

$$x = -3 \quad \Big| \quad x = 2 \qquad \text{Solve for } x.$$

The solutions are $x = -3$ or $x = 2$.

Check:

$$\frac{3}{x+1} - \frac{2}{x-1} = -1$$

$$\frac{3}{\boxed{-3}+1} - \frac{2}{\boxed{-3}-1} \stackrel{?}{=} -1$$

$$\frac{3}{-2} - \frac{2}{-4} \stackrel{?}{=} -1$$

$$\frac{6}{-4} - \frac{2}{-4} \stackrel{?}{=} -1$$

$$\frac{4}{-4} = -1 \checkmark$$

$$\frac{3}{x+1} - \frac{2}{x-1} = -1$$

$$\frac{3}{\boxed{2}+1} - \frac{2}{\boxed{2}-1} \stackrel{?}{=} -1$$

$$\frac{3}{3} - \frac{2}{1} \stackrel{?}{=} -1$$

$$1 - 2 = -1 \checkmark$$

Both solutions check.

Example 2: Solve $\frac{y^2}{y-2} = \frac{4}{y-2}$

Solution: Multiply both sides of the equation by the LCD, $y - 2$.

$$\boxed{(y-2)} \cdot \frac{y^2}{y-2} = \boxed{(y-2)} \cdot \frac{4}{y-2}$$

$$y^2 = 4$$

$$y = \pm 2 \qquad \text{Take the square root of both sides.}$$

The apparent solutions are $y = -2$ or $y = 2$.

Check:

$$\frac{y^2}{y-2} = \frac{4}{y-2} \qquad \bigg| \qquad \frac{y^2}{y-2} = \frac{4}{y-2}$$

$$\frac{(\boxed{-2})^2}{\boxed{-2}-2} \stackrel{?}{=} \frac{4}{\boxed{-2}-2} \qquad \bigg| \qquad \frac{\boxed{2}^2}{\boxed{2}-2} \stackrel{?}{=} \frac{4}{\boxed{2}-2}$$

$$\frac{4}{-4} = \frac{4}{-4} \checkmark \qquad \bigg| \qquad \frac{4}{0} \stackrel{?}{=} \frac{4}{0}$$

Since $\frac{4}{0}$ is undefined, $y = 2$ is not a solution. Therefore $y = -2$ is the only solution. This example illustrates the importance of checking all of your results in the original equation.

Example 3: Solve $x + 1 = \dfrac{6}{x}$

Solution: We multiply by the LCD, x, to clear of fractions.

$$\boxed{x}\,(x + 1) = \boxed{x}\left(\frac{6}{x}\right) \qquad \text{Multiply both sides by } x.$$

$$x^2 + x = 6$$

$$x^2 + x - 6 = 0 \qquad \text{Write in standard form.}$$

$$(x + 3)(x - 2) = 0 \qquad \text{Factor.}$$

$$x + 3 = 0 \quad \bigg| \quad x - 2 = 0 \qquad \text{Set equal to 0.}$$

$$x = -3 \quad \bigg| \quad x = 2$$

Check:

$$x = -3 \qquad \bigg| \qquad x = 2$$

$$x + 1 = \frac{6}{x} \qquad \bigg| \qquad x + 1 = \frac{6}{x}$$

$$\boxed{-3} + 1 \stackrel{?}{=} \frac{6}{\boxed{-3}} \qquad \bigg| \qquad \boxed{2} + 1 \stackrel{?}{=} \frac{6}{\boxed{2}}$$

$$-2 = -2 \checkmark \qquad \bigg| \qquad 3 = 3 \checkmark$$

Both solutions check.

EXERCISE 12.6

Solve. All of the apparent solutions must be checked in the original equations.

1. $x - 7 = \dfrac{-10}{x}$

2. $y - \dfrac{3}{y} = -2$

3. $x = \dfrac{9}{x}$

4. $y = \dfrac{16}{y}$

5. $x - 4 = \dfrac{12}{x}$

6. $\dfrac{-12}{x} = 1 - x$

7. $\dfrac{1}{x+2} + \dfrac{1}{x-2} = \dfrac{2}{x^2 - 4}$

8. $\dfrac{1}{2x} + \dfrac{1}{x} = 3x$

9. $\dfrac{12}{x^2 - 4} + 1 = \dfrac{3}{x-2}$

10. $\dfrac{1 - 2y}{y} + y = 0$

11. $\dfrac{14}{x^2 - 1} + \dfrac{x}{x - 1} = \dfrac{8}{x - 1}$

12. $\dfrac{1}{3y + 2} = 4y - 1$

13. $1 + \dfrac{2}{x - 5} = \dfrac{-7}{x^2 - 25}$

14. $\dfrac{y^2}{y - 1} = \dfrac{1}{y - 1}$

Answers to odd-numbered exercises on page 496.

12.7 RADICAL EQUATIONS RESULTING IN QUADRATIC EQUATIONS

Radical equations are generally solved by squaring both sides of the equation. Some of these resulting equations are quadratic equations.

This squaring process sometimes produces apparent solutions that will not check in the original equation. For example, the equation $x = 5$ has only one solution, the number 5. Squaring both sides ($x^2 = 25$) produces a new equation that has two solutions, 5 and -5.

All solutions must be checked in the *original equation*.

Example 1: $\sqrt{x + 12} = x$

Solution: Square both sides and solve for x.

$$\sqrt{x + 12} = x$$

$$(\sqrt{x + 12})^2 = x^2 \qquad \text{Square both sides.}$$

$$x + 12 = x^2$$

$$0 = x^2 - x - 12 \qquad \text{Put in standard form.}$$

$$0 = (x - 4)(x + 3) \qquad \text{Factor.}$$

$$x - 4 = 0 \quad \Big| \quad x + 3 = 0$$

$$x = 4 \quad \Big| \quad x = -3$$

$x = 4$ or $x = -3$ are the apparent solutions.

Check:

$\sqrt{x+12} = x$	$\sqrt{x+12} = x$
$\sqrt{\boxed{4}+12} \stackrel{?}{=} \boxed{4}$	$\sqrt{\boxed{-3}+12} \stackrel{?}{=} \boxed{-3}$
$\sqrt{16} \stackrel{?}{=} 4$	$\sqrt{9} \stackrel{?}{=} -3$ $\sqrt{9} = +3$ only.
$4 = 4$ ✓ Checks.	$3 \neq -3$ Does not check!

Therefore the *only* solution is $x = 4$.

Example 2: Solve $\sqrt{y-5} - y + 7 = 0$

Solution:

$\sqrt{y-5} = y - 7$	Isolate the radical on one side of the equation.
$(\sqrt{y-5})^2 = (y-7)^2$	Square both sides.
$y - 5 = y^2 - 14y + 49$	Remember the middle term on the right.
$0 = y^2 - 15y + 54$	
$0 = (y-9)(y-6)$	Factor.
$y - 9 = 0$ | $y - 6 = 0$	Solve for y.
$y = 9$ | $y = 6$	

$y = 9$ and $y = 6$ are the apparent solutions.

Check:

$\sqrt{y-5} = y - 7$	$\sqrt{y-5} = y - 7$
$\sqrt{\boxed{9}-5} \stackrel{?}{=} \boxed{9}-7$	$\sqrt{\boxed{6}-5} \stackrel{?}{=} \boxed{6}-7$
$\sqrt{4} \stackrel{?}{=} 2$	$\sqrt{1} \stackrel{?}{=} -1$
$2 = 2$ ✓ Checks.	$1 \neq -1$ Does not check.

The *only* solution is $y = 9$

Example 3: Solve $\sqrt{x^2+7} = 4$

Solution:

$(\sqrt{x^2+7})^2 = (4)^2$	Square both sides.
$x^2 + 7 = 16$	
$x^2 = 9$	Take the square root of both sides.
$x = \pm 3$	

$x = 3$ or $x = -3$ are the apparent solutions.

Check:

$\sqrt{x^2+7} = 4$	$\sqrt{x^2+7} = 4$
$\sqrt{\boxed{3}^2+7} \stackrel{?}{=} 4$	$\sqrt{(\boxed{-3})^2+7} \stackrel{?}{=} 4$
$\sqrt{9+7} \stackrel{?}{=} 4$	$\sqrt{9+7} \stackrel{?}{=} 4$
$\sqrt{16} \stackrel{?}{=} 4$	$\sqrt{16} \stackrel{?}{=} 4$
$4 = 4$ ✓	$4 = 4$ ✓

The solutions are $x = 3$ or -3.

EXERCISE 12.7

Solve and check:

1. $\sqrt{x^2+3} = 2$

2. $\sqrt{2x^2-7x} = 2$

3. $y = \sqrt{5y-6}$

4. $2x = \sqrt{5x-1}$

5. $\sqrt{x-3} = x-9$

6. $\sqrt{2y-5} = y+1$

7. $y+2 = \sqrt{y+2}$

8. $\sqrt{x^2+6} = x+3$

9. $\sqrt{2x + 7} = x + 4$

10. $\sqrt{6x - 8} = x$

11. $\sqrt{-4x - 2} = 2x + 1$

12. $2x = \sqrt{7x + 2}$

13. $3x = \sqrt{9x - 2}$

14. $y - 3 = \sqrt{27 - 3y}$

15. $y - 1 = \sqrt{2y + 1}$

16. $3y = \sqrt{6y - 1}$

17. $x - 3 = \sqrt{2x - 3}$

18. $\sqrt{-x - 4} = x + 4$

19. $2\sqrt{x-1} = x - 1$

20. $4\sqrt{y+1} = y + 4$

Answers to odd-numbered exercises on page 496.

12.8 APPLICATIONS OF QUADRATIC EQUATIONS

Many word problems result in quadratic equations. Be careful to see if all solutions make sense in the original problem.

Example 1: The product of two consecutive even whole numbers is 168. Find the numbers.

Solution: Let x = first even whole number; then $x + 2$ = next even whole number. Since their product is 168, we have the equation:

$$x(x+2) = 168$$

$$x^2 + 2x = 168$$

$$x^2 + 2x - 168 = 0$$

$$(x + 14)(x - 12) = 0$$

$$x + 14 = 0 \quad \Big| \quad x - 12 = 0$$

$$x = -14 \quad \Big| \quad x = 12$$

Since our problem asks for whole numbers, -14 is not a solution (whole numbers are not negative). The numbers are 12 and 14 (x and $x + 2$).

Check: 12 and 14 are consecutive even whole numbers, and $12 \cdot 14 = 168$.

Example 2: The altitude of a triangle is 3 cm greater than the base. If the area of the triangle is 54 cm², find the base and the altitude of the triangle.

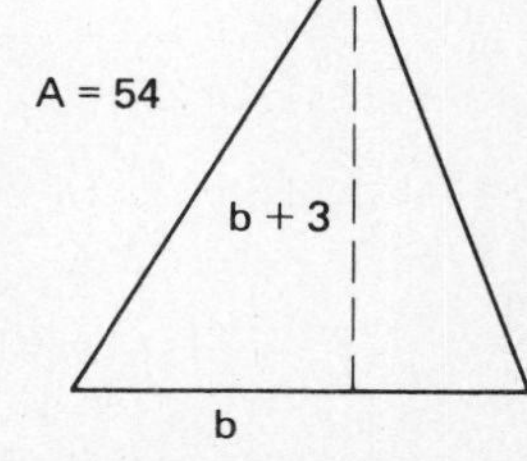

Solution: Let b = the base of the triangle, then $b + 3$ = the altitude of the triangle.

$$\frac{1}{2} \cdot \text{base} \cdot \text{height} = \text{area}$$

$$\frac{1}{2} \cdot b \cdot (b + 3) = 54 \qquad \text{Multiply by 2 to}$$

$$b(b + 3) = 108 \qquad \text{clear of fractions.}$$

$$b^2 + 3b = 108$$

$$b^2 + 3b - 108 = 0$$

$$(b + 12) \quad (b - 9) = 0$$

$$b + 12 = 0 \quad \Big| \quad b - 9 = 0$$

$$b = -12 \quad \Big| \quad b = 9$$

Since the base can't be negative, our only solution is $b = 9$ cm, and the altitude will be $b + 3 = 9 + 3 = 12$ cm.

Check: $a = 12$ cm is 3 cm more than $b = 9$ cm and the area is $\frac{1}{2} \cdot b \cdot h = \frac{1}{2} \cdot 12 \cdot 9 = 54 \text{ cm}^2$.

Example 3: A page in a book measures 7 inches by 9 inches. There is a blank border around the page leaving 35 square inches of print. How wide is the border?

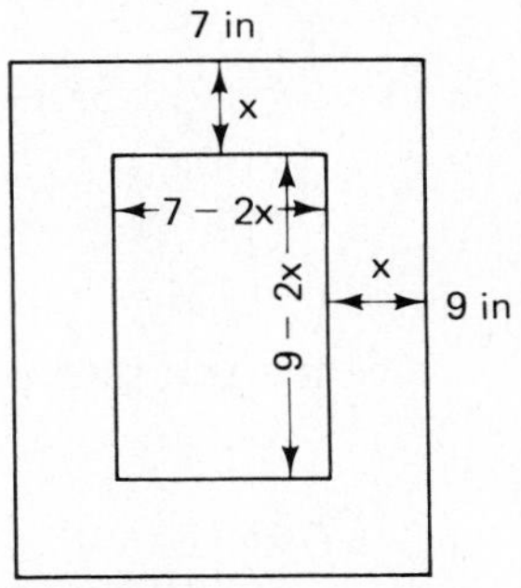

Solution: Let x = width of the border; then $9 - 2x$ = the length of the printed portion and $7 - 2x$ = the width of the printed portion.

$$\text{Area of print} = \ell \cdot \omega = 35$$

$$(9 - 2x)(7 - 2x) = 35$$

$$63 - 32x + 4x^2 = 35$$

$$4x^2 - 32x - 28 = 0 \qquad \text{Divide by 4.}$$

$$x^2 - 8x - 7 = 0$$

$$(x - 7)(x - 1) = 0 \qquad \text{Factor.}$$

$$x - 7 = 0 \quad \Big| \quad x - 1 = 0$$

$$x = 7 \quad \Big| \quad x = 1$$

Check: 7 cannot be a solution since a border of 7 inches would result in no print on the page. If the border is 1 inch, the length of the print $= 9 - 2x = 7$ inches, the width $= 7 - 2x = 5$ inches, and $7 \cdot 5 = 35$, which is the area of the printed portion of the page. Therefore, the width of the border is 1 inch.

Example 4: A canoeist travels 4 miles upstream, turns around, and travels 4 miles back downstream in a total time of 3 hours. If the speed of the current is 1 mile per hour, what is the speed of the canoe in still water?

Solution: Let r = the rate of speed of the canoe in still water; then $r - 1$ = the rate of the canoe upstream and $r + 1$ = the rate of the canoe downstream.

$$\text{distance upstream} = \text{distance downstream} = 4$$

Since $r \cdot t = d$, dividing by r gives us $t = \frac{d}{r}$

(time upstream)	+	(time downstream)	=	total time
$\frac{4}{r-1}$	+	$\frac{4}{r+1}$	=	3

To solve, multiply by the LCD = $(r-1)(r+1)$.

$$\boxed{\cancel{(r-1)}(r+1)}\cdot\frac{4}{\cancel{r-1}}+\boxed{(r-1)\cancel{(r+1)}}\cdot\frac{4}{\cancel{r+1}}=\boxed{(r-1)(r+1)}\cdot 3$$

$$4(r+1)+4(r-1)=3(r-1)(r+1)$$

$$4r+4+4r-4=3r^2-3$$

$$8r=3r^2-3$$

$$3r^2-8r-3=0$$ Put in standard form.

$$(3r+1)(r-3)=0$$ Factor.

$$3r+1=0 \quad\Big|\quad r-3=0$$

$$r=-\frac{1}{3} \quad\Big|\quad r=3$$

Since the rate of the boat is not negative we reject $r = -\frac{1}{3}$.

Check: $r = 3$

$$\frac{4}{r-1}+\frac{4}{r+1}=3$$

$$\frac{4}{\boxed{3}-1}+\frac{4}{\boxed{3}+1}\stackrel{?}{=}3$$

$$\frac{4}{2}+\frac{4}{4}=3\ \checkmark$$

The speed of the canoe in still water is 3 miles per hour.

The formula for finding the number of diagonals of an n-sided polygon is $\frac{n(n-3)}{2}$. For example, a 5-sided polygon (pentagon) would have

$$\frac{5(5-3)}{2}=\frac{5\cdot 2}{2}=5 \text{ diagonals}$$

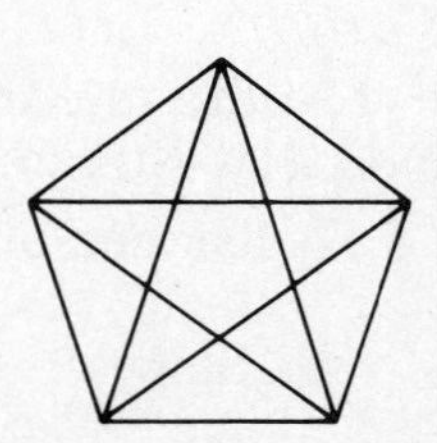

Example 5: How many sides does a polygon have if it has 20 diagonals?

Solution: $\dfrac{n(n-3)}{2} = 20$

$$n(n-3) = 40 \qquad \text{Multiply by 2.}$$

$$n^2 - 3n = 40$$

$$n^2 - 3n - 40 = 0 \qquad \text{Put in standard form.}$$

$$(n-8)(n+5) = 0 \qquad \text{Factor.}$$

$$n - 8 = 0 \quad \Big| \quad n + 5 = 0$$

$$n = 8 \quad \Big| \quad n = -5$$

We reject $n = -5$ and the polygon with 20 diagonals has 8 sides (octagon). Draw an octagon and see if you can draw 20 diagonals.

Example 6: The sum of a number and its reciprocal is $2\frac{1}{6}$. Find the number.

Solution: Let x = the desired number; then $\dfrac{1}{x}$ = the reciprocal of x.

Our equation is $$x + \frac{1}{x} = 2\frac{1}{6}$$

or $$x + \frac{1}{x} = \frac{13}{6}$$

$$\boxed{6x} \cdot x + \boxed{6x} \cdot \frac{1}{x} = \boxed{6x} \cdot \frac{13}{6} \qquad \text{Multiply by the LCD, } 6x.$$

$$6x^2 + 6 = 13x$$

$$6x^2 - 13x + 6 = 0 \qquad \text{Put in standard form.}$$

$$(3x-2)(2x-3) = 0 \qquad \text{Factor.}$$

$$3x - 2 = 0 \quad \Big| \quad 2x - 3 = 0$$

$$3x = 2 \quad \Big| \quad 2x = 3$$

$$x = \frac{2}{3} \quad \Big| \quad x = \frac{3}{2}$$

Each solution is the reciprocal of the other, and since

$$\frac{2}{3} + \frac{3}{2} = \frac{4}{6} + \frac{9}{6} = \frac{13}{6} = 2\frac{1}{6}$$

both solutions check.

The solutions are $\frac{2}{3}$ or $\frac{3}{2}$.

EXERCISE 12.8

1. The sum of the squares of two consecutive whole numbers is 113. Find the numbers.

2. Find a number so that the sum of the number and twice the square is 78.

3. The hypotenuse of a right triangle is 15 and one leg is 3 cm longer than the other. Find the length of each of the legs.

4. The hypotenuse of a right triangle is 1 cm greater than one of the legs. If the other leg is 5 cm, find the length of the other two sides.

5. The base of a triangle is 1 inch longer than its altitude. If the area of the triangle is 15 square inches, find the base and altitude of the triangle.

6. The length of a rectangle is one more than three times the width. If the area is 52, find its dimensions.

7. The diagonal of a rectangle is 20. If the length of the rectangle is 4 more than its width, find the dimensions of the rectangle.

8. The product of two consecutive odd whole numbers is 143. Find the numbers.

9. The area of a square is equal to its perimeter. Find its dimensions.

10. The outside dimensions of a frame around a painting are 22 inches by 16 inches. If the area of the painting inside the frame is 247 square inches, how wide is the frame?

11. The page of a book measures 25 cm by 20 cm. There is a blank border of equal width on all sides surrounding the printed material. Find the width of the border if the area of the printed portion of the page is 336 cm^2.

12. A photograph measures 6 inches by 9 inches. The frame around the photograph is of equal width and has an area equal to the area of the photograph. Find the width of the frame.

13. A boat travels upstream a distance of 15 miles and downstream the same distance in a total time of 4 hours. If the speed of the current is 5 miles per hour, what is the speed of the boat in still water?

14. A boat takes a 10-hour trip traveling 40 miles upstream and 40 miles downstream. If the current is moving at a rate of 3 miles per hour, what is the speed of the boat in still water?

15. How many sides does a polygon have if it has 9 diagonals? (See example 5.)

16. How many sides does a polygon have if it has 27 diagonals? (See example 5.)

17. The sum of a number and its reciprocal is $\frac{25}{12}$. What is the number?

18. The sum of a number and its reciprocal is $3\frac{1}{24}$. What is the number?

19. The sum of a number and its reciprocal is 4. Find the number.

20. A box is 4 inches longer than it is wide. If the box is 5 inches high and has a volume of 300 in^3, find the length and width of the box.

21. A rectangular sheet of cardboard is three times as long as it is wide. If 3-inch squares are cut from the corners, and the sides and ends folded up, a box having a volume of 288 in^3 is formed. What are the length and width of the piece of cardboard?

22. Find the width of an L-shaped corridor that measures 24 feet by 9 feet if the total floor area is 116 ft^2.

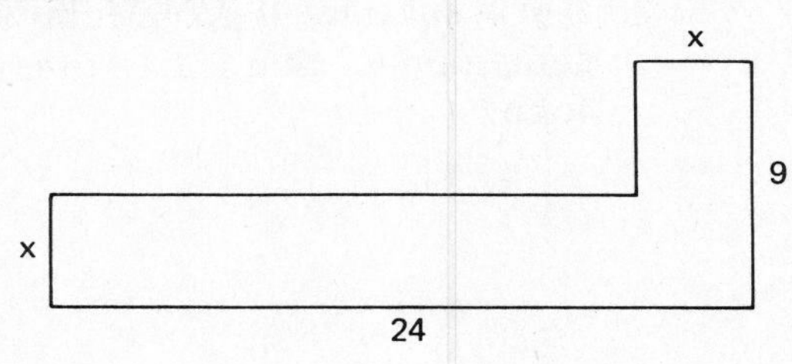

23. A rectangular garden is 20 feet longer than it is wide. If the area is 300 square feet, find the length and width of the garden.

24. Twice the square of a number is equal to twenty times the number. What is the number?

25. The length of a rectangular table is 2 feet more than its width. If the area of the table is 4 more than its perimeter, find the dimensions of the table.

26. Twice the square of two consecutive even integers is 200. Find the numbers.

27. 480 square tiles are required to cover a floor. If tiles that are 2 inches longer on a side are used, only 270 tiles are required to cover the same floor. Find the size of the smaller tiles.

28. If you subtract 10 from John's age and square the result, you will get a number that is five times as large as John's age. How old is John?

29. With the current moving at a rate of 3 miles per hour, a boat travels 6 miles upstream and 12 miles downstream in a total time of 2 hours. What is the rate of speed of the boat in still water?

30. The sides of a right triangle are $x, x + 2$, and $x + 4$. Find the length of each side.

Answers to odd-numbered exercises on page 497.

12.9 Chapter Summary

Examples

(12.1) The **standard form** of a quadratic equation is

$$ax^2 + bx + c = 0$$

where a, b and c are real numbers and $a > 0$

(12.1) $6x^2 - 8x + 1 = 0$ is in standard form.

(12.2) To solve a quadratic equation by factoring:

1. Put the equation in standard form $ax^2 + bx + c = 0$.
2. Factor the left side of the equation.
3. Set each factor equal to zero and solve for x.
4. Check each solution in the original equation.

(12.2) Solve $x^2 - 7x = -10$ by factoring.

1. $x^2 - 7x + 10 = 0$
2. $(x - 5)(x - 2) = 0$
3. $x - 5 = 0$, $x = 5$ | $x - 2 = 0$, $x = 2$
4. $5^2 - 7 \cdot 5 + 10 \stackrel{?}{=} 0$, $25 - 35 + 10 \stackrel{?}{=} 0$, $0 = 0 \checkmark$ | $2^2 - 7 \cdot 2 + 10 \stackrel{?}{=} 0$, $4 - 14 + 10 \stackrel{?}{=} 0$, $0 = 0 \checkmark$

(12.3) To solve a quadratic equation of the form $ax^2 - c = 0$:

1. Add c to both sides to obtain $ax^2 = c$.
2. Divide both sides by a, yielding $x^2 = \frac{c}{a}$.
3. Take the square root of both sides, obtaining $x = \pm\sqrt{\frac{c}{a}}$.
4. Check the solutions in the original equation.

(12.3) Solve $2x^2 - 18 = 0$

1. $2x^2 = 18$
2. $x^2 = 9$
3. $x = \pm\sqrt{9} = \pm 3$
4. $2(3)^2 \stackrel{?}{=} 18$, $2 \cdot 9 \stackrel{?}{=} 18$, $18 = 18$ | $2(-3)^2 \stackrel{?}{=} 18$, $2 \cdot 9 \stackrel{?}{=} 18$, $18 = 18$

(12.4) To solve a quadratic equation $ax^2 + bx + c = 0$ by completing the square:

1. Subtract c from both sides of the equation.
2. If $a \neq 1$, divide both sides by a.
3. Complete the square by adding the square of half the coefficient of x, $\left(\frac{b}{a}\right)^2$, to both sides.
4. Factor the left side of the equation.
5. Take the square root of both sides and solve for x.

(12.4) Solve $x^2 - 4x - 9 = 0$ by completing the square.

1. $x^2 - 4x + \_\_\_\_ = 9$
2. Coefficient of x^2 is 1
3. $x^2 - 4x \boxed{+ 4} = 9 \boxed{+ 4}$
4. $(x - 2)^2 = 13$
5. $x - 2 = \pm\sqrt{13}$

   $x = 2 \pm \sqrt{13}$

(12.5) To solve a quadratic equation $ax^2 + bx + c = 0$ using the quadratic formula

(12.5) Solve $3x^2 + 2x - 5 = 0$ using the quadratic formula.

$$x = \frac{-b \pm \sqrt{b^2 - 4ac}}{2a}$$

1. Write the equation in standard form $ax^2 + bx + c = 0$.
2. Identify the constants a, b, and c.
3. Substitute the values for a, b, and c into the quadratic formula.
4. Evaluate to find the solutions.

1. $3x^2 + 2x - 5 = 0$
2. $a = 3, b = 2, c = -5$
3. $x = \frac{-2 \pm \sqrt{2^2 - 4 \cdot 3 \cdot (-5)}}{2 \cdot 3}$

$x = \frac{-2 \pm \sqrt{4 + 60}}{6}$

$x = \frac{-2 \pm \sqrt{64}}{6} = \frac{-2 \pm 8}{6}$

$x = \frac{-2 \boxed{+} 8}{6} = \frac{6}{6} = 1$

or

$x = \frac{-2 \boxed{-} 8}{6} = \frac{-10}{6} = \frac{-5}{3}$

The solutions are $x = 1$ or

or $x = \frac{-5}{3}$.

Exercise 12.9 Chapter Review

(12.1) Write the following in standard form:

1. $3x^2 = 9 - 2x$

2. $-7 = 5x\ (2 - x)$

3. $\dfrac{x^2}{3} - 4 = \dfrac{3x}{2}$

(12.2) Solve by factoring:

4. $x^2 - 49 = 0$

5. $x^2 + 2x - 15 = 0$

6. $6x^2 + 7x - 3 = 0$

7. $8x = 5 - 4x^2$

8. $x^2 - 2x = 9x + 3x^2 - 21$

(12.3) Solve using the method described in Section 12.3:

9. $a^2 = 81$

10. $x^2 = 12$

11. $3y^2 = 60$

12. $2x^2 - \frac{8}{9} = 0$

13. $\frac{t^2}{4} - 7 = 0$

(12.4) Complete the square:

14. $x^2 + 6x +$ ________ $= (x +$ ________ $)^2$

15. $y^2 - 5x +$ ________ $= (y -$ ________ $)^2$

16. $t - \frac{1}{2}t +$ ________ $= (t -$ ________ $)^2$

(12.4) Solve by completing the square:

17. $x^2 - 4x - 5 = 0$

18. $x^2 - 6x + 5 = 0$

19. $y^2 - y - 1 = 0$

20. $3t^2 - 3t - 2 = 0$

21. $2m^2 - 3m - 2 = 0$

22. $3x^2 + 6x - 9 = 0$

(12.5) Solve using the quadratic formula:

23. $x^2 - x - 12 = 0$

24. $m^2 - 3m - 5 = 0$

25. $y^2 + 9y + 1 = 0$

26. $2t^2 - 7t - 15 = 0$

27. $x^2 - 10 = 0$

28. $y^2 = 5y - 2$

29. $x(x - 1) + 2(x + 5) = 12$

30. $(x + 3)(x - 1) = 2(x + 1)$

(12.2–12.7) Solve by any method:

31. $x^2 - 6x - 7 = 0$

32. $2x^2 + 11x + 5 = 0$

33. $5a^2 - 60 = 0$

34. $t^2 - 3t - 6 = 0$

35. $t^2 - 7t + 2 = 0$

36. $15x^2 - 2x - 1 = 0$

37. $m^2 = 8$

38. $3x^2 - 19x + 20 = 0$

39. $(x - 3)(x + 5) = 4(x + 1)$

40. $(x - 2)^2 = x(3x - 2)$

41. $\sqrt{x^2 - 7} = 3$

42. $\frac{3}{y^2} - 1 = \frac{2}{y}$

43. $2x - 8 = 2\sqrt{3x - 2}$

44. $\frac{3}{2y + 5} + \frac{x}{4} = \frac{3}{4}$

45. $\sqrt{6x - 1} = 3x$

(12.8) Solve the following word problems:

46. Five times the square of a whole number minus that number is equal to 18. Find the number.

47. The length of a rectangle is 3 inches more than its width. If the diagonal of the rectangle is 15, find the length and width of the rectangle.

48. A boat travels 30 miles upstream and back in a total of 8 hours. If the speed of the current is 2 miles per hour, what is the speed of the boat in still water?

49. How many sides does a polygon have if it has 35 diagonals?

50. An L-shaped kitchen counter has a working area that measures 8 feet by $1\frac{1}{2}$ feet. How wide is the counter if it has a total area of 30 ft^2?

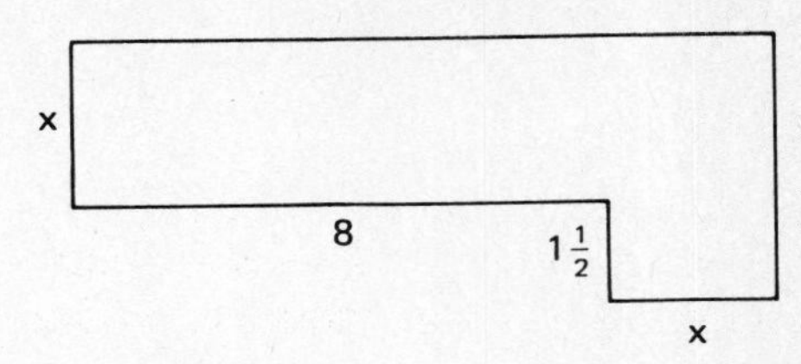

Answers to odd-numbered exercises on page 497.

Chapter 12 Achievement Test

Name ______________________

Class ______________________

This test should be taken before you are tested in class on the material in Chapter 12. Solutions to each problem and the section where the type of problem is found are given on page 498.

Write in standard form:

1. $2 + 3x^2 = -9x$

2. $\frac{x^2}{3} - \frac{5}{2} = \frac{x}{4}$

Solve by factoring:

3. $x^2 + x - 12 = 0$

4. $6x^2 + 13x = 5$

Solve by the method described in Section 12.3:

5. $x^2 = 100$

6. $4x^2 = 80$

Complete the square:

7. $x^2 - 4x +$ ________ $= (x -$ ________ $)^2$

1. ______________________

2. ______________________

3. ______________________

4. ______________________

5. ______________________

6. ______________________

7. ______________________

8. $x^2 + x +$ ________ $= (x +$ ________ $)^2$

Solve by completing the square:

9. $x^2 - x - 6 = 0$

10. $x^2 - 8x + 1 = 0$

Solve using the quadratic formula:

11. $x^2 - 2x - 8 = 0$

12. $2x^2 - 3x - 1 = 0$

Solve by any method:

13. $6x^2 - 36 = 0$

14. $6x^2 = 1 - x$

15. $x - 1 = 6\sqrt{x - 9}$

8. ________

9. ________

10. ________

11. ________

12. ________

13. ________

14. ________

15. ________

Solve the following word problems:

16. The length of a rectangle is 2 inches more than twice its width. What are the dimensions of the rectangle if the area is 60 in^2?

16. ____________________

17. The sum of a number and its reciprocal is $4\frac{13}{18}$. Find the number.

17. ____________________

18. A boat travels upstream 8 miles and downstream 20 miles. The total trip takes 4 hours. If the current has a speed of 3 miles per hour, find the speed of the boat in still water.

18. ____________________

Appendix
Arithmetic Review

A.1 FRACTIONS

To reduce a fraction to its lowest terms, divide the numerator and denominator by the *same* non-zero number.

$$\frac{8}{12}=\frac{\overset{2}{\cancel{8}}}{\underset{3}{\cancel{12}}}=\frac{2}{3} \qquad \frac{8}{12} \text{ and } \frac{2}{3} \text{ are called } \textbf{equivalent fractions.}$$

To raise a fraction to higher terms, multiply the numerator and denominator by the *same* non-zero number.

$$\frac{3}{4}=\frac{3\cdot 5}{4\cdot 5}=\frac{15}{20} \qquad \frac{3}{4} \text{ and } \frac{15}{20} \text{ are called } \textbf{equivalent fractions.}$$

To multiply fractions, cancel any numerator with any denominator. Then multiply the numerators together and multiply the denominators together.

$$\frac{5}{8}\cdot\frac{12}{25}=\frac{\overset{1}{\cancel{5}}}{\underset{2}{\cancel{8}}}\cdot\frac{\overset{3}{\cancel{12}}}{\underset{5}{\cancel{25}}}=\frac{1\cdot 3}{2\cdot 5}=\frac{3}{10}$$

To divide fractions, invert the second fraction and multiply.

$$\frac{3}{8}\div\frac{9}{16}=\frac{\overset{1}{\cancel{3}}}{\underset{1}{\cancel{8}}}\cdot\frac{\overset{2}{\cancel{16}}}{\underset{3}{\cancel{9}}}=\frac{2}{3}$$

To add (or subtract) fractions, the denominators must be the same. If they are not, change each given fraction to an equivalent fraction with a denominator equal to the LCD (least common denominator). Then add (or subtract) the numerators and keep the denominator.

$$\frac{3}{8} + \frac{5}{12} \qquad \text{LCD} = 24 \text{ (by inspection).}$$

$$\frac{9}{24} + \frac{10}{24} = \frac{19}{24}$$

A.2 MIXED NUMBERS

To change an improper fraction to a mixed number, divide the numerator by the denominator and write the remainder as a fraction.

$$\frac{22}{7} = 3\frac{1}{7}$$

To change a mixed number to an improper fraction, multiply the whole number by the denominator and add the numerator. Keep the same denominator.

$$3\frac{5}{8} = \frac{3 \cdot 8 + 5}{8} = \frac{29}{8}$$

To multiply (or divide) mixed numbers, convert them to improper fractions and follow the rules for multiplying (or dividing) fractions.

$$4\frac{2}{3} \cdot 3\frac{3}{8} = \frac{\overset{7}{\cancel{14}}}{\underset{1}{\cancel{3}}} \cdot \frac{\overset{9}{\cancel{27}}}{\underset{4}{\cancel{8}}} = \frac{63}{4} \text{ or } 15\frac{3}{4}$$

$$2\frac{3}{4} \div 5\frac{1}{2} = \frac{11}{4} \div \frac{11}{2} = \frac{\overset{1}{\cancel{11}}}{\underset{2}{\cancel{4}}} \cdot \frac{\overset{1}{\cancel{2}}}{\underset{1}{\cancel{11}}} = \frac{1}{2}$$

To add mixed numbers, add the fractions separately from the whole numbers.

$$\begin{array}{rl} 11\frac{1}{2} = & \frac{3}{6} \\ + 15\frac{2}{3} = & \frac{4}{6} \\ \hline 26 & \frac{7}{6} = \boxed{1}\frac{1}{6} \end{array}$$

↖ Add this 1 with the whole numbers.

The answer is $27\frac{1}{6}$.

To subtract mixed numbers, subtract the fractions separately from the whole numbers. If necessary, borrow from the whole number.

$$
\begin{array}{r}
\overset{7}{\cancel{18}}\frac{1}{3} = \frac{2}{6} + \frac{6}{6} = \frac{8}{6} \\
-\ 12\frac{1}{2} = \frac{3}{6} \qquad\qquad \\
\hline
5\frac{5}{6} \qquad\qquad\qquad
\end{array}
$$

A.3 DECIMALS

Every whole number has a decimal point implied (but not written) at the right of the number.

26 = 26. ↖

The number of decimal places in a number is the number of digits which come after the decimal point.

3.061 (3 decimal places under "061")

3 decimal places

To add or subtract decimals, line up the decimal points vertically.

Add:

$$
\begin{array}{r}
3.126 \\
20.04 \\
.003 \\
11.4 \\
\hline
34.569
\end{array}
$$

Subtract: 3.029 from 6.7

$$
\begin{array}{r}
6.700 \\
-\ 3.029 \\
\hline
3.671
\end{array}
$$

6.700 ← add zeros

To multiply decimals, the number of decimal places in the answer is equal to the sum of the number of decimal places in the numbers being multiplied.

$$
\begin{array}{r}
2.04 \\
\times\ 1.8 \\
\hline
1632 \\
204\ \\
\hline
3.672
\end{array}
$$

2.04 ← 2 decimal places
× 1.8 ← 1 decimal place
3.672 ← 3 decimal places

$$
\begin{array}{r}
1.003 \\
\times\ .06 \\
\hline
.06018
\end{array}
$$

1.003 ← 3 decimal places
× .06 ← 2 decimal places
.06018 ← 5 decimal places
↖ must add a zero

To divide decimals, move the decimal point a sufficient number of decimal places in both the divisor and dividend so that you will be dividing by a *whole number*.

$$\begin{array}{r} .52 \\ 4.1_{\smile})\overline{2.1_{\smile}32} \\ 2\,0\,5 \\ \hline 82 \\ 82 \\ \hline 0 \end{array}$$

To write a fraction as a decimal, divide the numerator by the denominator.

$$\frac{3}{8} = \frac{3}{8} \qquad 8\overline{)3.000}^{\,.375} \qquad \frac{7}{9} = \frac{7}{9} \qquad 9\overline{)7.000}^{\,.777} \approx .78$$

is approximately equal to

rounded to the nearest hundredth

To write a decimal as a fraction, just read the decimal and write it in fraction form.

Reduce to lowest terms:

$$.26 = 26 \text{ hundredths} = \frac{26}{100} = \frac{13}{50}$$

For a more comprehensive review of arithmetic, see Derek I. Bloomfield, *From Arithmetic to Algebra,* 2nd Edition, Reston, VA: Reston Publishing Co., 1982.

EXERCISE A.1

Reduce to lowest terms if possible:

1. $\frac{12}{18}$
2. $\frac{15}{21}$
3. $\frac{28}{87}$
4. $\frac{24}{32}$

Multiply:

5. $\frac{14}{15} \cdot \frac{10}{21}$
6. $\frac{5}{7} \cdot \frac{14}{15}$
7. $\frac{2}{3} \cdot \frac{12}{15}$
8. $\frac{2}{7} \cdot \frac{7}{16} \cdot \frac{4}{15}$
9. $\frac{3}{5} \cdot 6 \cdot \frac{5}{8} \cdot \frac{8}{12}$

Divide:

10. $\frac{6}{7} \div \frac{3}{5}$

11. $\frac{1}{2} \div \frac{5}{8}$

12. $\frac{3}{8} \div \frac{3}{8}$

Add or subtract as indicated:

13. $\frac{9}{16} - \frac{5}{16}$

14. $\frac{2}{3} + \frac{3}{8} + \frac{1}{6}$

15. $\frac{1}{8} + \frac{5}{16} + \frac{1}{4}$

16. $\frac{7}{16} - \frac{5}{12}$

Write as mixed numbers:

17. $\frac{23}{6}$

18. $\frac{29}{3}$

19. $\frac{57}{4}$

Write as improper fractions:

20. $5\frac{1}{3}$

21. $9\frac{3}{5}$

22. $2\frac{1}{5}$

Multiply or divide as indicated:

23. $2\frac{1}{3} \div 3\frac{3}{4}$

24. $3\frac{3}{5} \div 5\frac{1}{4}$

25. $4\frac{1}{6} \cdot 2\frac{4}{5}$

26. $3\frac{1}{4} \div 2\frac{2}{3}$

27. $2\frac{1}{3} \div 3\frac{8}{9}$

28. $5\frac{1}{6} \cdot \frac{7}{18} \cdot 4$

Add or subtract as indicated:

29. $\begin{array}{r} 4\frac{1}{6} \\ +\ 3\frac{1}{3} \\ \hline \end{array}$

30. $\begin{array}{r} 14\frac{2}{3} \\ +\ 22\frac{7}{8} \\ \hline \end{array}$

31. $\begin{array}{r} 5\frac{5}{8} \\ -\ 3\frac{1}{3} \\ \hline \end{array}$

32. $\begin{array}{r} 9\frac{1}{5} \\ -\ 7\frac{2}{3} \\ \hline \end{array}$

33. $\begin{array}{r} 3\frac{7}{8} \\ 4\frac{1}{2} \\ +\ 7\frac{3}{4} \\ \hline \end{array}$

34. $\begin{array}{r} 13 \\ -\ 3\frac{1}{3} \\ \hline \end{array}$

Add or subtract as indicated:

35. 3.041 + .003 + 20.09 + .041

36. .024 + 6 + 15.01 + 2.14 + .006

37. 6.824 − 1.956

38. 17 − 11.41

Multiply:

39. $\begin{array}{r} 4.12 \\ \times\ 1.5 \\ \hline \end{array}$

40. $\begin{array}{r} .215 \\ \times\ 2.3 \\ \hline \end{array}$

41. $\begin{array}{r} .0029 \\ \times\ .07 \\ \hline \end{array}$

Divide:

42. $.06\,\overline{)1.284}$

43. $2.3\,\overline{)9.453}$

44. $.002\,\overline{)16.8}$

Write as a decimal (round off to the nearest hundredth if necessary):

45. $\frac{3}{4}$

46. $\frac{5}{8}$

47. $\frac{3}{7}$

Write as a fraction and reduce to lowest terms:

48. .25

49. .56

50. .034

Answers to odd-numbered exercises on page 498.

Answers to Odd-Numbered Exercises

CHAPTER 1

EXERCISE 1.1, PAGE 2

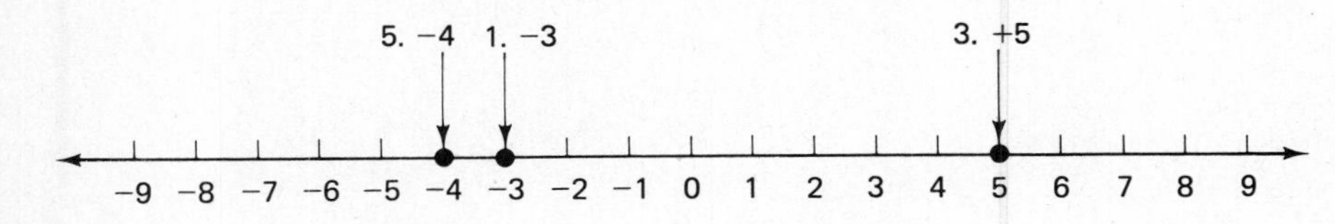

7. −$5.00 9. +7288 11. +1990 13. −15 15. −$25 17. $-2\frac{1}{4}$

EXERCISE 1.2, PAGE 3

1. 8 3. 4 5. 1462 7. 0 9. 6

EXERCISE 1.3, PAGE 4

1. true 3. false 5. true 7. false

9. true 11. false 13. $17 > 14$ 15. $23 > -5$

17. $46 < 49$ 19. $0 < 14$ 21. $586 > -12$ 23. $314 > 8.9$

EXERCISE 1.4, PAGE 6

1. 14 3. -3 5. 2 7. 7

9. -1 11. -6 13. -10 15. -8

17. -17 19. -30 21. $-\frac{4}{8} = -\frac{1}{2}$ 23. $-\frac{2}{7}$

25. lost \$.82 or $-$ \$.82 27. down $4\frac{1}{3}$ or $-4\frac{1}{3}$

29.

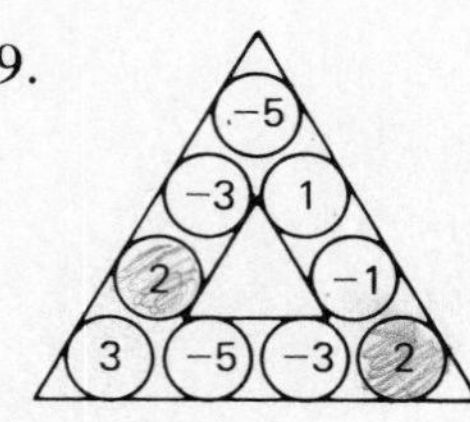

EXERCISE 1.5, PAGE 10

1. -2 3. -5 5. 9

7. -11 9. -19 11. -30

13. 0 15. 10 17. 1

19. 16 21. -4 23. 4

25. 1 27. $-8\frac{4}{8} = -8\frac{1}{2}$ 29. $\frac{1}{7}$

31. -9 33. -29 35. -4

37. $-\frac{1}{10}$ 39. 7.05 41. 10

43. (a) loss of $1\frac{1}{4}$ or $-1\frac{1}{4}$
(b) $63\frac{1}{4}$

EXERCISE 1.6, PAGE 16

1. 24 3. -28 5. -16

7. 66 9. -60 11. 132

13. -528 15. -98 17. -4.08

19. -54 21. -4 23. -38.7

25. -144 27. -120 29. -1

31. 17.64 33. 81 35. 1

37. $-\$11{,}200$

EXERCISE 1.7, PAGE 19

1. 3 3. 1 5. -7

7. -9 9. $-\frac{1}{6}$ 11. 0

13. undefined 15. 4 17. $\frac{3}{2}$

19. $-\frac{5}{2}$ 21. $-.4$ 23. undefined

25. lost \$6 per race 27. lost \$1.75 per share

EXERCISE 1.8, PAGE 24

1. true 3. false 5. true

7. true 9. false 11. false

13. true

15.
$$6(3 + 5) \stackrel{?}{=} 6 \cdot 3 + 6 \cdot 5$$
$$6(8) \stackrel{?}{=} 18 + 30$$
$$48 = 48$$

17.
$$-8(4 + 7) \stackrel{?}{=} (-8)(4) + (-8)(7)$$
$$-8(11) \stackrel{?}{=} (-32) + (-56)$$
$$-88 = -88$$

19.
$$4.3(1.2 + .3) \stackrel{?}{=} (4.3)(1.2) + (4.3)(.3)$$
$$4.3(1.5) \stackrel{?}{=} (5.16) + 1.29$$
$$6.45 = 6.45$$

EXERCISE 1.9, PAGE 26

1. 16 3. 4 5. 81

7. 64 9. 256 11. 1000

13. 1 15. -1000 17. $\frac{1}{4}$

19. .0009 21. $-\frac{8}{27}$ 23. 1

25. $10^n = 1$ followed by n zeros

EXERCISE 1.10, PAGE 28

1. 5 3. −6 5. 7
7. −3 9. 3 11. 1
13. 20 15. 6 17. 2

EXERCISE 1.11, PAGE 30

1. 7 3. 1 5. 19
7. −7 9. 6 11. 5
13. 38 15. −9 17. −16
19. 100 21. −71

EXERCISE 1.12, PAGE 32

1. −30 3. 15 5. −23
7. 42 9. −7 11. 15
13. 209

EXERCISE 1.13, CHAPTER REVIEW, PAGE 37

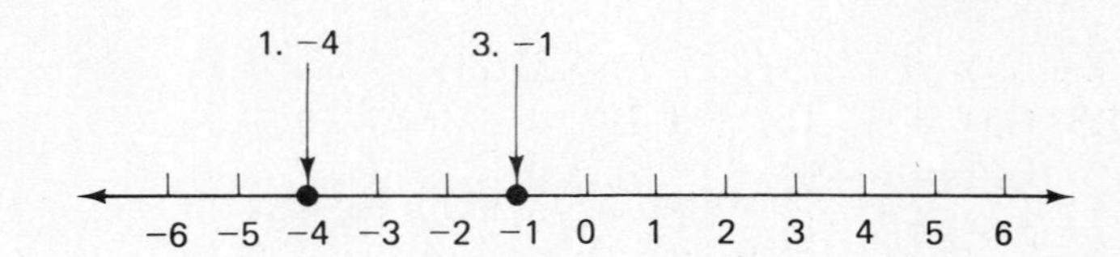

5. +8 7. 4 9. 12
11. $-3 < 5$ 13. $3 > -6$ 15. −4
17. 2 19. −1 21. −3
23. 9 25. undefined 27. −10

29. $-.06$ 31. 72 33. no

35. 27 37. 81 39. 1,000,000

41. 6 43. 2 45. 10

47. -11 49. -6 51. 4

53. -9

CHAPTER 1 ACHIEVEMENT TEST, PAGE 39

(1.1)

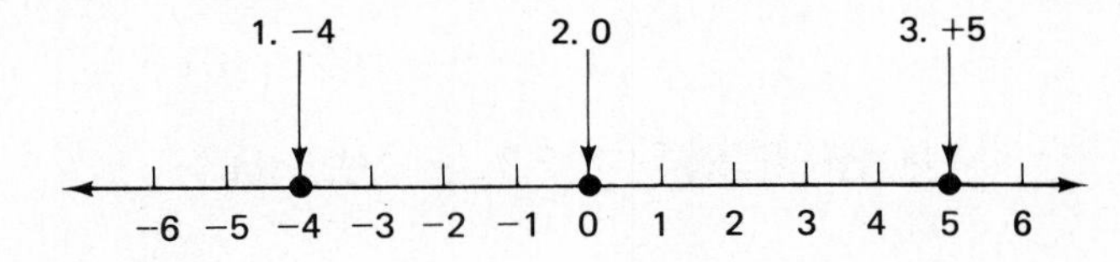

4. $-\$24$ 5. $+\$50$

(1.2) 6. 0 7. 8 8. 13

(1.3) 9. $-8 < -4$ 10. $6 > -5$ 11. $-4 < 0$

(1.4—1.10) 12. -1 13. 1 14. 1 15. 20 16. 4

17. undefined 18. 0 19. -9 20. -4 21. -48

22. -3.16 23. $5\frac{2}{5}$ 24. 64 25. 16 26. -8

27. 100,000 28. 3

(1.11) 29. 31 30. 9

(1.12) 31. 5 32. -8

CHAPTER 2

EXERCISE 2.1, PAGE 45

1. 108 sq ft 3. \$1500 5. 176 miles

7. 48 sq units 9. 50.24 sq meters 11. 19

13. 320 15. 16 17. 900

19. 6.28 21. 5 23. 942

25. -13

27. (a) \$21.08
(b) \$ 6.08
(c) \$181.12—much less than half.

29. (a) $T = \frac{1}{2} \cdot 2 \cdot 3 = 3$
$T = \frac{1}{2} \cdot 4 \cdot 5 = 10$
$T = \frac{1}{2} \cdot 6 \cdot 7 = 21$
(b) 55
(c) 5050

EXERCISE 2.2, PAGE 49

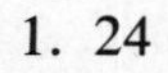

1. 24   3. -4 5. -19 7. 190

9. 0   11. -39 13. 76   15. 72

17. 16   19. 16   21. 14   23. 60

25. -8 27. -146 29. 18

EXERCISE 2.3, PAGE 53

1. terms are $2x$ and $3y$
numerical coefficients are 2 and 3
literal parts are x and y
3. terms are $-6x^3y$ and $2x^2y^3$
numerical coefficients are -6 and 2
literal parts are x^3y and x^2y^3
5. terms are $7x^3y$, $-14xy^3$, $2z$ and -5
numerical coefficients are 7, -14, 2 and -5
literal parts are x^3y, xy^3, and z
7. terms are $-7xy^4$, $2x^2$ and $-3xyz$
numerical coefficients are -7, 2 and -3
literal parts are xy^4, x^2 and xyz
9. terms are $-11a^3$, $-14b^3$ and $17ac^4$
numerical coefficients are -11, -14 and 17
literal parts are a^3, b^3 and ac^4

11. $3x$ 13. $5a$

15. $17x$ 17. $3x$

19. $4a^2b$ 21. $-2x^2y^2$

23. $-8x^3$ 25. $10x^3 + 7x$

27. $14mn$ 29. $8ab - 8b + 3$

31. $-11cd - 5c^2d + 4cd^2$

33. $9x^2y - 9$

35. $5m + 6n + 2$

37. $8a - 11b - 3c$

39. cannot be combined further

EXERCISE 2.4, PAGE 56

1. a^5
3. x^{12}
5. y^5
7. 10^7 or 10,000,000
9. 3^5 or 243
11. x^6
13. x^9
15. cannot be combined
17. 10^6 or 1,000,000
19. cannot be combined
21. x^{a+b}
23. a^{x+2}
25. cannot be combined
27. x^{2a}
29. x^{4a+b}

EXERCISE 2.5, PAGE 57

1. $15x^2$
3. $35x^7$
5. $-10a^5b^5$
7. $-12x^3y^2z^5$
9. $18a^2b^2$
11. $-3a^4$
13. $12x^3y^3$
15. $-6r^2s^2t^2$
17. $-3a^6b^6c^4$
19. $12x^2yz$

EXERCISE 2.6, PAGE 59

1. $6x - 6y$
3. $3x + 15$
5. $3a - 21$
7. $6x^3 + 9x^2$
9. $12x^2 + 10x$
11. $6x - 9y$
13. $-42x^4 - 14x^3 + 35x^2$
15. $-28x^2 - 12x + 4$
17. $-3x^2y + 4xy^4 - 7xy^8 - xy$
19. $-7x - 5 + 6y^3$
21. $-6a^3b + 18a^2b^2 + 21a^2b^3 - 6ab^5$
23. $-21x^5y^2z - 28x^6y^8 + 21x^7y^2$
25. $30a^2x^2 - 20abcx^2 + 10ax^2y^4$
27. $-12x^6 + 18x^7 - 21x^3 + 3x^2$
29. $18a^3b - 12ab^2 + 6a^2b - 6ab^2 + 6ab$

EXERCISE 2.7, PAGE 61

1. $-a-2b$
3. $-8m + 15$
5. $2x - 7y$
7. $-4x$
9. $-12x^2 + 3xy + 4y$
11. $-2a$
13. $2m + 8n$
15. $-3x^3 - 15x^2 + 10x$
17. $3x - 3$
19. $49x - 52$
21. $n - 5$
23. $118x - 320$
25. $x + 6$
27. $22x + 172$

EXERCISE 2.8, CHAPTER REVIEW, PAGE 66

1. $I = 2040$
3. $A = 49$
5. -2
7. 32
9. terms are $3a$ and $2b$; numerical coefficients are 3 and 2; literal parts are a and b
11. terms are $-11m^2n, 2mn^2$ and -1; numerical coefficients are -11, 2 and -1; literal parts are m^2n and mn^2
13. $-2a - 5b$
15. cannot be combined
17. h^7
19. x^{13}
21. $-6x^4y^5$
23. a^5b^5
25. $8x^4 + 12x^3$
27. $-4a^3 - 3a^2 + 2a + 7$
29. $a - 5b$
31. $-8a^2 - 2ab$

CHAPTER 2 ACHIEVEMENT TEST, PAGE 69

(2.1) 1. $\frac{32}{3}$
2. $A = 27$

(2.2) 3. 35
4. -48

(2.3) 5. (a) terms: $-2ab^3$, $6a^2b^2$, 2
(b) numerical coefficients: -2, 6, 2
(c) literal parts: ab^3, a^2b^2

6. cannot be combined

7. $-3a^2 - 7b^2$

8. $2rst - 2r + 1$

(2.4) 9. x^{13}

10. h^{14}

11. cannot be simplified

12. x^{m+n}

(2.5) 13. $-42a^5b^5$

14. $3a^3b^8$

(2.6) 15. $8x^2y + 12xy^3$

16. $-6x^5 - 4x^4 + 10x^3 + 2x^2$

17. $-4x^5 - 3x^3 + 7x + 5$

18. $21x^3y^2 - 6x^2y^3 + 12xy^2$

19. $2x - 15$

20. $a^2b + ab^2$

21. $x + 6$

22. $14 + 13x$

CHAPTER 3

EXERCISE 3.1, PAGE 73

1. a. yes b. no c. yes d. no e. yes

3. a. no b. yes c. yes d. yes e. yes

EXERCISE 3.2, PAGE 76

1. $x = 11$

3. $y = 11$

5. $x = -2$

7. $h = 10$

9. $x = -1$

11. $x = -9$

13. $y = -15$

15. $m = -2$

17. $x = 4.25$

19. $x = 7$

21. $x = 8.53$

23. $x = .3$

25. $y = -7$

EXERCISE 3.3, PAGE 80

1. $x = 4$

3. $x = 8$

5. $h = -8$

7. $y = 6$

9. $x = -35$

11. $t = 18$

13. $t = \frac{6}{15} = \frac{2}{5}$

15. $y = -25$

17. $x = -6$

19. $x = 48$

EXERCISE 3.4, PAGE 83

1. $x = 7$

3. $x = 8$

5. $a = -4$

7. $y = -6$

9. $t = -7$

11. $y = 2$

13. $x = 8$

15. $n = 7$

17. $P = 10$

19. $x = -32$

21. $x = -\frac{1}{2}$

EXERCISE 3.5, PAGE 88

1. $x = 4$

3. $x = 8$

5. $y = -3$

7. $x = 4$

9. $x = \frac{14}{5}$

11. $t = 0$

13. $x = 150$

15. $x = .2$ or $\frac{1}{5}$

17. $x = 6$

19. $x = 6$

21. $y = 21$

23. $y = 27$

25. $t = -4$

27. $x = \frac{-11}{13}$

29. $w = \frac{8}{3}$

31. $x = \frac{29}{2}$

33. $x = -\frac{38}{5}$

EXERCISE 3.6, PAGE 91

1. contradiction

3. identity

5. contradiction

7. identity

9. contradiction

EXERCISE 3.7, PAGE 94

1. $y = b - a$

3. $y = x - z$

5. $y = \frac{z - x}{a}$

7. $b = P - a - c$

9. $y = \frac{z - 5x}{5}$

11. $b = \frac{2A}{h}$

13. $b = \frac{1 - 6a}{3}$

15. $x = -2y$

17. $P = \frac{CT}{V}$

19. $g = \frac{2S}{t^2}$

21. $y = \frac{4z - 2x}{3}$

23. $F = \frac{9C + 160}{5}$

25. $R = \frac{E}{I} - r$ or $\frac{E - Ir}{I}$

EXERCISE 3.8, PAGE 98

1. $x > -3$

3. $y \geqslant -13$

5. $x > 6$

7. $x > -5$

9. $x > 2$

11. $x > -3$

13. $x > 3$

15. $x \geqslant \frac{-9}{2}$ ✓

17. $x \geqslant 3$

19. $y > 6$

21. $a \leq -15$

23. $x < 16$

25. $y < 40$

27. $x \leq -6$

29. $x > -\frac{33}{2}$

EXERCISE 3.9, CHAPTER REVIEW, PAGE 102

1. yes

3. no

5. no

7. $x = 11$

9. $x = -1$

11. $h = -12$

13. $y = -\frac{1}{13}$

15. $a = 6$

17. $x = 6$

19. $y = \frac{1}{3}$

21. $x = -36$

23. $x = -30$

25. $x = \frac{27}{5}$

27. identity

29. contradiction

31. conditional, $x = 4$

33. $x = 2b - a$

35. $h = \frac{v}{\ell w}$

37. $y = \frac{bx}{a}$

39. $x = 6y + 4$

41. $y = \frac{5z - 3x}{4}$

43. $x > 2$

45. $y \geqslant 1$

47. $z < \frac{13}{8}$

49. $y < -3$

51. $x < -36$

CHAPTER 3 ACHIEVEMENT TEST, PAGE 105

(3.1) 1. no 2. yes

(3.2–3.4) 3. $x = 11$ 4. $x = -6$ 5. $x = 5$ 6. $x = -7$

(3.5) 7. $x = 11$ 8. $x = 3$ 9. $x = 8$ 10. $x = \frac{17}{7}$

11.
$$\begin{aligned} 3(-7) - 4 &\stackrel{?}{=} 5(-7) + 10 \\ -21 - 4 &\stackrel{?}{=} -35 + 10 \\ -25 &= -25 \end{aligned}$$

(3.6) 12. identity 13. contradiction 14. conditional, $x = 0$

(3.7) 15. $y = \frac{2x + 4z}{3}$

(3.8) 16. $x > 3$ 17. $x \leq \frac{16}{3}$

CHAPTER 4

EXERCISE 4.1, PAGE 109

1. $x + 6 = 12$

3. $17 = 4 + x$

5. $\frac{x}{-4} = 5$

7. $x = 8 + 5$

9. $x - 5 = 12$

11. $A + 4 = 2(A - 1)$

13. $x - 1 = \frac{1}{2}(x + 18)$

15. $x = 2x - 6$

17. $x + 4 = 2(x - 8)$

19. $8 - x + 6 = 10$ 21. $x + 3 = 2x - 2$ 23. $\frac{16 - 3x}{5} = x$

25. $2x - 7 = \frac{x}{3}$

EXERCISE 4.2, PAGE 116

1. 8
3. 24
5. 7
7. 36
9. 5,7,9
11. 24,100
13. physics book costs $24
    math book costs $18
15. 36
17. 9,11,13
19. 21,23,25
21. 7,8,9,10
23. 8,10,12
25. Mr. Meyers received 12 votes
    Mrs. Meyers received 486 votes
27. identity, all solutions will work
29. 21,42
31. 12 ft, 28 ft
33. 18 ✓
35. −39,−37,−35
37. 88
39. −24,−22,−20,−18
41. 3 calories
43. 4 hours
45. identify, all solutions will work
47. $180
49. chairs cost $52 each
    table costs $415

EXERCISE 4.3, PAGE 125

1. 144
3. 98
5. 19.78
7. 70
9. 18.2%
11. .3%
13. 23
15. 1066.7
17. 2880
19. $157,692.31
21. 20% wrong, 80% correct
23. $1680
25. 40%
27. $7.42
29. 574

EXERCISE 4.4, PAGE 134

1. a. $\frac{18}{7}$ b. $\frac{7}{18}$ c. $\frac{18}{25}$ d. $\frac{7}{25}$

3. a. $\frac{3}{2}$ b. $\frac{3}{6} = \frac{1}{2}$ c. $\frac{1}{2}$ d. $\frac{2}{6} = \frac{1}{3}$

5. $\frac{2}{5}$

7. $\frac{26}{36} = \frac{13}{18}$

9. $\frac{8}{24} = \frac{1}{3}$

11. $x = 25$

13. $x = \frac{28}{3}$

15. $a = 11.1$

17. $m = 50$

19. $z = 200$

21. $y = \frac{45}{8}$

23. $n = 1\frac{29}{51}$

25. $16\frac{2}{3}$ milligrams

27. 1050 miles

29. $1010.23

31. $26\frac{2}{3}$ cups

33. 126 items

35. 72 min or 1 hr 12 min

37. 475.2 or 475 points

39. $1200

41. $c = 5, a = 7\frac{1}{7}$

43. $37\frac{1}{2}$ feet

EXERCISE 4.5, PAGE 143

1. 216 in

3. 14.1 ft

5. 172,800 sec

7. $19.4 \frac{\text{m}}{\text{sec}}$

9. $62.1 \frac{\text{miles}}{\text{hr}}$

11. 929 cm^2

13. 102,400 rod^2

15. 2 miles

17. 2.6 yd^3

19. $750 \frac{\text{miles}}{\text{hr}}$

21. $22.4 \frac{\text{miles}}{\text{hr}}$

23. 7568 cc

25. .036 lbs

27. $9.8 \frac{\text{meters}}{\text{sec}^2}$

29. 24.4 in^3

EXERCISE 4.6, CHAPTER REVIEW, PAGE 149

1. $6 = x + 4$

3. $x + 8 = 2x$

5. $x - 8 = \frac{3}{4}(x + 6)$

7. 11, 28

9. 39, 81

11. 98

13. 50

15. $10.74

17. (a) $\frac{12}{20}=\frac{3}{5}$ (b) $\frac{8}{20}=\frac{2}{5}$ (c) $\frac{12}{8}=\frac{3}{2}$

19. $x = \frac{5}{12}$ 21. $16\frac{1}{2}$ feet by 21 feet 23. 4.5 lbs.

25. 16.4 gal

27. 1000 meters is longer by 640.8 ft.

29. $88\frac{\text{km}}{\text{hr}}$

CHAPTER 4 ACHIEVEMENT TEST, PAGE 153

(4.1) 1. $2(x + 5) = 18$ 2. $3x - 4 = 2(x + 5)$

(4.2) 3. 32 4. 16, 18, 20 5. 88

(4.3) 6. 80% 7. 5.28 8. $75,000

(4.4) 9. $\frac{1}{8}$ 10. $\frac{6}{5}$ 11. 180 items 12. 150 feet

(4.5) 13. $95.5\frac{\text{miles}}{\text{hr}}$ 14. 3500 cc 15. 19 sq yd

CHAPTER 5

EXERCISE 5.1, PAGE 160

1. x^8 3. y^8 5. x^{15}

7. x^3 9. y^2 11. cannot be simplified

13. x^9y^9 15. $27x^3$ 17. 4

19. $16x^4y^4$ 21. x^{h+5} 23. x^{h-5}

25. $5^h x^h$ 27. 3^{3a} 29. 6^{2+x}

31. x^{12a} 33. 6^{xy} 35. 3^{x^2}

EXERCISE 5.2, PAGE 165

1. $\frac{1}{x^7}$ 3. $\frac{1}{8}$ 5. y

7. $\frac{1}{10}$ 9. $\frac{1}{x^2y^2}$ 11. $\frac{x^2}{y^2}$

13. $\frac{y^5}{x^5}$ 15. $\frac{b^x}{a^x}$ 17. m^5n^3

19. 1 21. 1 23. 1

25. $\frac{8}{125}$ 27. $\frac{q}{m^2n^5}$

EXERCISE 5.3, PAGE 168

1. $\frac{y^2}{x^2}$ 3. $125x^6$ 5. $\frac{1}{125x^6}$

7. $6y^4$ 9. x^3y^3 11. x^4y^4

13. $\frac{y^4}{16x^8}$ 15. $\frac{y^6}{36x^4}$ 17. 1

19. $\frac{x^6}{4}$ 21. $5x^2y^4$ 23. $\frac{1}{y^3}$

EXERCISE 5.4, PAGE 173

	ORDINARY NOTATION	*SCIENTIFIC NOTATION*
1.	3480	3.48×10^3
3.	.0000061	6.1×10^{-6}
5.	.0000031	3.1×10^{-6}
7.	84.26	8.426×10^1
9.	467,000,000	4.67×10^8
11.	.000000000466	4.66×10^{-10}
13.	74.2	7.42×10^1
15.	.014	1.4×10^{-2}

17. 602,000,000,000,000,000,000,000 19. 3.6×10^4 21. 4.8×10^{12}

23. 4.8×10^{-22} 25. 5×10^5 27. 9×10^6

29. 5×10^7

EXERCISE 5.5, CHAPTER REVIEW, PAGE 176

1. a^{11} 3. x^{21} 5. x^{20}

7. y^6 9. $\frac{x^8}{y^8}$ 11. cannot be simplified

13. $\frac{1}{x^5}$ 15. x^4 17. $\frac{1}{7}$

19. $\frac{n^5}{m^5}$ 21. $\frac{y^5}{x^3}$ 23. $\frac{9}{4}$

25. x^2y^6 27. $9x^4$ 29. $49y^{10}$

31. cannot be simplified 33. $\frac{25}{4x^{10}}$ 35. $\frac{9b^2}{25}$

37. 7.6×10^5 39. 1.5×10^{-5} 41. .000000028

43. 1.5×10^{-2} 45. 8.88×10^0 47. 2.66×10^{12}

49. 6.48×10^{-2} 51. 3×10^{-8}

CHAPTER 5 ACHIEVEMENT TEST, PAGE 179

(5.1) 1. $\frac{1}{a^2}$ 2. $\frac{1}{x^5}$ 3. $8x^3y^3$ 4. $4x^6y^4$ 5. x^8

(5.2) 6. $\frac{b^5}{a^5}$ 7. $\frac{1}{x^6y^2}$ 8. -6 9. h^7

(5.3) 10. 1 11. $\frac{9}{4x^{14}}$ 12. $\frac{64}{27}$

(5.4) 13. 7.4×10^{-7} 14. 5.63×10^9 15. 3140000

16. .0000752 17. 1.84×10^1 18. 1.1×10^{-4}

CHAPTER 6

EXERCISE 6.1, PAGE 183

	Polynominal	*Descending Order*	*No. Of Terms*	*Name*	*Degree*
1.	$2x - 7 + 3x^2$	$3x^2 + 2x - 7$	3	Trinominal	2
3.	$-124x + 6$	$-124x + 6$	2	Binomial	1

5.	$x^3 - 2x + \frac{1}{x}$	not a polynomial	—	—	—
7.	-4	-4	1	Monomial	0
9.	$6x$	$6x$	1	Monomial	1
11.	$6x^4 - 7x^{-3} + 2$	not a polynomial	—	—	—
13.	$-23x^5$	$-23x^5$	1	Monomial	5
15.	$-6 + 14z - 52z^7$	$-52z^7 + 14z - 6$	3	Trinomial	7

EXERCISE 6.2, PAGE 186

1. $10x^2 - 3x - 6$
3. $13y^2 + 11y + 5$
5. $2a^4 - 3a^3 - 24a + 2$
7. $7y^5 + 12y^3 + 11y - 7$
9. $13x^3 + 7x^2 - 8x + 4$
11. $-23a^4 + 5a^3 + 2a^2 - 6$
13. $6x^2 - 11x + 11$
15. $8y^3 + 12y^2 + 2y + 8$
17. $-7x^4 + 2x^3 - 2x^2 + 8x - 5$
19. $-9a^3 + 13a^2 - 9a + 6$
21. $-2y^3 - 22y^2 - 3y + 1$
23. $-11x^3 - 2x^2 + x - 4$
25. $4x^2 + 8x - 5$
27. $7x^4 + 5x^3 + 13$
29. $2x^3 - 2x^2 - 2x + 3$
31. $-9y^4 - 7y^3 + y^2 - 16y - 10$
33. $12x^3 + 2x^2 - 26x + 1$
35. $7y^3 + y^2 + 12y - 9$

EXERCISE 6.3, PAGE 192

1. $-30x^3$
3. $16y^4$
5. $6x^3y^2z^5$
7. $6x^2 + 10x$
9. $3x^4 + 9x^3 - 27x^2$
11. $24x^5 + 12x^4 + 18x^3 + 12x^2 - 42x$
13. $-60a^5 + 70a^4 + 30a^3 - 90a^2$
15. $2x^2 + 8x - 24$

17. $x^3 + 7x^2 + 17x + 35$

19. $6x^3 - 49x^2 - 40x + 8$

21. $21x^3 + 55x^2 + 2x - 28$

23. $x^4 - 3x^3 - 3x^2 + 7x + 6$

25. $x^5 + 4x^4 - 2x^3 - 13x^2 - 29x - 3$

27. $4x^5 + 3x^4 + 6x^3 + 3x^2 - 18x - 18$

29. $21a^5 - 14a^4 + 25a^3 - 51a^2 - 10a - 25$

EXERCISE 6.4, PAGE 198

1. $2x^3 + 3x^2 - 4x + 8$

3. $2x^3 + x^2 - 3 + \frac{5}{x}$

5. $-2 - \frac{3}{a} + \frac{4}{a^2} - \frac{6}{a^4}$

7. $\frac{3x^3}{2} + x^2 - 4x + 11 - \frac{11}{2x}$

9. $\frac{3}{x^2} + \frac{2}{x^3} - \frac{2}{x^4} - \frac{1}{x^5}$

11. $x - 8$

13. $a - 3$

15. $2y - 3$

17. $2n + 3$

19. $2x - 5, R\ 4$

21. $4x - 1, R\ 6$

23. $x^2 - 7x + 28, R - 12$

25. $x + 7, R\ 7$

27. $x^2 + 2x + 1$

29. $y^3 + 2y + 4$

31. $x^2 - 2x - 4$

EXERCISE 6.5, CHAPTER REVIEW, PAGE 203

	Polynominal	Descending Order	No. Of Terms	Name	Degree
1.	$3x + 4x^2 - 8$	$4x^2 + 3x - 8$	3	Trinomial	2
3.	$y^4 - 3y + 2y^2 - 1$	$y^4 + 2y^2 - 3y - 1$	4	Polynomial	4
5.	$-14x$	$-14x$	1	Monomial	1

7. $7y^4 + 4y^3 - 2y^2 - 3y - 7$

9. $-11x^3 + 2x^2 - 10x + 3$

11. $y^3 + 6y^2 + 10y - 20$

13. $5y^3 - 2y^2 + 7y - 10$

15. $12x^3 - 4x^2 + 3x + 4$

17. $-4x^4 + 3x^3 + 7x^2 - 5$

19. $-a^4 + 4a^3 + 2a + 5$

21. $4y^4 + 4y^3 - 12y^2 + 9y - 1$

23. $16a^7y^5$

25. $18x^2 - 33x - 40$

27. $2x^3 + 11x^2 + 8x - 16$

29. $8a^3 + 27$

31. $2x^3 - 3x^2 - 5x - 1$

33. $-4a^3 - 3a^2 + 4a - 1 - \frac{3}{a}$

35. $x + 3$

37. $3x - 1$

39. $2y^2 - 3y + 5$, R 12

CHAPTER 6 ACHIEVEMENT TEST, PAGE 207

(6.1)

	Polynominal	*Descending Order*	*No. Of Terms*	*Name*	*Degree*
1.	$-6x^2 + 4x - 3x^3$	$-3x^3 - 6x^2 + 4x$	3	Trinomial	3
2.	$-26y$	$-26y$	1	Monomial	1
3.	$-7 - 3x$	$-3x - 7$	2	Binomial	1

(6.2) 4. $10x^3 + 6x^2 - 3x - 4$ 5. $7a^4 + 6a^3 + 8a^2 - 6a - 11$

6. $x^4 - 3x^2 + 6x + 11$ 7. $-3m^4 - 3m^3 - 5m^2 + 9m - 1$

8. $-4y^3 + 8y^2 - 19$

(6.3) 9. $60x^5y^5$ 10. $-12x^4 - 6x^3 + 15x^2$

11. $21x^2 + 22x - 8$ 12. $x^3 - 2x^2 + x - 36$ 13. $8x^3 + 9x^2 - 13x + 3$

(6.4) 14. $-9a^2 + a - 2 + \frac{6}{a}$ 15. $\frac{3x^3}{2} - \frac{7x^2}{2} + 3x - \frac{9}{4x^2}$

16. $x + 2$ 17. $x^2 + 3x + 1$ 18. $x^2 + 2x - 3$ 19. $x^3 + 3x^2 + 2x + 6$

CHAPTER 7

EXERCISE 7.1, PAGE 212

1. $3(x + 3)$

3. $2(6x - 7)$

5. $6a^2(a + 2)$

7. $3x(6x - 1)$

9. $9x^4(3x^4 - 4)$

11. $3(a^2b + 2x^2y)$

13. $8m(2m^2 + m - 3)$

15. $y^2(y^4 - 3y^3 + 1)$

17. $a(2a^5 + 7a^3 - 3)$

19. prime

21. $-4x(3x^2 + 4x + 6)$

23. $9m^2nq(6 - 4m^2n + 2nq)$

25. $-9xy^4z^3(3x^2y^4 + z^9 + 5y^2z^6)$

27. $8r^3(6t^2 - 4rt^4 + 3r^3)$

EXERCISE 7.2, PAGE 216

1. $x^2 + 6x + 8$

3. $a^2 + 7a + 10$

5. $x^2 - 2x - 15$

7. $n^2 - 9n + 14$

9. $x^2 - 4$

11. $6x^2 - x - 12$

13. $9x^2 - 4$

15. $x^2 + 6x + 9$

17. $x^2 - 14x + 49$

19. $y^2 - 18y + 81$

21. $12y^2 - 7y - 12$

23. $49x^2 - 4$

25. $9x^2 - 12x + 4$

27. $25n^2 - 9$

29. $49n^2 - 14n + 1$

31. $49x^2 + 28x + 4$

33. $x^2 + 4xy + 3y^2$

35. $9a^2 - b^2$

37. $50m^2 + 55mn - 21n^2$

39. $16a^2 - 40ab + 25b^2$

EXERCISE 7.3, PAGE 219

1. $(x + 4)(x + 1)$

3. $(x + 2)(x + 2)$

5. $(x - 3)(x - 1)$

7. $(y + 5)(y + 7)$

9. prime

11. $(a + 14)(a - 1)$

13. $(a - 7)(a - 2)$

15. $(a - 8)(a - 2)$

17. $(b - 6)(b + 1)$

19. $(b - 2)(b - 3)$

21. $(h - 3)(h - 9)$

23. $(h + 5)(h + 2)$

25. $(h - 6)(h - 3)$

27. prime

29. $(u - 9)(u - 2)$

31. prime

33. $(t + 15)(t - 2)$

35. $(t - 6)(t - 5)$

37. $(z - 7)(z - 7)$

39. prime

41. $(m - 6)(m - 4)$

43. $(m + 3)(m + 8)$

45. $(m - 8)(m + 3)$

47. $(y + 7)(y + 8)$

49. $(a - 14)(a + 3)$

EXERCISE 7.4, PAGE 223

1. $(2x + 3)(x + 1)$
3. $(3x + 5)(x - 1)$
5. $(2y + 3)(y - 1)$
7. $(3y - 2)(y - 2)$
9. $(2h - 5)(2h - 1)$
11. $(3a + 1)(a + 3)$
13. $(5t + 1)(t + 4)$
15. prime
17. $(3y + 2)(y - 3)$
19. $(5x - 7)(x - 1)$
21. $(2y + 1)(y + 1)$
23. $(2b - 3)(b - 2)$
25. $(2t + 1)(2t + 1)$
27. prime
29. $(3x - 2)(3x + 4)$
31. $(4x + 3)(x + 2)$
33. $(6d - 1)(2d + 1)$
35. prime
37. prime
39. $(7w + 1)(w - 3)$
41. $(3x - 2)(2x - 3)$
43. $(5b - 4)(5b + 1)$
45. $(4a + 3)(2a - 7)$
47. $(3x + 4)(3x + 2)$
49. $(4y - 9)(2y + 1)$
51. prime
53. $(5x + 6)(3x - 7)$
55. prime
57. $(9x + 2)(x - 6)$
59. $(2t - 9)(4t + 3)$

EXERCISE 7.5, PAGE 227

1. $(x - 6)(x + 6)$
3. $(t - 7)(t + 7)$
5. $(4x - 3)(4x + 3)$
7. $(3y - 2)(3y + 2)$
9. prime
11. $(7m - 2n)(7m + 2n)$
13. $(9a - 7b)(9a + 7b)$
15. prime
17. $(y^3 - 5)(y^3 + 5)$
19. $(2x^6 - y^3)(2x^6 + y^3)$
21. $(5x - 6x^3y^2)(5x + 6x^3y^2)$
23. $(12x^3 - 7m^3n^9)(12x^3 + 7m^3n^9)$
25. prime

EXERCISE 7.6, PAGE 230

1. $3(x - y)(x + y)$
3. $2(x + 2)(x + 1)$
5. $3(2a - b)(2a + b)$
7. $4(2y + 1)(y - 2)$
9. $3(x - 4)(x + 2)$
11. $(3x - 5)(2x - 3)$
13. $(t - 2)(t + 2)(t^2 + 4)$
15. $a(a - 2)(a + 2)$
17. $4c(c - 5)(c + 5)$
19. $6(d^2 + 2c^2)$

21. $2h^2(3h + 1)(2h - 3)$

23. $(x^2 - 3)(x^2 + 3)(x^4 + 9)$

25. $5(x^{13} - 1)$

27. $2x(4x - 5y)(4x + 5y)$

29. $7(2x + 5)(x - 1)$

31. $-(2x + 7)(x - 3)$

EXERCISE 7.7, CHAPTER REVIEW, PAGE 234

1. $6(x - 5)$

3. $3x(2x + 3)$

5. prime

7. prime

9. $-2y^2(18y^2 + 8y + 9)$

11. $x^2 + x - 12$

13. $6x^2 + 7x - 3$

15. $10x^2 - 21x + 9$

17. $x^2 + 8x + 16$

19. $4x^2 - 20x + 25$

21. $(y + 6)(y + 2)$

23. $(x - 5)(x + 2)$

25. $(x + 7)(x - 5)$

27. $(b - 6)(b - 7)$

29. $(y + 12)(y + 5)$

31. $(2x + 3)(x - 4)$

33. $(3b + 7)(b - 1)$

35. $(2y - 3)(y - 5)$

37. $(5c - 6)(7c - 2)$

39. $(4x + 7)(2x + 7)$

41. $(x - 4)(x + 4)$

43. $(5c - 1)(5c + 1)$

45. $(2a - 5b)(2a + 5b)$

47. $(x^4 - 7)(x^4 + 7)$

49. $(5x^2 - 6y)(5x^2 + 6y)$

51. $2(a - 2b)(a + 2b)$

53. $3(2x + 1)(x - 2)$

55. $2a(2a + 5)(a + 2)$

57. $y(2x - 5)(x + 1)$

59. $-(2x - 7)(x - 2)$

CHAPTER 7 ACHIEVEMENT TEST, PAGE 237

(7.2)
1. $x^2 - 9x + 20$
2. $12x^2 + 5x - 2$
3. $18x^2 - 9x - 35$
4. $9x^2 - 30x + 25$

(7.1–7.5)
5. $x^3(6 + 5x^3)$
6. $3a^3b^2(4 + 3a)$
7. $3x^2y(3x - 5y + 3x^2)$
8. $(y - 3)(y + 2)$

9. prime

10. $(x - 7)(x - 6)$

11. $(4a - 1)(2a + 1)$

12. $(4t + 3)(t + 2)$

13. $(6x + 1)(3x - 5)$

14. $(4h - 3)(2h - 3)$

15. $(3y - 5)(3y + 5)$

16. $(2a - 7b)(2a + 7b)$

17. prime

18. $2(y^3 - 2)(y^3 + 2)(y^6 + 4)$

19. $2y(4x - 3)(2x + 1)$

20. $3(4t - 5)(3t - 1)$

CHAPTER 8

EXERCISE 8.1, PAGE 241

1. $x = 3$

3. $x = -3$

5. $x = 4$

7. $x = -3, 1$

9. $x = 0$

11. $x = 2, -5$

13. -3

15. 4

17. -5

19. $y - 2$

21. $x - 1$

EXERCISE 8.2, PAGE 244

1. $\frac{2}{3}$

3. $\frac{y^3}{2x^2}$

5. $\frac{1}{-3y}$

7. $\frac{1}{2}$

9. cannot be simplified further

11. $\frac{2}{x+3}$

13. $\frac{1}{x+2}$

15. 3

17. 1

19. $\frac{x+1}{x-1}$

21. $\frac{1}{x+6}$

23. $\frac{2x-5}{x-4}$

25. $\frac{t+5}{t+3}$

27. $\frac{x-3}{x+5}$

29. $-\frac{3a}{a+b}$

EXERCISE 8.3, PAGE 249

1. $\frac{3}{8}$

3. x^3y

5. 1

7. $\frac{b}{3}$

9. 2

11. $\frac{a}{a-1}$

13. $\frac{2}{x-2}$

15. $\frac{x+7}{x-4}$

17. $-\frac{1}{2x+5}$

19. $\frac{x+3}{x+1}$

21. $\frac{1}{x-3}$

23. -1

25. $\frac{x-3}{x+3}$

EXERCISE 8.4, PAGE 256

1. $\frac{1}{2}$

3. $\frac{2}{x}$

5. $\frac{2}{x}$

7. 4

9. $\frac{a+b+c}{x}$

11. 1

13. 120

15. $24x^2y^3$

17. $x(x-4)$

19. $(x+1)(x+1)$

21. $7x(7x+1)(7x-1)$

23. $a(a-2)(a+2)(a+2)$

25. The LCD is equal to the product of the denominators when no factor of any denominator appears as a factor in any other denominator.

27. 6

29. $25x^2y^3$

31. $49bc^2$

33. $4(x-4)$

35. $(x-2)(x+2)$

37. $4(x-4)(x+1)$

39. $(x-4)(x-4)$

EXERCISE 8.5, PAGE 261

1. $\frac{35}{36}$

3. $\frac{2y-x}{x^2y^2}$

5. $\frac{10+27y}{12y^2}$

7. $\frac{a^2+b^2+2c^2}{abc}$

9. $\frac{x^2+3x+9}{x(x+3)}$

11. $\frac{x^2}{y(x-y)}$

13. $\dfrac{a}{5(a+2)}$

15. $\dfrac{-2x-25}{(x+2)(x-5)}$

17. $\dfrac{2y^2-10y+20}{5(y+2)(y-5)}$

19. $\dfrac{2}{x-1}$

21. $-\dfrac{x^2+4x+7}{(x-3)(x+3)}$

23. $\dfrac{3x^2+4x-3}{(x+1)(x-1)(x+2)}$

25. $\dfrac{2x^2+11x+3}{2(x-3)(x+3)}$

27. $\dfrac{8}{a}$

29. $\dfrac{2}{x+2}$

31. $\dfrac{x-10}{(x-2)(x+2)}$

EXERCISE 8.6, PAGE 266

1. $\dfrac{b}{a^2}$

3. $\dfrac{1}{b}$

5. $\dfrac{3}{29}$

7. $\dfrac{xy^2}{2}$

9. $\dfrac{b+a}{a^2b}$

11. $\dfrac{x^2-2xy}{2xy-y^2}$

13. $\dfrac{-3}{2x}$

15. $-2y$

17. $\dfrac{a+2}{3+4a^2}$

19. x^2+y^2

21. $\dfrac{2x-3}{3x}$

23. $\dfrac{x-1}{x+2}$

25. $\dfrac{a^2-a^3+2}{4a-a^3}$

27. $\dfrac{-x+5}{x^2-2x}$

29. $\dfrac{-x}{x^2+x-2}$

EXERCISE 8.7, PAGE 272

1. $x = 30$

3. $y = 20$

5. $a = \dfrac{3}{2}$

7. no solution

9. $x = -1$

11. $x = -\dfrac{2}{3}$

13. $x = 1$

15. $n = 14$

17. no solution

19. no solution

21. $x = \dfrac{1}{2}$

23. no solution

25. $x = 1$

27. $x = -\dfrac{1}{3}$

29. no solution

31. 3

33. 5

35. $\dfrac{3}{8}$ and $\dfrac{3}{2}$

37. 3 and 5

EXERCISE 8.8, CHAPTER REVIEW, PAGE 280

1. 1
3. 2, −2
5. $-y - 4$
7. $\frac{3}{8}$
9. cannot be reduced further
11. $\frac{x+3}{x-3}$
13. $\frac{6}{x^2y^2}$
15. $\frac{(x+4)(x-5)}{3x^2}$
17. −1
19. $\frac{3}{y}$
21. 40
23. $(x-5)(x-3)$
25. $\frac{10x+27}{6x^3}$
27. $\frac{x-3}{(x-2)(x+1)}$
29. $\frac{3}{x-3}$
31. $\frac{a^2-3ab}{3ab-b^2}$
33. $x = \frac{27}{7}$
35. no solution
37. $x = -1$

CHAPTER 8 ACHIEVEMENT TEST, PAGE 283

(8.1) 1. −1 2. −5, −1 3. −3 4. −2

(8.2) 5. $\frac{x+1}{2}$ 6. cannot be reduced 7. $\frac{x-1}{x+4}$

(8.3) 8. $\frac{x+y}{5xy+1}$ 9. 1

(8.4–8.5) 10. 2 11. $\frac{3x^2+4}{x}$ 12. $\frac{3-a}{a-2}$

13. $\frac{2x-10}{(x-1)(x+1)(x+1)}$ 14. $\frac{-6}{x+1}$

(8.6) 15. $\frac{-x+3}{2x}$ 16. $\frac{2a^2+a}{2a-4}$

(8.7) 17. $x = 1$ 18. $x = 2$ 19. $a = -5$ 20. −36, −30

CHAPTER 9

EXERCISE 9.1, PAGE 289

1.

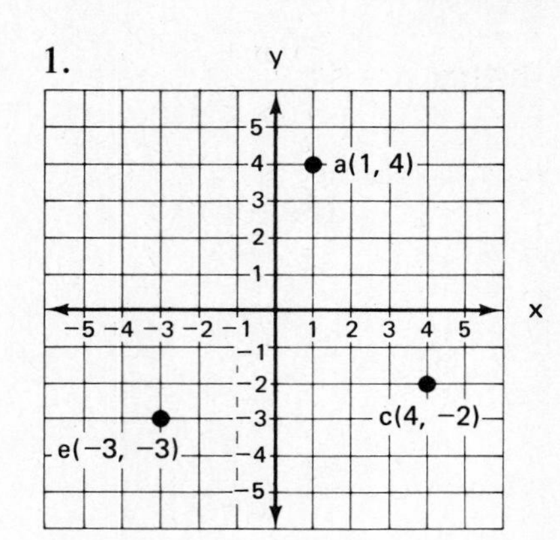

3.

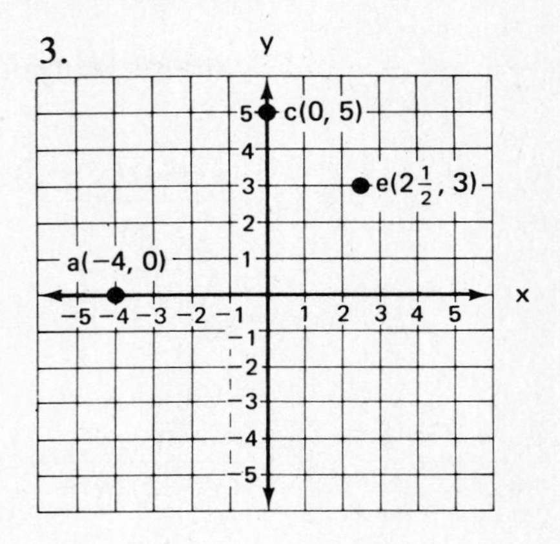

5. A(3,5) C(−4,3) E(4,0) G(0,0)

EXERCISE 9.2, PAGE 295

1.

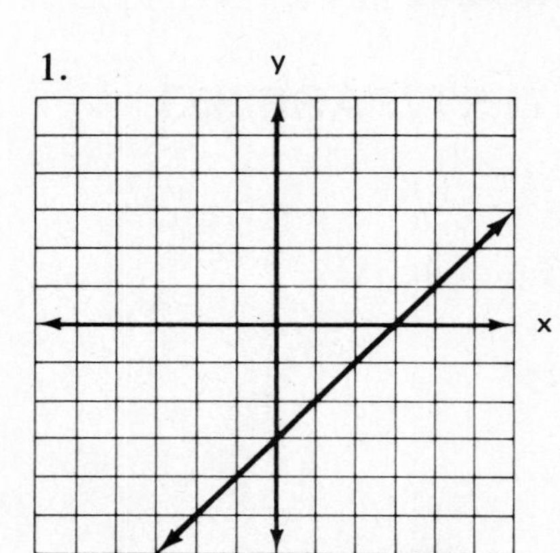

3.

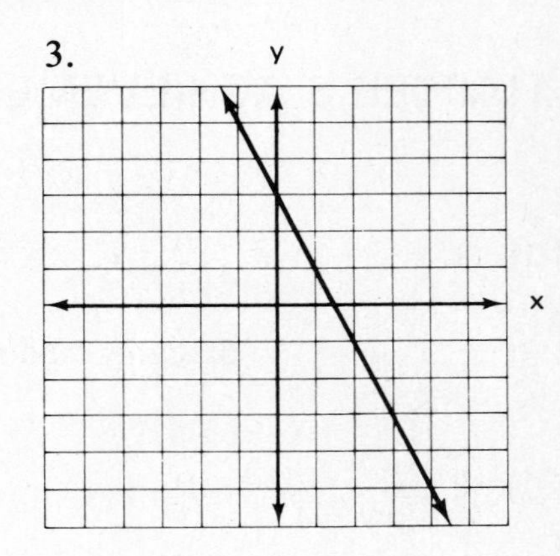

5.

y

x

7.

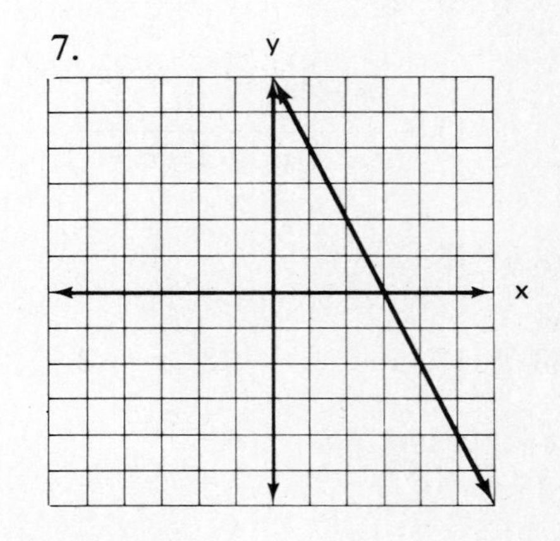

9.

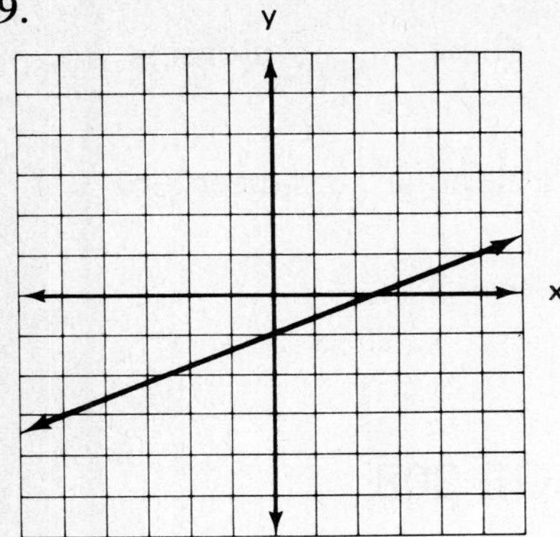

11.

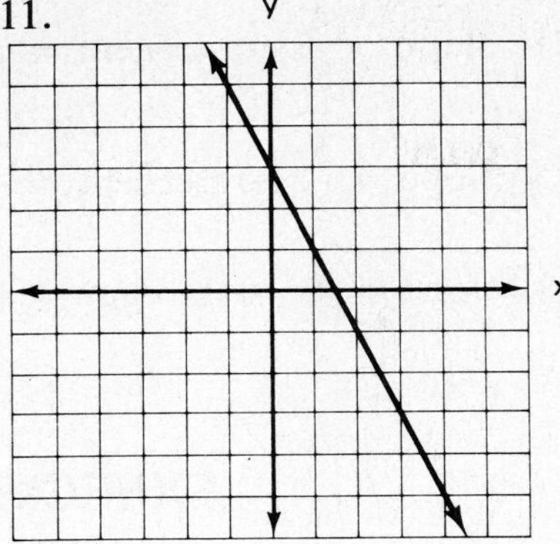

13.

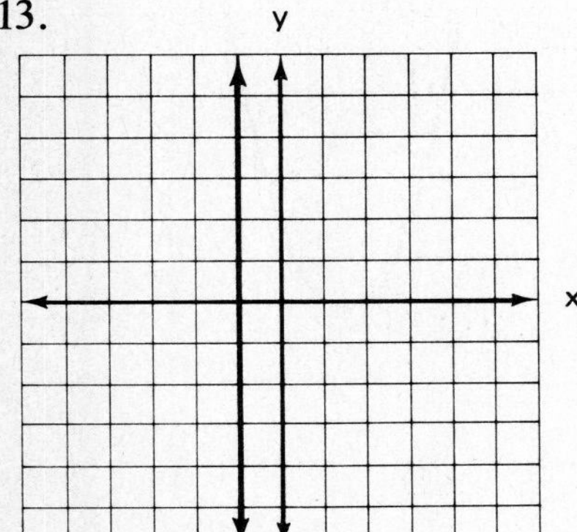

EXERCISE 9.3, PAGE 300

1. $\frac{1}{2}$
3. $-\frac{1}{2}$
5. 1
7. 2
9. undefined
11. 0
13. 0
15. $\frac{5}{3}$
17. 0
19. -2
21. -1

EXERCISE 9.4, PAGE 305

1. $2x - y + 1 = 0$
3. $3x + y + 11 = 0$
5. $x - y - 5 = 0$
7. $3x - 4y - 13 = 0$
9. $y = 4$
11. $4x - 3y = 0$
13. $x + y - 5 = 0$
15. $x + 2y - 6 = 0$
17. $6x - 5y - 34 = 0$
19. undefined

21. slope $= \frac{4}{3}$, y-intercept $= 2$

23. slope $= \frac{1}{3}$, y-intercept $= 2$

25. slope $= \frac{5}{3}$, y-intercept $= -3$

27. slope $= \frac{6}{5}$, y-intercept $= 7$

29. slope $= 0$, y-intercept $= -4$

31. $y = 3$

33. $x = -2$

EXERCISE 9.5, PAGE 308

1.

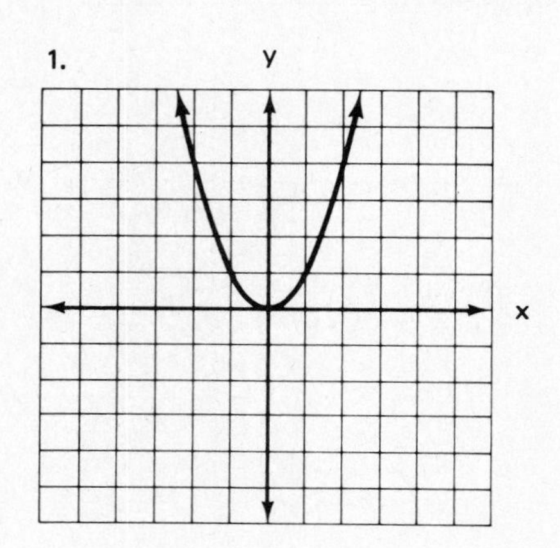

3.

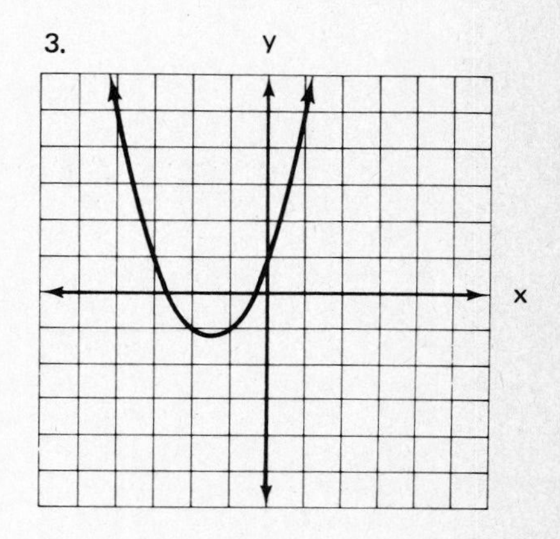

5.

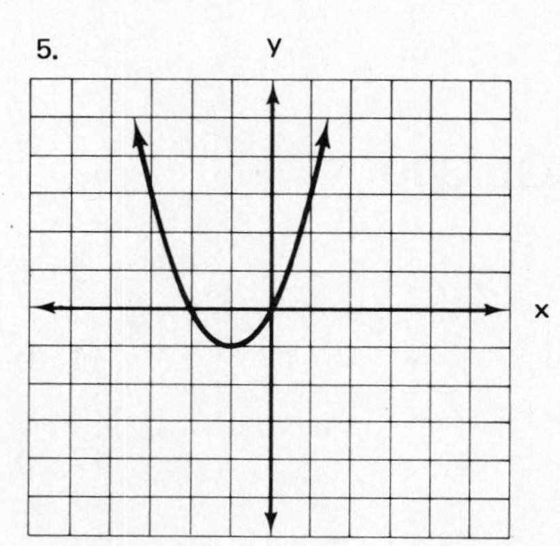

7.

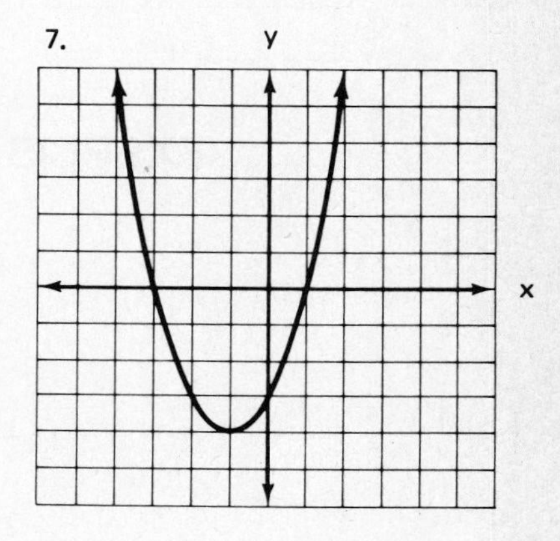

9.

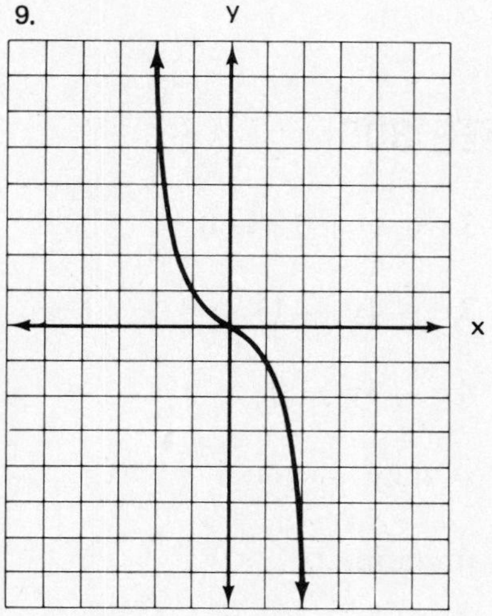

y

x

EXERCISE 9.6, PAGE 312

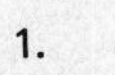

1.

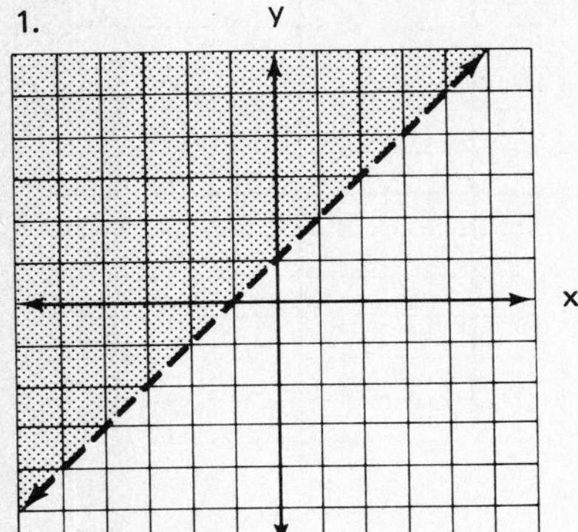

3.

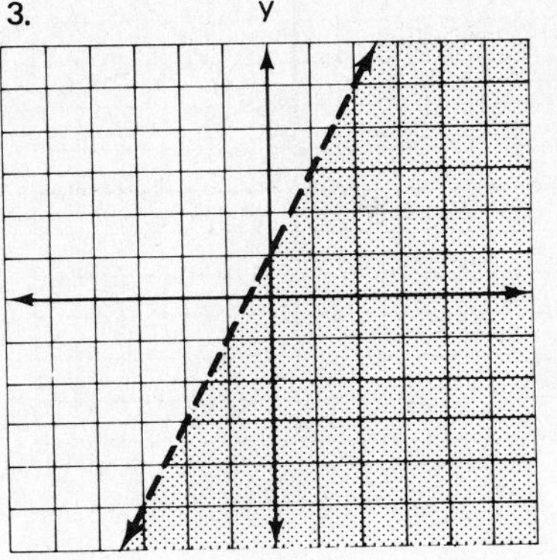

5.

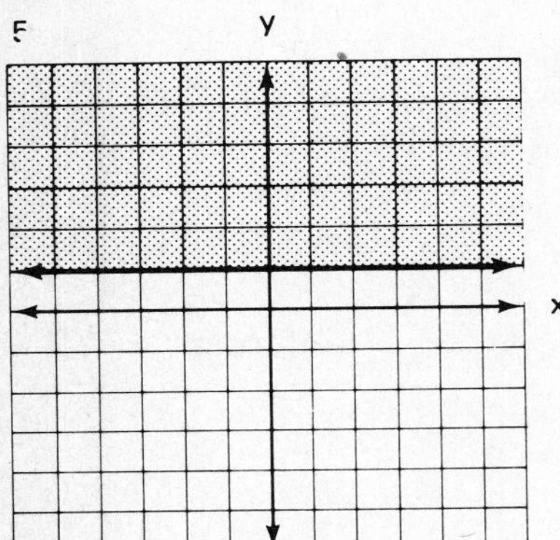

7.

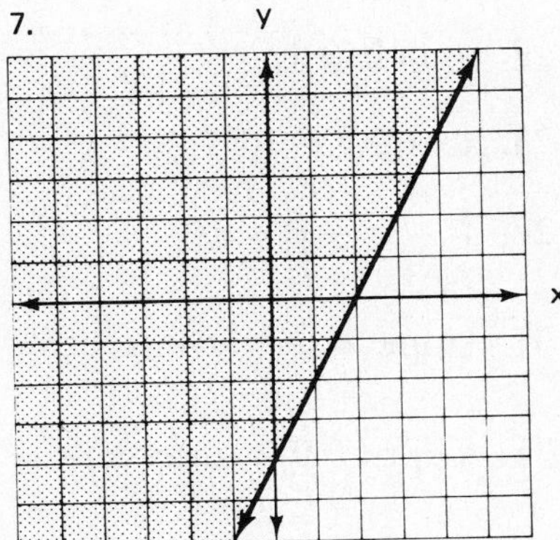

9.

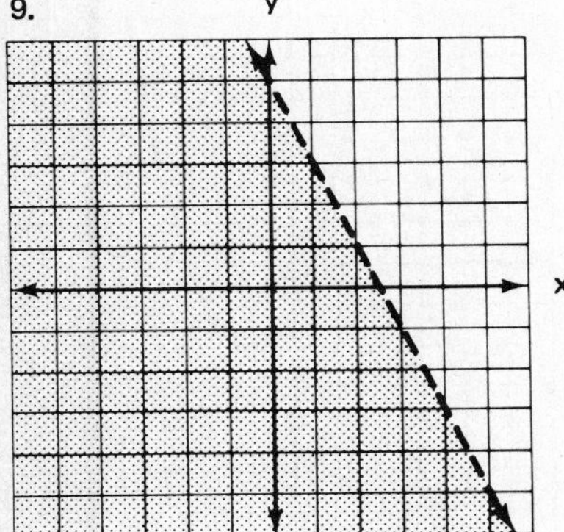

EXERCISE 9.7, CHAPTER REVIEW, PAGE 317

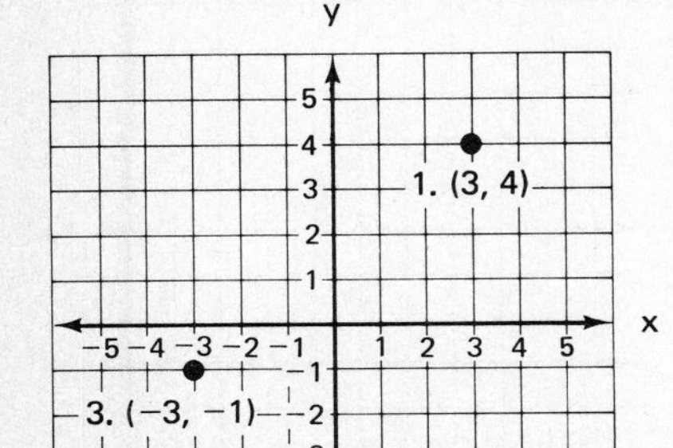

7. B(−2, −2)

9. D(−2, 0)

11.

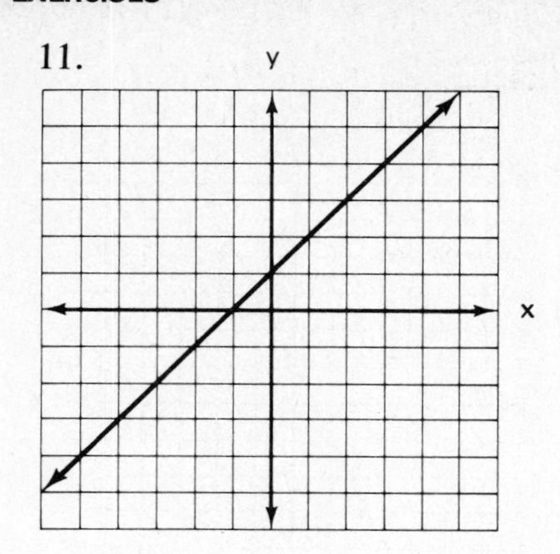

13.

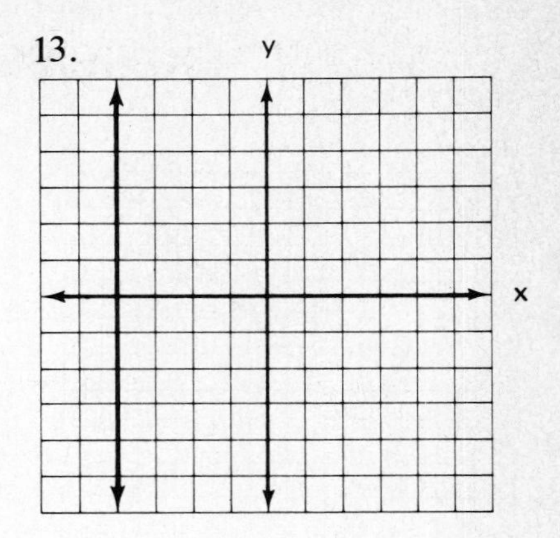

15. -4

17. 1

19. undefined

21. $4x - y - 7 = 0$

23. $x - 2y + 10 = 0$

25. $y + 3 = 0$

27. $x + y - 6 = 0$

29. $x = -3$

31. slope $= -3$, y-intercept $= -4$

33. slope $= \frac{3}{2}$, y-intercept $= -2$

35. slope $= 0$, y-intercept $= -2$

37. $y = 1$

39.

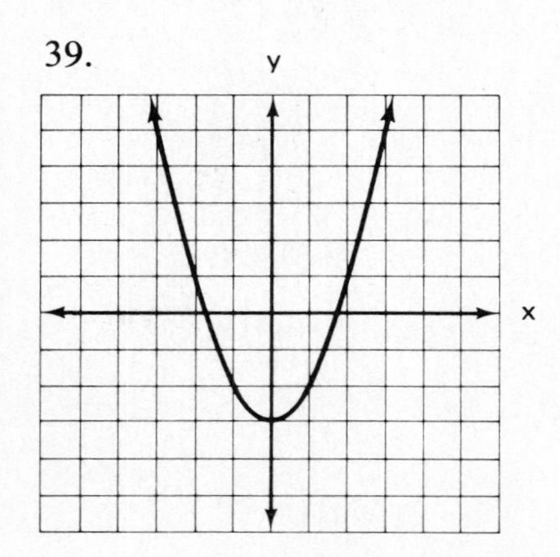

41.

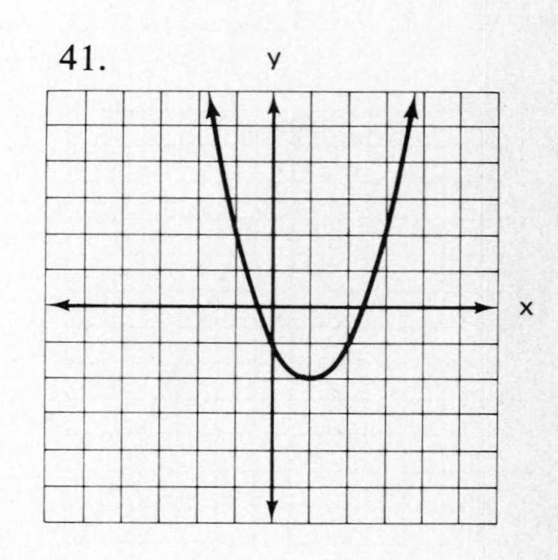

43.

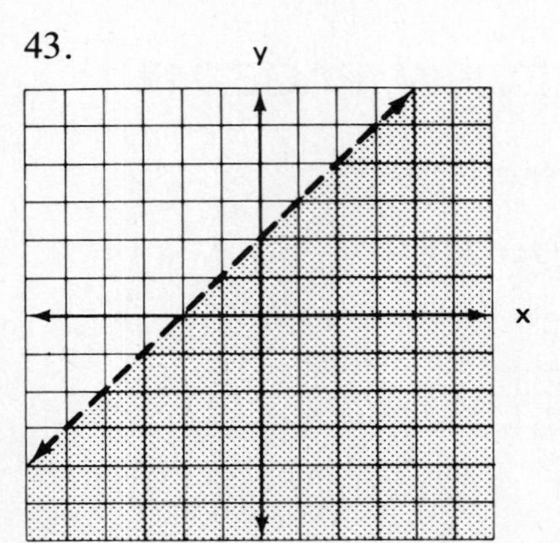

45.

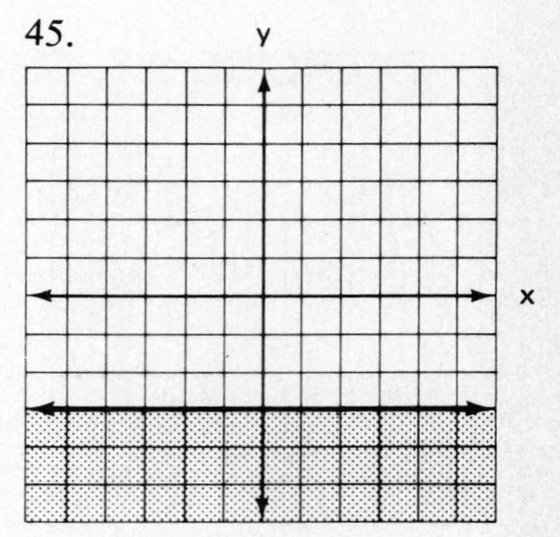

CHAPTER 9 ACHIEVEMENT TEST, PAGE 321

(9.1)

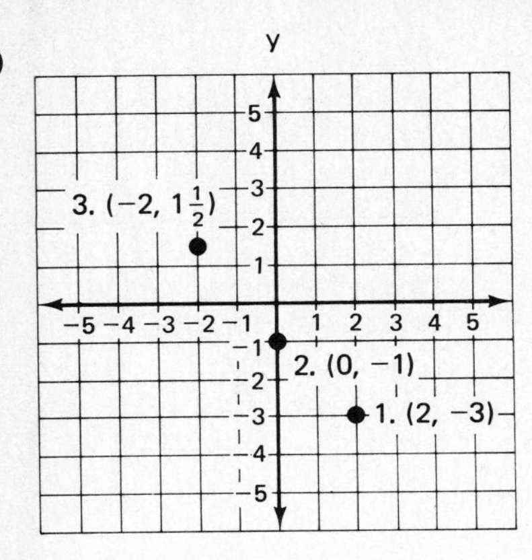

(9.1) 4. A(2,3)

5. B(3,0)

6. C(−1,−1)

(9.2) 7.

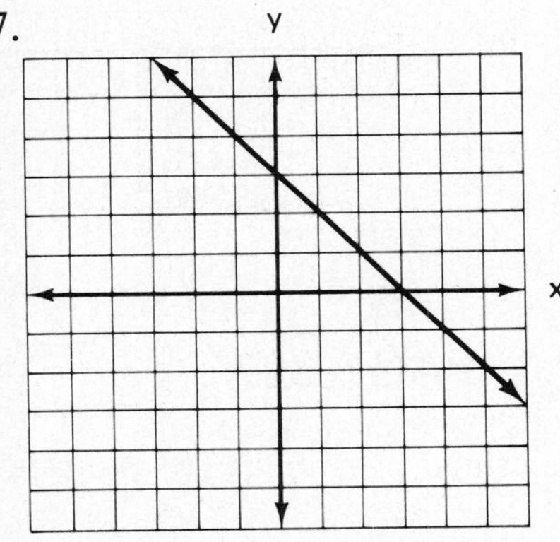

8.

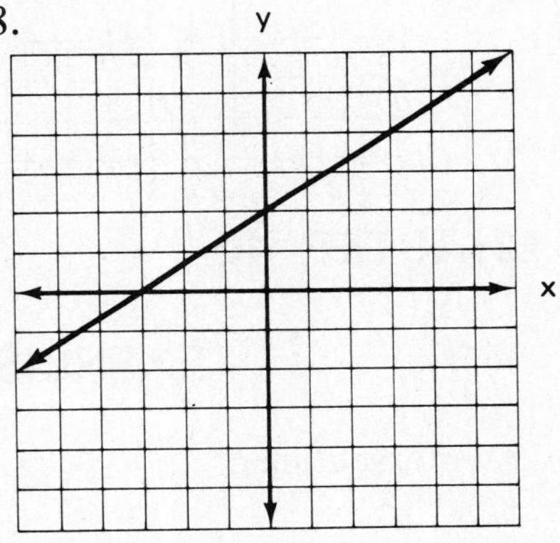

(9.3) 9. $-\frac{2}{3}$ 10. 0 11. undefined

(9.4) 12. $4x - y - 11 = 0$ 13. $x - 2y - 7 = 0$ 14. $x - 2 = 0$

15. $3x + 7y - 5 = 0$ 16. $x + 2 = 0$ 17. $y + 1 = 0$

18. slope $= -5$, y-intercept $= -3$

19. slope $= \frac{3}{4}$, y-intercept $= 3$

20. slope $= 0$, y-intercept $= 3$

21. slope is undefined, no y-intercept 22. $y = -4$

23. $x = 3$

(9.5) 24.

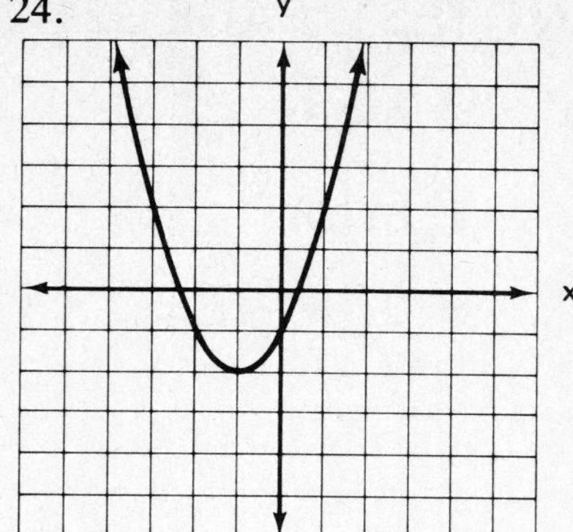

25.

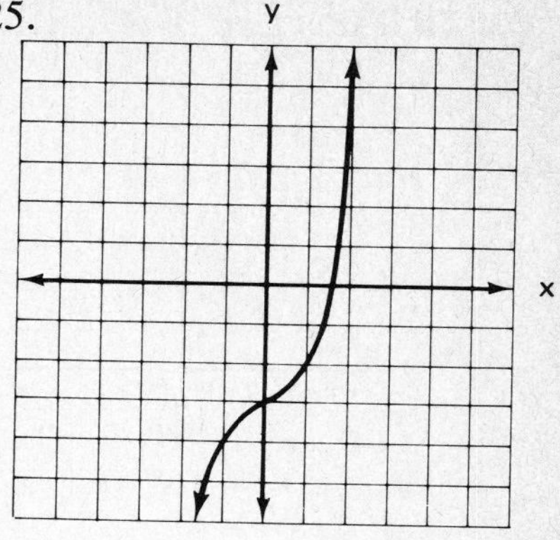

(9.6) 26.

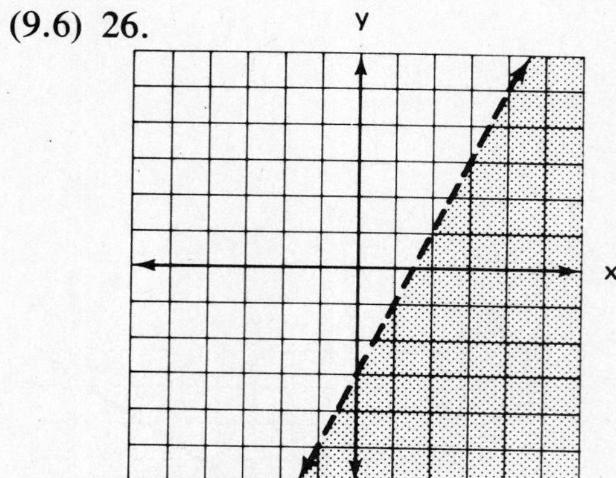

27.

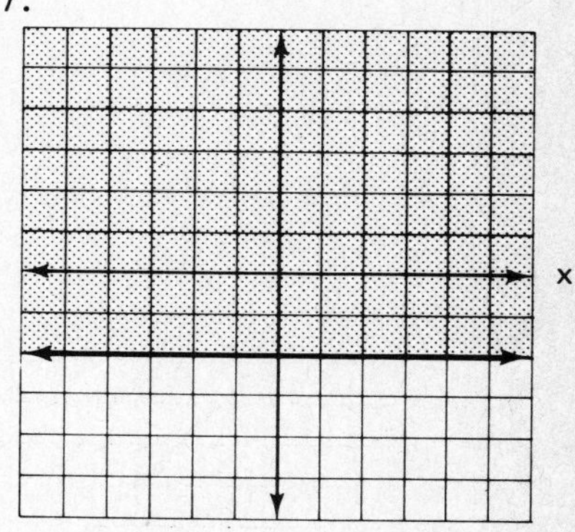

CHAPTER 10

EXERCISE 10.1, PAGE 327

1. intersecting
3. intersecting
5. coincident
7. parallel
9. intersecting
11. intersecting

EXERCISE 10.2, PAGE 330

1.

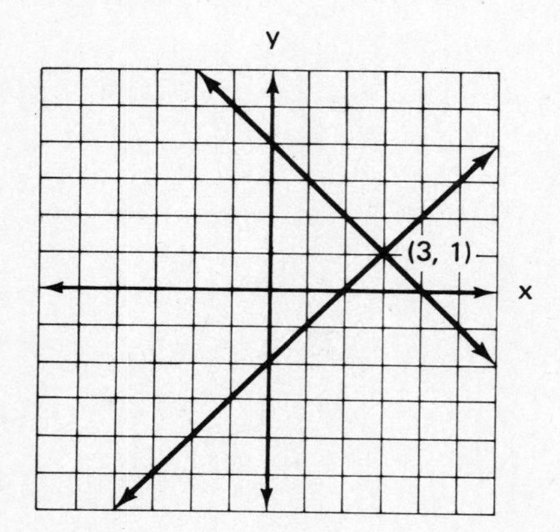

3. parallel lines, no solution

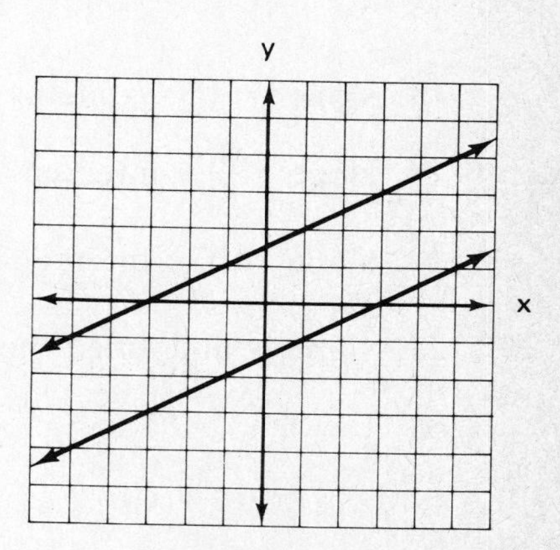

5.

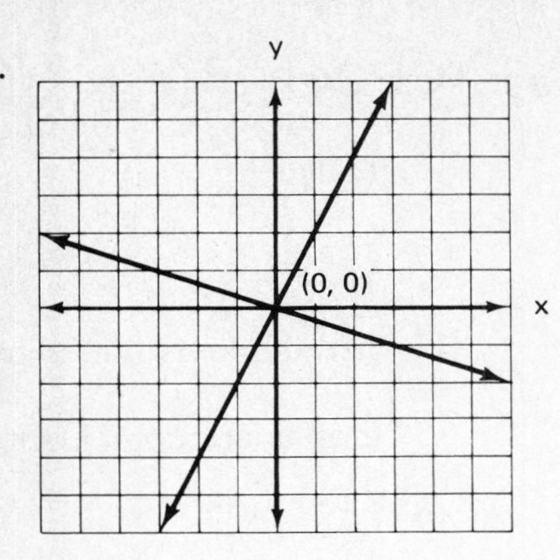

7.

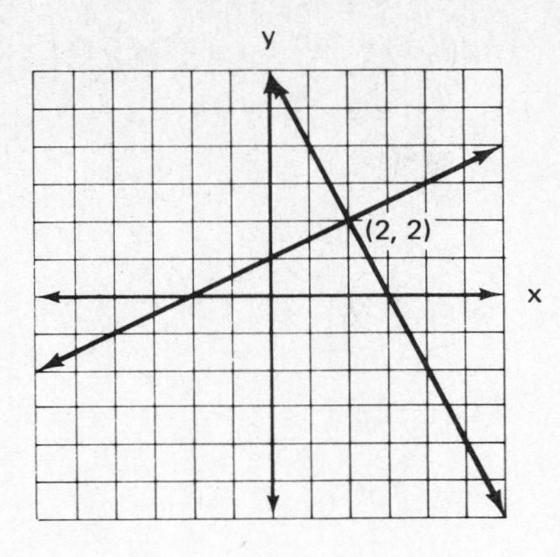

9. parallel lines, no solution

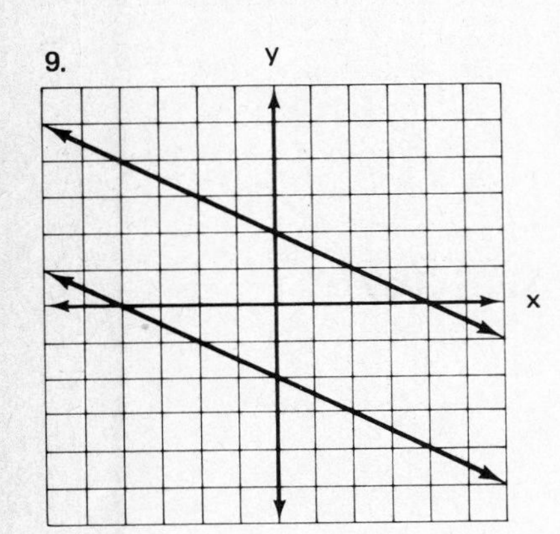

EXERCISE 10.3, PAGE 336

1. $(2,-2)$
3. $(5,-2)$
5. $(-3,-5)$
7. $(-4,1)$
9. $\left(\frac{1}{2},\frac{1}{2}\right)$
11. parallel lines, no solution
13. $\left(\frac{5}{2},1\right)$
15. parallel lines, no solution
17. $(2,0)$
19. $(3,-8)$
21. parallel lines, no solution
23. Graphs are coincident. Any ordered pair that satisfies either equation is a solution to the system.
25. $(34,-24)$

EXERCISE 10.4, PAGE 341

1. (1,2)
3. (2,1)
5. (3, −1)
7. (−1,4)
9. $\left(\frac{3}{4}, 0\right)$
11. parallel lines, no solution
13. (−2,3)
15. graphs are coincident
17. $\left(\frac{1}{2}, 0\right)$
19. $\left(\frac{1}{2}, -1\right)$
21. $\left(\frac{3}{2}, 1\right)$
23. graphs are coincident
25. (0,2)

EXERCISE 10.5, PAGE 347

1. 29,12
3. width is 5 inches, length is 24 inches
5. 7 five-dollar bills and 6 twenty-dollar bills
7. 23 hours, 29 hours
9. 4.5 pounds of chocolates, 1.5 pounds of caramels
11. 20 nickels, 25 dimes
13. 126°, 54°
15. Fred is 36, Freddie is 12
17. 182 students, 65 non-students
19. cookies cost $1.30 per dozen and bread costs $1.25 per loaf
21. 20, 32
23. length is 9, width is 4
25. 42 dogs, 36 owners
27. 40,120

EXERCISE 10.6, CHAPTER REVIEW, PAGE 355

1. intersecting   3. coincident   5. intersecting

7.

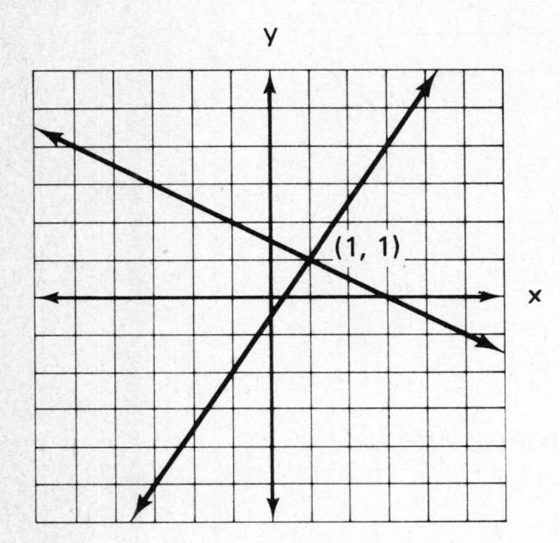

9. graphs coincide

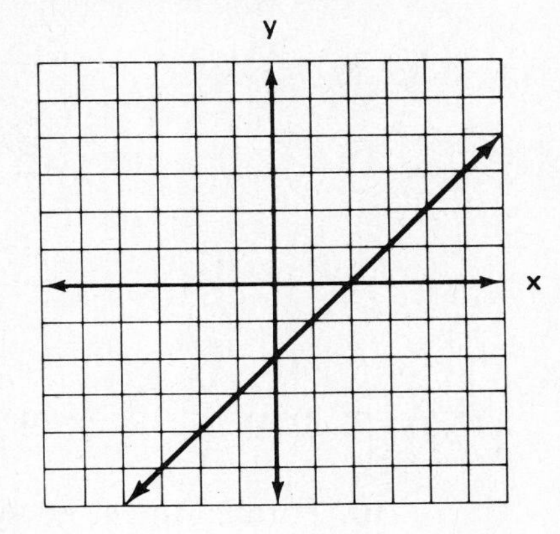

11. (1,1)

13. $\left(4, \frac{1}{3}\right)$

15. coincident lines

17. (3,2)

19. (−3,0)

21. parallel lines, no solution

23. 32, 24

25. 87 nickels, 39 pennies

27. 15 thirty-cent and 13 twenty-cent stamps

29. 19°, 71°

CHAPTER 10 ACHIEVEMENT TEST, PAGE 359

(10.1) 1. coincident   2. intersecting   3. parallel

(10.2) 4. parallel lines, no solution

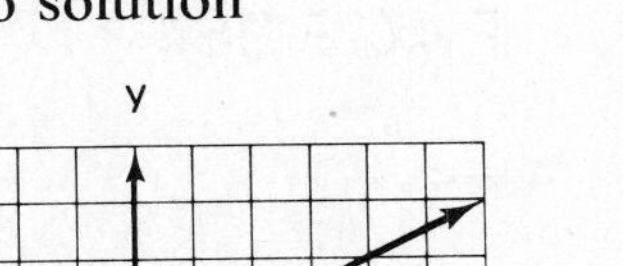

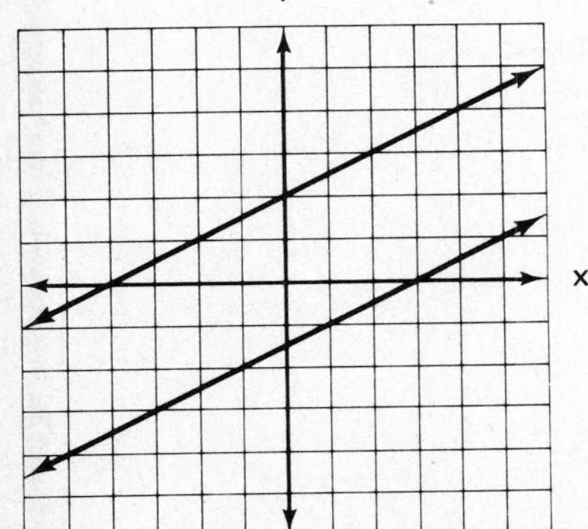

5.

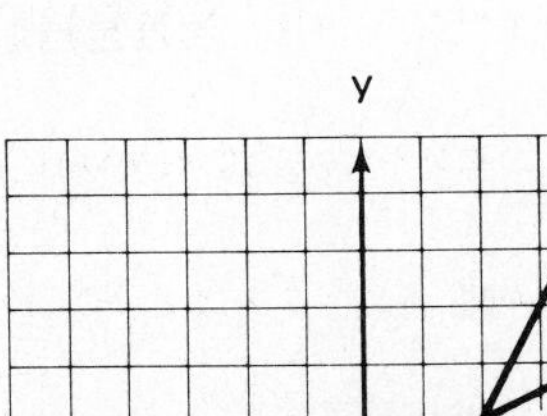

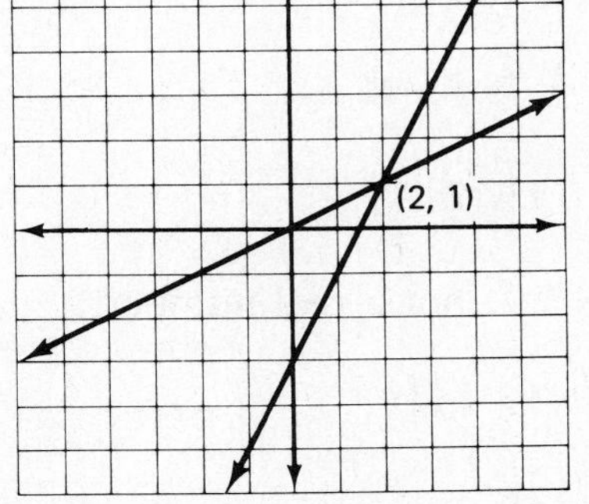

6. coincident lines

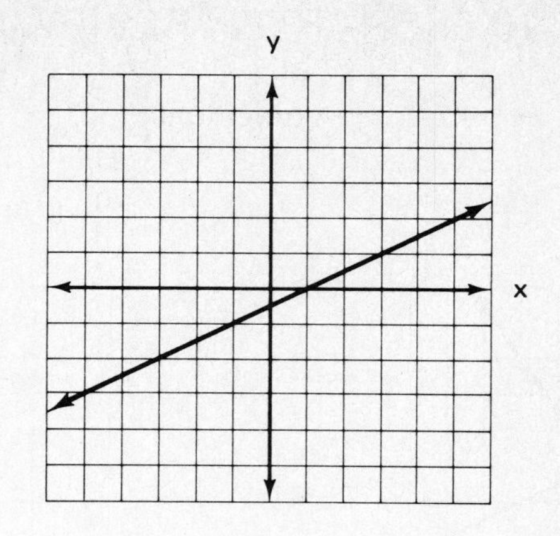

(10.3) 7. (1,1) 8. (4,0) 9. coincident lines

10. parallel lines, no solution

(10.4) 11. (3,5) 12. coincident lines 13. (4,3)

14. parallel lines, no solution

(10.5) 15. 50, 20 16. length is 52, width is 13

17. 14 quarters, 23 dimes

18. 12 pounds of $2.30 candy, 18 pounds of $1.80 candy

19. grandfather is 60, granddaughter is 7

CHAPTER 11

EXERCISE 11.1, PAGE 364

1. $5, -5$ 3. $4, -4$ 5. $10, -10$

7. $8, -8$ 9. $11, -11$ 11. $12, -12$

13. 2 15. 1 17. 7

19. not a real number 21. -9 23. -13

25. $|a|$ 27. $|3x|$ 29. $|-x|$

31. $|x - 2|$ 33. $-|x - 2|$

EXERCISE 11.2, PAGE 367

	NUMBER	RATIONAL	IRRATIONAL	REAL
1.	3	x		x
2.	$\frac{0}{4}$	x		x
3.	$\sqrt{8}$		x	x
4.	$\frac{22}{7}$	x		x
5.	π		x	x
6.	6.11111...	x		x
7.	.1262626...	x		x
8.	$\sqrt{9}$	x		x
9.	$-\sqrt{9}$	x		x
10.	$\sqrt{-9}$			
11.	$-\sqrt{-9}$			
12.	6.301425187...		x	x
13.	0	x		x
14.	−6	x		x
15.	$17\frac{4}{5}$	x		x

EXERCISE 11.3, PAGE 370

1. $\sqrt{6}$
3. 6
5. $2\sqrt{10}$
7. $6\sqrt{2}$
9. $2\sqrt{3}$
11. y^2

13. $4x^3$
15. $3z^2\sqrt{6z}$
17. $t^{12}\sqrt{t}$
19. $x - 1$
21. $5\sqrt{2}$
23. $5\sqrt{3}$
25. $3\sqrt{2}$
27. a^2
29. $4x\sqrt{3y}$
31. $5x^7\sqrt{2x}$
33. $4x^4\sqrt{3x}$
35. $6x^2y^3$

EXERCISE 11.4, PAGE 372

1. $5\sqrt{5}$
3. $10\sqrt{x}$
5. cannot be combined
7. $\sqrt{2}$
9. $16\sqrt{3}$
11. $12\sqrt{5}$
13. $3\sqrt{2}$
15. $-12\sqrt{2}$
17. $7\sqrt{3x}$
19. $17\sqrt{7a}$

EXERCISE 11.5, PAGE 375

1. $\sqrt{3}$
3. 2
5. $\sqrt{6}$
7. 4
9. $\frac{7}{2}$
11. $\frac{5}{3}$
13. 2
15. $x\sqrt{5}$
17. $x^2\sqrt{6}$
19. $\frac{\sqrt{6}}{2}$
21. $\frac{2\sqrt{3}}{3}$
23. $\frac{\sqrt{30}}{3}$
25. $2\sqrt{10}$
27. $\frac{\sqrt{7}}{7}$
29. $\frac{2\sqrt{y}}{y}$
31. $\frac{\sqrt{15}}{6}$
33. $\frac{\sqrt{6x}}{9x}$

EXERCISE 11.6, PAGE 379

1. $12 + 2\sqrt{3}$
3. $2\sqrt{21} - 14$
5. $6 - 6\sqrt{3}$
7. $3\sqrt{y} + 2y$
9. $16 + 8\sqrt{3}$
11. $-1 + 2\sqrt{2}$

13. $7 - 3\sqrt{6}$

15. $16 + 7\sqrt{6}$

17. -23

19. $3 + \sqrt{6} - 3\sqrt{2} - 2\sqrt{3}$

21. $28 + 10\sqrt{3}$

23. $7 + 4\sqrt{3}$

25. $3 + \sqrt{7}$

27. $\sqrt{5} - 3\sqrt{6}$

29. $x + \sqrt{5}$

31. $2\sqrt{2} - 2$

33. $-3\sqrt{3} + 3\sqrt{5}$

35. $1 + \sqrt{3}$

37. $6 + 3\sqrt{3}$

39. $2 - \sqrt{3}$

41. $\frac{13 + 9\sqrt{2}}{7}$

43. $\frac{23 - 14\sqrt{5}}{11}$

EXERCISE 11.7, PAGE 383

1. $x = 16$

3. $x = 25$

5. $x = 11$

7. $x = 39$

9. $y = 2$

11. $x = 66$

13. $x = 6$

15. $x = 2$

17. no solution

19. $x = 10$

EXERCISE 11.8, PAGE 386

1. 5

3. $\sqrt{77}$

5. 12

7. $2\sqrt{34}$

9. 3

11. $3\sqrt{2}$

13. $5\sqrt{3}$

15. $3\sqrt{3}$

17. yes

19. yes

21. no

23. 13 cm

25. 10 inches

27. 30 miles

29. 52 feet

EXERCISE 11.9, CHAPTER REVIEW, PAGE 392

1. $7, -7$

3. $11, -11$

5. 5

7. ± 5

9. 0

11. $|y|$

13. $|h - 3|$

15. rational, real

17. irrational, real

19. not a real number

21. rational, real

23. $4\sqrt{5}$

25. y^4

27. $(y - 3)^3$

29. $3\sqrt{5}$

31. y^6

33. $4xy^2\sqrt{3}$

35. $-7\sqrt{x}$

37. cannot be combined

39. $11\sqrt{2}$

41. $9\sqrt{3x} - 10\sqrt{3y}$ cannot be combined

43. 2

45. $x\sqrt{6}$

47. $\dfrac{\sqrt{15}}{5}$

49. $\dfrac{\sqrt{6y}}{y}$

51. $2\sqrt{15} - \sqrt{5}$

53. $2x\sqrt{3} - 2\sqrt{2x}$

55. $8 + 3\sqrt{21}$

57. $4 + \sqrt{5}$

59. $-2\sqrt{2} - 4$

61. $\dfrac{1 - \sqrt{6}}{5}$

63. no solution

65. $y = 71$

67. $3\sqrt{5}$

69. no

71. $3\sqrt{111}$ or approximately 31.6 feet

CHAPTER 11 ACHIEVEMENT TEST, PAGE 397

(11.1) 1. 8, −8 2. 6 3. −6 4. not a real number

(11.2) 5. rational 6. irrational 7. rational 8. irrational

(11.3) 9. $3\sqrt{5}$ 10. $3x^2\sqrt{2x}$ 11. $5\sqrt{3}$ 12. y^6

(11.4) 13. $-3\sqrt{5}$ 14. $7\sqrt{3}$ 15. cannot be combined

(11.5) 16. $\sqrt{3}$ 17. $\frac{3}{4}$ 18. $x\sqrt{2}$ 19. $\frac{\sqrt{30}}{10}$ 20. $\sqrt{15}$

(11.6) 21. $2\sqrt{15}-3\sqrt{2}$ 22. -6 23. $-1-7\sqrt{7}$

24. $\sqrt{2}+3\sqrt{5}$ 25. $3\sqrt{2}+3$ 26. $\frac{17-7\sqrt{7}}{9}$

(11.7) 27. $y = 6$ 28. no solution

(11.8) 29. $4\sqrt{2}$ 30. yes 31. 75 feet

CHAPTER 12

EXERCISE 12.1, PAGE 402

1. $4x^2 - 3x + 2 = 0$
3. $3x^2 - x + 1 = 0$
5. $x^2 - 2x + 6 = 0$
7. $5x^2 + x + 2 = 0$
9. $4x^2 + 30x - 5 = 0$

EXERCISE 12.2, PAGE 405

1. $4, -4$
3. $-2, -8$
5. $-4, 3$
7. $0, 14$
9. $-1, -6$
11. $\frac{1}{2}, -\frac{3}{2}$
13. $\frac{1}{3}, -2$
15. $-\frac{5}{2}, 1$
17. $3, 2$
19. $4, 2$

EXERCISE 12.3, PAGE 407

1. ± 4
3. ± 5
5. ± 3
7. ± 3
9. $\pm 2\sqrt{3}$
11. $\pm\frac{3}{2}$
13. ± 5
15. $\pm\sqrt{5}$
17. $\pm\frac{6}{\sqrt{5}}$ or $\frac{6\sqrt{5}}{5}$
19. ± 3

EXERCISE 12.4, PAGE 411

1. $x^2 + 4x + 4 = (x + 2)^2$
3. $y^2 - 8y + 16 = (y - 4)^2$
5. $x^2 - 7x + \frac{49}{4} = \left(x - \frac{7}{2}\right)^2$
7. $x^2 + \frac{1}{4}x + \frac{1}{64} = \left(x + \frac{1}{8}\right)^2$
9. $-2, -4$
11. $3, 5$
13. $-3, -7$
15. $5, -2$
17. $-4, 3$
19. $-1 \pm \sqrt{6}$
21. $\frac{-3 \pm \sqrt{17}}{2}$
23. $\frac{5 \pm \sqrt{65}}{4}$

EXERCISE 12.5, PAGE 415

1. $-4, -1$
3. $\frac{4 \pm \sqrt{10}}{2}$
5. $-3 \pm \sqrt{6}$
7. $\frac{-1 \pm \sqrt{5}}{2}$
9. $\frac{3}{5}, 1$
11. $3, -3$
13. $-\frac{2}{3}, \frac{3}{4}$
15. $-2 \pm \sqrt{5}$
17. $3 \pm 2\sqrt{3}$
19. $\frac{2 \pm \sqrt{3}}{3}$
21. $\frac{-5 \pm 3\sqrt{5}}{2}$
23. $\frac{-7 \pm \sqrt{61}}{2}$

EXERCISE 12.6, PAGE 419

1. $5, 2$
3. $-3, 3$
5. $6, 2$
7. 1
9. 1
11. 6
13. $-4, 2$

EXERCISE 12.7, PAGE 422

1. $-1, 1$
3. $2, 3$
5. 12, 7 does not check
7. $-2, -1$
9. -3
11. $-\frac{1}{2}$, $-\frac{3}{2}$ does not check
13. $\frac{2}{3}, \frac{1}{3}$
15. 4, 0 does not check
17. 6, 2 does not check
19. $5, 1$

EXERCISE 12.8, PAGE 428

1. 7,8
3. 9,12
5. base is 6, altitude is 5
7. width is 12, length is 16
9. 4 on a side
11. 4 cm wide
13. 8 mph
15. 6 sides (hexagon)
17. $\frac{3}{4}$ or $\frac{4}{3}$
19. $2 \pm \sqrt{3}$
21. 10 inches long, 6 inches wide
23. 30 feet long, 10 feet wide
25. 4 feet wide, 6 feet long
27. 6 inches square
29. 9 mph

EXERCISE 12.9, CHAPTER REVIEW, PAGE 435

1. $3x^2 + 2x - 9 = 0$
3. $2x^2 - 9x - 24 = 0$
5. 3, −5
7. $\frac{1}{2}, -\frac{5}{2}$
9. ± 9
11. $\pm 2\sqrt{5}$
13. $\pm 2\sqrt{7}$
15. $y^2 - 5x + \frac{25}{4} = \left(y - \frac{5}{2}\right)^2$
17. 5, −1
19. $\frac{1 \pm \sqrt{5}}{2}$
21. $-\frac{1}{2}, 2$
23. 4, −3
25. $\frac{-9 \pm \sqrt{77}}{2}$
27. $\pm \sqrt{10}$
29. −2, 1
31. −1, 7
33. $\pm 2\sqrt{3}$
35. $\frac{7 \pm \sqrt{41}}{2}$
37. $\pm 2\sqrt{2}$
39. $1 \pm 2\sqrt{5}$
41. ± 4
43. 9, 2 doesn't check
45. $\frac{1}{3}$
47. length is 12, width is 9
49. 10

CHAPTER 12 ACHIEVEMENT TEST, PAGE 441

(12.1) 1. $3x^2 + 9x + 2 = 0$ 2. $4x^2 - 3x - 30 = 0$

(12.2) 3. $(-4,3)$ 4. $\frac{1}{3}, -\frac{5}{2}$

(12.3) 5. ± 10 6. $\pm 2\sqrt{5}$

(12.4) 7. $x^2 - 4x + 4 = (x - 2)^2$ 8. $x^2 + x + \frac{1}{4} = (x + \frac{1}{2})^2$

9. $3, -2$ 10. $4 \pm \sqrt{15}$

(12.5) 11. $4, -2$ 12. $\frac{3 \pm \sqrt{17}}{4}$

(12.2–12.7) 13. $\pm\sqrt{6}$ 14. $\frac{1}{3}, -\frac{1}{2}$ 15. 13, 25

(12.8) 16. length is 12 inches, width is 5 inches 17. $\frac{2}{9}$ or $\frac{9}{2}$

18. 7 mph

APPENDIX I, ARITHMETIC REVIEW

EXERCISE A.1, PAGE 448

1. $\frac{2}{3}$ 3. in lowest terms 5. $\frac{4}{9}$

7. $\frac{8}{15}$ 9. $\frac{3}{2}$ 11. $\frac{4}{5}$

13. $\frac{1}{4}$ 15. $\frac{11}{16}$ 17. $3\frac{5}{6}$

19. $14\frac{1}{4}$ 21. $\frac{48}{5}$ 23. $\frac{28}{45}$

25. $\frac{35}{3}$ 27. $\frac{3}{5}$ 29. $7\frac{1}{2}$

31. $2\frac{7}{24}$ 33. $16\frac{1}{8}$ 35. 23.175

37. 4.868 39. 6.18 41. .000203

43. 4.11 45. .75 47. .43

49. $\frac{14}{25}$

Index